# FOOD PROCESS ENGINEERING
## SECOND EDITION

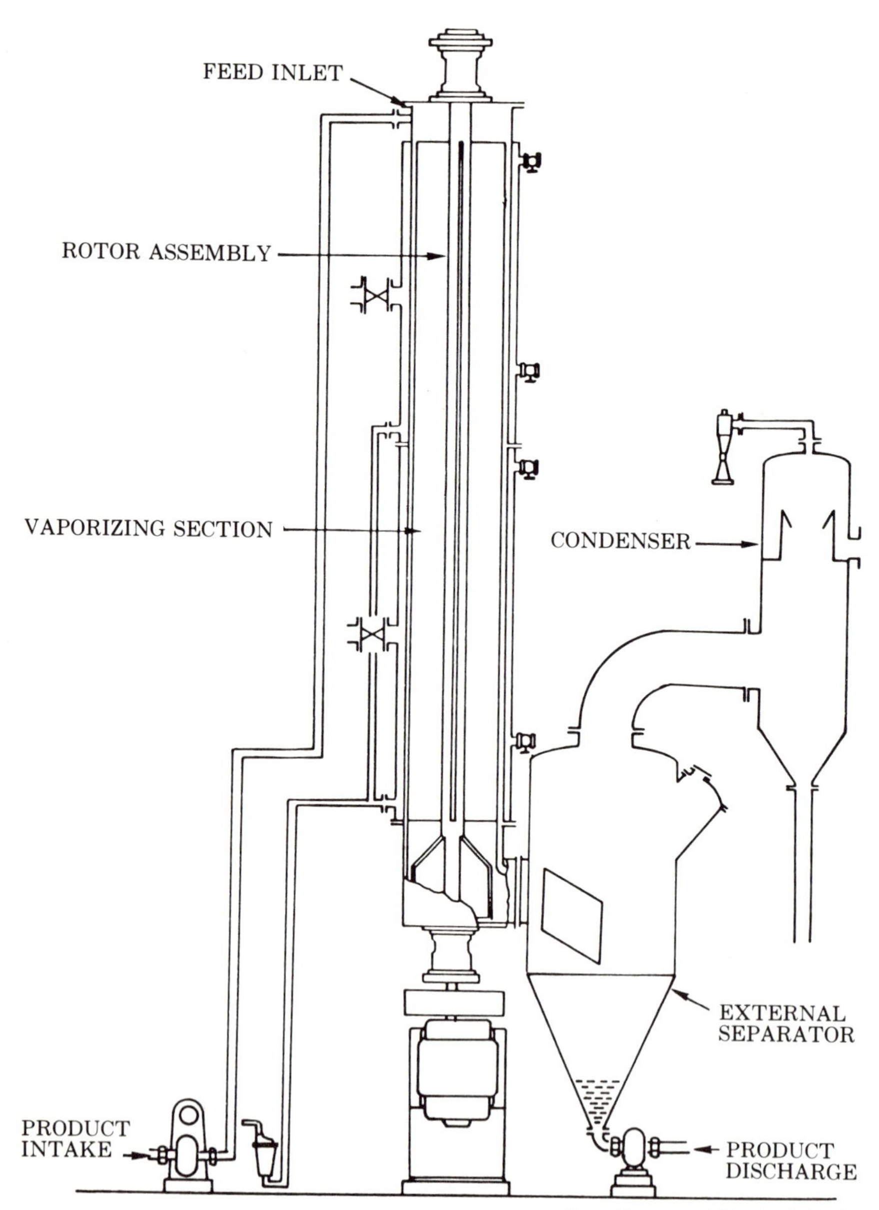


*From Moore and Hessler (1963)*

# FOOD PROCESS ENGINEERING

## SECOND EDITION



by
Dennis R. Heldman
Professor of Food Engineering
Michigan State University

and

R. Paul Singh
Associate Professor of Food Engineering
University of California, Davis

AVI PUBLISHING COMPANY, INC.
Westport, Connecticut

Library of Congress Cataloging in Publication Data

Heldman, Dennis R
Food process engineering.

Includes bibliographies and index.
1. Food industry and trade. I. Singh, R. Paul, joint author. II. Title.
TP370.H44 1981 664 80-27689
ISBN 0-87055-380-1

Printed in the United States of America
by The Saybrook Press, Inc.

# Foreword

The Second Edition of *Food Process Engineering* by Dr. Dennis Heldman, my former student, and co-author Paul Singh, his former student, attests to the importance of the previous edition. In the Foreword to the First Edition, I noted the need for people in all facets of the food processing industry to consider those variables of design of particular importance in engineering for the food processing field.

In addition to recognizing the many variables involved in the biological food product being handled from production to consumption, the engineer must oftentimes adapt equations developed for non-biological materials. As more and more research is done, those equations are appropriately modified to be more accurate or new equations are developed specifically for designing to process foods. This Edition updates equations used. This book serves a very important need in acquainting engineers and technologists, particularly those with a mathematics and physics background, with the information necessary to provide a more efficient design to accomplish the objectives. Of prime importance, at present and in the future, is to design for efficient use of energy. Now, it is often economical to put considerably more money into first costs for an efficient design than previously, when energy costs were a much smaller proportion of the total cost of process engineering.

All chapters in the new edition relate in some manner to efficiency of operations: either efficiency of energy utilization for a particular process or overall efficiency for carrying out a series of operations, including the materials for accomplishing the design.

A major development since the last edition is the gradual adoption of SI (the International System) metric units of measurement. Although final agreement has not been obtained for several areas of applications of SI units in the food processing industry, these are evolving and the latest units used are included herein. Students, faculty, and industry people can benefit from this book.

CARL W. HALL, P.E., Dean
Washington State University
Pullman, Washington

*September 1980*

# Foreword

The availability of the second edition of this book in the area of food engineering will be most significant in the field of food science and technology, particularly for advanced undergraduates and graduate students who wish to develop competency in the area of food engineering, and for professional personnel in the food and allied industries, government and universities. I have had the opportunity to work closely with Professor Dennis Heldman during the course of his doctoral studies and his work on the faculty of the Departments of Food Science and Agricultural Engineering at Michigan State University, as well as during two periods when he was a visiting faculty member with the Departments of Food Science and Technology and Agricultural Engineering here at the University of California at Davis in 1970 and again in 1980. His unique experience in engineering and in food science and technology provides an excellent set of capabilities and competencies for developing revision of this book. I also am equally proud to work with Professor Paul Singh, who holds a joint faculty appointment in the Departments of Agricultural Engineering and Food Science and Technology here. Since joining the faculty, he has made major contributions to the food engineering program in teaching, research and public service.

The area of food science and technology with the greatest need for further development is, in my opinion, that of food engineering. Scientists with the capability of bridging these fields, such as Professors Heldman and Singh, are rare, and we are very fortunate to have scientists with such insights and capabilities develop this book. This material should be of special timeliness as the food and allied industries evolve and become increasingly scientific, particularly in quantifying process and product procedures and in evolving from "batch processes" to "continuous processes." Current issues of energy and water conservation in food processing are good illustrations.

As we look ahead, we see growing emphasis on the nutritive value and sensory properties of processed foods, on the utilization of physical and chemical characteristics of food constituents in the development of new products, and on safety and environmental protection programs. These activities will require increasing attention to the food engineering concepts for the product and process innovations. I am confident that the information provided in this text will form the basis for the educational development of many young food scientists and food engineers in this country and abroad. It is hoped that such an awareness and capability will augment the changes occurring within the food and allied industries outlined above. It is for these reasons I commend to the reader careful attention to the material developed in this Second Edition.

B. S. SCHWEIGERT, Chairman
Department of Food Science & Technology
University of California
Davis, California

*September 1980*

# Preface to the Second Edition

The role of processes and equipment in the ability of the Food Industry to supply nutritious and healthful foods to the consumer is increasing in importance. As costs of energy required for all aspects of food processing continue to increase, the need to improve efficiency of processes becomes more evident. As analysis of these processes is initiated, the need for understanding and use of engineering principles becomes more obvious.

Food engineering education is very complex and involves development of courses and curricula for engineering as well as non-engineering students. It follows that textbooks must be designed to meet the unique needs of the students enrolled in food engineering courses. Textbooks for non-engineering students must deal with introductory engineering concepts and mathematics, and limited emphasis should be placed on engineering design. The food engineering textbook for the engineering student should emphasize design of processes and equipment and can build on the student's background in mathematics, basic engineering and thermodynamics. This textbook on *Food Process Engineering* is written for the undergraduate engineering student with particular interest in operations and processes used in the food processing plant.

Several developments which occurred during the five years since publication of the First Edition make the availability of this Second Edition very timely. The influence of increasing energy costs has become a significant factor in the handling, processing, packaging and distribution of food. The need to improve the efficiency of existing processes and to develop new, more efficient processes is more evident than at any time during the evolution of the modern food industry. Unfortunately, the design of existing processes and equipment has been based on availability of inexpensive energy and may require significant revision in order to achieve optimum energy efficiency. This textbook deals with process design concepts in a manner that will allow the student to gain an understanding of the process sufficient to improve or redesign the process or equipment in response to energy efficiency requirements.

The second development of significance is one of units of measurements. The First Edition utilized English units in a manner consistent with practice in the food industry. Although the units used in the industry have not changed, it is evident that the change to the *Système International d'Unités* (SI units) will occur, and new graduates accepting positions in the industry must provide leadership in the conversion process. With this in mind, the Second Edition is presented entirely with SI units.

The authors wish to express appreciation to numerous individuals and groups who gave significant input to the Second Edition. Major portions of the Edition were completed by the senior author during a sabbatical from Michigan State University. Since this sabbatical leave was completed at the University of California at Davis, both institutions have made significant contributions. In particular, the support and encouragement of Dr. B.S. Schweigert, Chairman of Food Science and Techology, and Dr. R.E. Garrett, Chairman of Agricultural Engineering, at the University of California at Davis, is acknowledged. Throughout the preparation of the Second Edition, the authors have enjoyed significant support from students with interest in the subject of the textbook. Finally, the authors are indebted to Karin Clawson and Kathy Adamski for the typing of the manuscript drafts for the Second Edition.

There are similarities between the format followed in this new edition and others on the subject of food engineering, including Fundamentals of Food Engineering by S.E. Charm, Unit Operations in Food Processing by R.L. Earle, Elements of Food Engineering by J.C. Harper and Fundamentals of Food Processing Engineering by R.T. Toledo. This book is based extensively on The Fundamentals of Food Engineering, First, Second and Third Editions by Dr. S.E. Charm. The basis of presentation of many of the topics was established by Dr. Charm and in this book the authors have attempted to select those topics of particular significance to the engineering student and to emphasize design-oriented problems. Most food engineering students will find the use of all these books of particular value.

The authors hope that this textbook will be useful to students with interest in Food Process Engineering. Their contribution has been invaluable and will continue to stimulate future developments.

Dennis R. Heldman
R. Paul Singh

*November 1980*

# Preface to the First Edition

The ability of a food industry to provide a continuous supply of nutritious and healthful food to the consumer is highly dependent on the processes and equipment used throughout all stages of handling, processing and distribution. The efficiency involved in accomplishing these stages is a function of the process and/or equipment design. The role of engineering in all aspects of process and equipment design is very evident.

As the demand for large quantities of high-quality food increases, the need for individuals with specific competencies becomes greater. The design of highly complex processes and equipment that will be required in the future will demand individuals who have a clear understanding of the processes and equipment utilized in handling, processing and distribution of food, as well as adequate knowledge of the chemical, microbiological and biochemical characteristics of food products. It seems evident that that education of the food engineer can be accomplished most efficiently by providing areas of instruction specifically designed to illustrate applications of engineering principles and concepts to handling, processing and distribution of food. The information presented in this book is intended to provide the engineering student with an opportunity to apply engineering principles in the design of processes and systems for the food industry.

This volume is design-oriented, and assumes a background of coursework in calculus and differential equations, heat transfer, thermodynamics or physical chemistry, and fluid mechanics. It would seem most logical for this material to be presented late in the curriculum of the engineering student and in the graduate program of the food science student with engineering interests.

There are obvious similarities between the format used in this book and those of *Fundamentals of Food Engineering,* Editions 1 and 2, by S. E. Charm and *Unit Operations in Food Processing* by R. L. Earle. Certainly, the approaches used previously in presentation of engineering concepts of food-processing systems had a significant influence on the style used for presentation in this volume. The author's primary objective has been to bring together those topics of primary significance to the engineering student preparing for employment in some phase of the food industry.

There are several areas which have been omitted or presented in only a brief manner. The intent of the author is not to deemphasize any important area in favor of another. In general, it is hoped that the information presented will provide sufficient background for the student to handle through self-study the material which is omitted. One obvious area is design of thermal processes, which would require considerable space for adequate coverage. The basic tools for describing survivor curves and heating or cooling curves are presented for the student to use in his study of thermal processing.

The author's stimulation for preparation of this book was first provided by an advisor, colleague and friend, Dr. C. W. Hall, Dean of Engineering at Washington State University. During the years of preparation, encouragement was provided by Dr. B. S. Schweigert, Chairman of Food Science and Technology, University of California, Davis, and Dr. B. A. Stout, Chairman of Agricultural Engineering, Michigan State University.

The following colleagues and associates have made a significant contribution by reviewing various chapters of the manuscript.

Dr. R. L. Merson, Assistant Professor, University of California, Davis.

Dr. D. D. Lund, Associate Professor, University of Wisconsin, Madison.

Dr. J. C. Harper, Professor, University of California, Davis.

Dr. C. G. Haugh, Professor, Purdue University, Lafayette, Indiana.

Dr. B. L. Clary, Associate Professor, Oklahoma State University, Stillwater.

Mr. R. W. Dickerson, Chief, Food Process, Evaluation Branch, FDA, Cincinnati, OH.

Dr. D. R. Thompson, Assistant Professor, University of Minnesota, St. Paul.

Dr. J. L. Blaisdell, Associate Professor, Ohio State University, Columbus.

Dr. S. W. Fletcher, Associate Professor, University of Massachusetts, Amherst.

Dr. D. F. Farkas, Chemical Engineer, Engineering and Development Laboratory, Western Regional Research Lab., USDA, Albany, California.

Dr. F. W. Bakker-Arkema, Professor, Michigan State University, East Lansing.

Dr. J. M. Harper, Professor, Colorado State University, Fort Collins.

Dr. T. P. Labuza, Professor, University of Minnesota, St. Paul.

Dr. V. A. Jones, Associate Professor, North Carolina State University, Raleigh.

Dr. J. R. Rosenau, Associate Professor, University of Massachusetts, Amherst.

Dr. R. Holmes, Assistant Professor, North Carolina State University, Raleigh.

Professor E. Seltzer, Rutgers University, New Brunswick, New Jersey.

The suggestions, constructive criticism and encouragement provided by these authorities have resulted in a significantly improved book. In addition, the author is indebted to Carole Steinberg for the many hours devoted to typing several drafts.

The author wishes to express gratitude to all students who have devoted considerable time to study of the manuscript in draft form. The comments and suggestions have been received with enthusiasm, since the information presented is intended for all students with an interest in Food Engineering.

DENNIS R. HELDMAN

*March 15, 1974*

# Dedication

Dedicated
to
Our Wonderful Families
Joyce
Cindy, Doug, Kristi
Candy, Craig
Anila
Raj

# Contents

# 1

# Introduction

The modern supermarket contains a large variety of food items which have been processed by different methods in an effort to obtain products that are most acceptable to the consumer. The industry which provides this great variety of products relies on basic concepts of chemistry, microbiology, and engineering to process and package them. The engineering phase of the food industry deals primarily with the processes involved in changing or altering a raw food material. There are two distinct parts of the discipline which can be called Food Engineering: (a) the descriptive part dealing with the detailed description of the equipment and the processes involved in the processing of food, and (b) the theoretical part, which comprises a mathematical description of the processing equipment and the changes that may occur within the products during handling, processing, and storage. Both parts are important, and in many cases the two supplement or support each other. The subject matter to be presented will be more closely related to the theoretical aspects of food processing. Every effort will be made, however, to illustrate how these mathematical operations are utilized in arriving at useful computations which lead to the sizing of equipment and description of the processes involved.

It is almost impossible to describe a single type of food-processing plant which would incorporate all processes involved for all products; there are too many variations between the types of materials, the types of processes involved, and the types of products produced by the modern processing plant. An attempt will be made, however, to describe a hypothetical plant which contains the most important unit operations involved in the modern food-processing industry. In addition to introducing these more important unit operations, several auxiliary factors and subjects will be introduced. Without these related areas the modern plant could not operate.

By referring to Fig. 1.1, it should be possible to visualize that the raw food material may arrive at the processing plant in one of many forms,

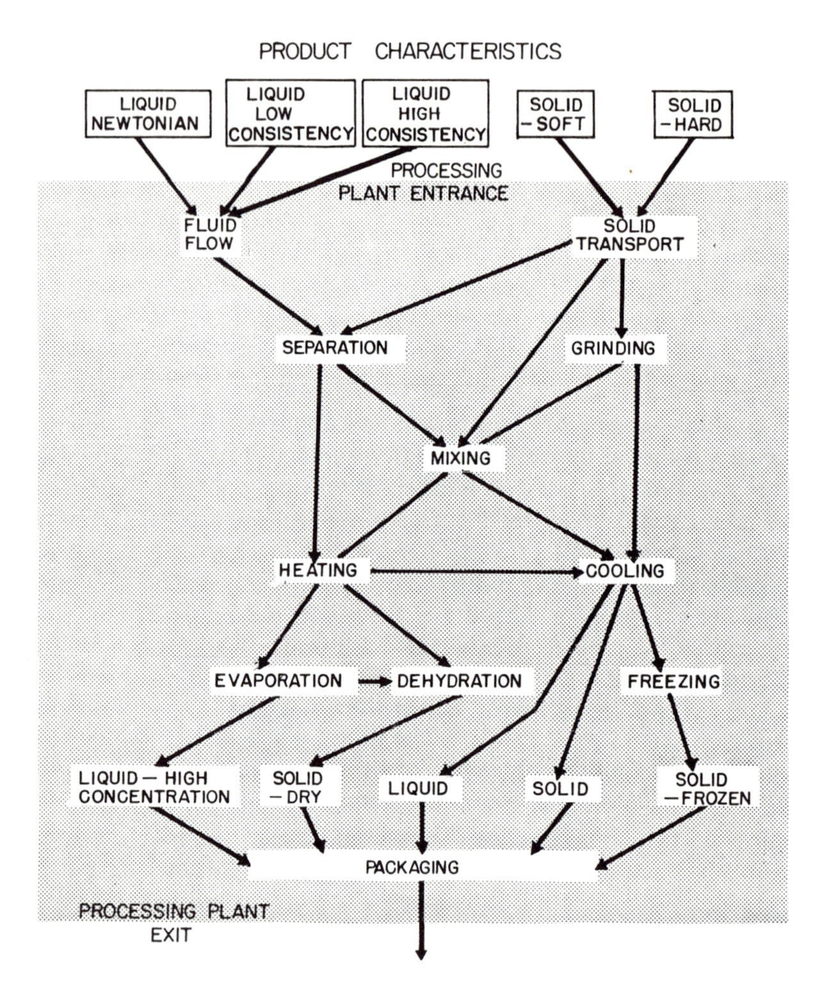


FIG. 1.1. SCHEMATIC ILLUSTRATION OF FLOW IN GENERALIZED FOOD PROCESSING PLANT

which may range from low viscosity liquids to hard, solid materials. These materials must be transported into and throughout the plant by some means. These product transport operations introduce the first of the important unit operations that must be performed in the modern food processing plant. The subject of fluid flow or hydraulics is of considerable importance in most food processing plants. Although this is a discipline of its own, a considerable number of the basic concepts must be known and understood in order to design a transport system required for liquid food products. The subject matter area is further complicated by the fact that many food products have non-Newtonian flow characteristics and cannot be treated as fluids similar to water. The transport of solid materials may take so many different forms that it is impossible to develop any general description.

The next logical group of unit operations is that which deals with separation of the product into components based on either physical or chemical characteristics. These operations include (a) mechanical separation by either centrifugal force of filtration; (b) distillation based on the volatility of the product; (c) extraction, involving either a solvent with solid materials or liquid-liquid extraction; and (d) sedimentation based on the influence of gravity on components of the product. A similar operation that might be involved at this point in the processing plant is grinding, which is applied only to solid materials. After the raw materials are separated into components or ground into smaller parts, the next logical operation is that of mixing or blending of some components into an acceptable form which the consumer may desire. The type of mixing operation required will vary considerably depending on the kind of product being made.

The single unit operation which is nearly universal to all processed food products is heat transfer. Almost every processed food product is either heated and/or cooled at some point between the time it reaches the processing plant and the time it reaches the consumer. A basic understanding of the heating and cooling characteristics of food products is of utmost importance. The specific operations that involve heat transfer include the various types of heat exchangers, the pasteurization processes, and the sterilization processes. At this point and after completing this phase of the processing, many products will go directly to the packaging operation, where they are placed in the container which ultimately reaches the consumer.

There are still other unit operations that deserve the attention of the student interested in theoretical aspects of food process engineering. The first of these is freezing, because of the many products which reach the consumer in the frozen state. The second operation which cannot be overlooked is evaporation, due to its importance in reducing the volume of many of the high-moisture liquid foods either for transport purposes or before a dehydration process. Food dehydration is the third operation which must receive some attention because of its significance in the production of the many dried foods that are available in the modern supermarket.

The unit operations described up to this point represent the major processes involved in food manufacturing. Several auxiliary or related topics, however, must be mentioned in order to ensure that the student is not misled about the importance of any one operation. These related topics include (a) instrumentation and process control, (b) cleaning operations, (c) materials handling, (d) electric power, (e) equipment maintenance and (f) plant layout and design. Each of these rather unrelated topics is very important to the successful operation of the processing plant; they are described in detail by Brennan *et al.* (1976), Earle (1966), Farrall (1976, 1979), Joslyn and Heid (1963), Harper (1976), Toledo (1980), Slade (1967, 1971), Charm (1978) or Loncin and Merson (1979).

Processed food products leave the modern processing plant in the same variety of forms in which they entered the plant. These forms include liquids, solids, high concentrated liquids, dry solids and semi-solids. With this introduction to the complex type of operation with which we are dealing, we will proceed to introduce several basic areas in a rather brief manner.

## 1.1 THERMODYNAMICS APPLIED TO FOOD PROCESSING

It is not the intent of this section to present an exhaustive coverage of thermodynamics. The objective will be to present the more important parameters in the area of food processing and illustrate their applications. One of the most basic parameters is heat content or enthalpy. In thermodynamics this particular parameter is defined by the following equation:

$$H = E + PV \tag{1.1}$$

where $E$ is the internal energy of a system, $P$ is the absolute pressure, and $V$ is the volume of the system. Any change in the enthalpy will be due to a change in heat content of the system when the change occurs at a constant pressure. The enthalpy ($H$) then becomes what is defined as a state function along with other basic parameters such as internal energy ($E$), temperature ($T$), pressure ($P$), and volume ($V$).

One of the basic thermodynamic parameters used considerably in the description of heating and cooling of food products is specific heat. The parameter which is referred to in almost every case is the specific heat at a constant pressure, as defined by the following equation:

$$c_p = \left(\frac{\partial H}{\partial T}\right)_p \tag{1.2}$$

As indicated by equation (1.2), the specific heat at a constant pressure represents the change in enthalpy or heat content for a given change in temperature when the change occurs at a constant pressure. This parameter will take on different values depending on the product considered. Both predicted and experimental values for different food products are given in appendix A.9. If the values of this thermodynamic parameter are known, it is almost exclusively used in the computation of the heat content of enthalpy of a product above a given reference level.

EXAMPLE 1.1 Determine the heat content of tomato soup concentrate at 25 C above a reference of 0 C.

SOLUTION

From the appendix (Table A.9) the specific heat of tomato soup concentrate is found to be 3.676 kJ/kg C. The enthalpy ($\Delta H$) will be:

$$\Delta H = c_p \, \Delta T = 3.676 \; (25-0) = 91.9 \text{ kJ/kg}$$

Another state function that must be introduced at this point is entropy. The change in entropy is defined by the following equation:

$$dS = \frac{dq}{T} \tag{1.3}$$

Although it is very difficult to attach physical significance to the value of entropy, it should be pointed out that for any real process the change in entropy will always be positive.

If a system is held at a constant temperature and constant pressure, as is the case of many food products which are in storage, a rather special type of equilibrium condition is attained. The equilibrium condition results in the defining of a new thermodynamic function called free energy and is described by the following equation:

$$G = H - TS \tag{1.4}$$

As indicated by equation (1.4) the free energy ($G$) is equal to the difference between the enthalpy ($H$) and the product of the temperature and entropy. A change in free energy during any kind of reaction is frequently equated to the net work done on a system in a reversible process. For isothermal processes, it is obvious from the equation (1.4) that the change in free energy can be defined by the following equation:

$$\Delta G = \Delta H - T \, \Delta S \tag{1.5}$$

The value and use of these equations will become more obvious in later sections.

Many processes used in the food industry involve a change of state within the product. Thermodynamics can be utilized to describe or explain the conditions which exist. If this change occurs at a constant temperature and pressure, as is the case in most phase changes of concern, the system changes from a state of high free energy to a state of low free energy. This change in free energy is normally considered to be the thermodynamic potential—the difference between the free energy levels of the two phases involved in a phase change.

This definition of thermodynamic potential leads to the definition of chemical potential which is stated by the following equation:

$$\eta_i = \left[\frac{\partial G}{\partial n_i}\right]_{T,\, p,\, n_j} \tag{1.6}$$

where $n$ represents the number of moles of component $(i)$ present in the system. The chemical potential represents the change in free energy in the system with the change in number of moles of the given component $(i)$ while the temperature, pressure and number of moles of all other components, indicated by $j$, are kept constant. In other words, chemical potential is a measure of how the free energy of a change in phase depends on changes in composition. In later chapters, chemical potential will be utilized in the derivation of expressions describing freezing-point depression and boiling-point rise that can be applied to food products.

The factors involved in the phase change can be most easily understood by study of the phase diagram for a pure substance. Figure 1.2 illustrates the phase diagram for water on pressure-temperature coordinates. Lines A-C and A-D represent the equilibrium conditions between liquid and vapor, and solid and liquid, respectively. Curve A-B represents the conditions under which sublimation may occur, while point A is the triple point at which liquid, solid and vapor co-exist at the same temperature and pressure. Point C is referred to as the critical point above which the liquid phase can no longer be distinguished.

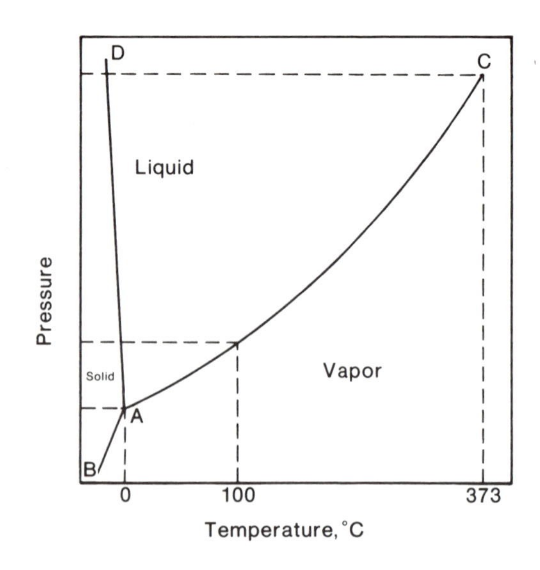


FIG. 1.2. PHASE DIAGRAM FOR WATER (NOT DRAWN TO SCALE)

One of the fundamental theoretical equations describing the conditions that exist at equilibrium between phases is the Clausius-Clapeyron equation. This equation, as derived from considerations of chemical potential, can be stated as follows:

$$\frac{dp}{dT} = \frac{\lambda}{T \Delta \overline{V}} \tag{1.7}$$

indicating that the change in pressure with the temperature is equal to the latent heat of the phase transformation divided by the product temperature and change in volume. The most frequent use of the Clausius-Clapeyron equation when applied to food products is in the determination of the effective latent heat of vaporization required during dehydration of the product. In this application, the logarithm of the equilibrium vapor pressure is plotted against the inverse of temperature so that the following expression is satisfied:

$$\frac{d(\ln p)}{dT} = \frac{\lambda}{RT^2} \tag{1.8}$$

Equation (1.8) can be derived from equation (1.7) by neglecting the volume of liquid compared to that of the vapor and assuming ideal gas behavior which introduces the universal gas constant ($R$).

One of the frequently used applications of the Clausius-Clapeyron equation was presented by Othmer (1940). By equating the vapor pressure conditions for a food product and water vapor using equation (1.8), the following expression is obtained:

$$\frac{\lambda}{\lambda'} = \frac{\log p_2 - \log p_1}{\log p_2' - \log p_1'} \tag{1.9}$$

In equation (1.9), $\lambda$ and $p$ represent latent heat and vapor pressure for the product and $\lambda^1$ and $p^1$ are latent heat and vapor pressure for water vapor, respectively.

EXAMPLE 1.2. The equilibrium relative humidity for precooked freeze-dried beef during moisture removal was 0.065, 0.1, 0.38 and 0.44 at 10% moisture content (nonfat dry basis) and 4, 10, 20, 40 C, respectively. Compute the latent heat of vaporization for the product at 38 C.

SOLUTION

(1) Using equation (1.9), a plot of the logarithm of product vapor pressure versus the logarithm of water-vapor pressure can be used to obtain the ratio of product latent heat to latent heat of water vapor (Othmer plot).

(2) The Othmer plot is obtained using the following information.

| $p/p_s$ | Temperature (C) | $p_s$ (kPa) | $p$ (kPa) |
|---|---|---|---|
| 0.065 | 3 | 0.7577 | 0.049 |
| 0.1 | 9 | 1.1477 | 0.115 |
| 0.38 | 21 | 2.487 | 0.945 |
| 0.44 | 36 | 5.947 | 2.617 |

(3) By plotting product vapor pressure versus vapor pressure of water vapor on log - log coordinates, as illustrated in Fig. 1.3, the latent heat ratio of 2.169 is obtained from the slope of the curve.

(4) Since the latent heat of vaporization for water at 38°C is 2412 kJ/kg:

$$\lambda = 2.169\ (2412) = 5231.6\ \text{kJ/kg}$$

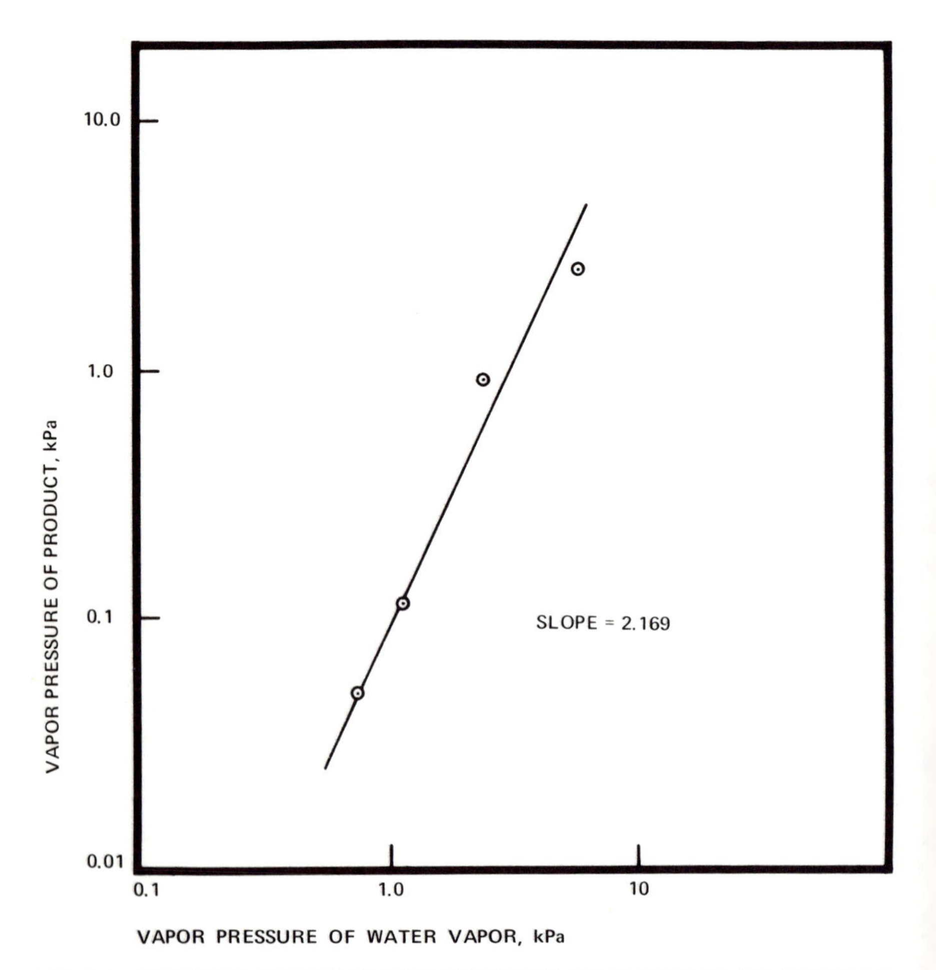

FIG. 1.3. OTHMER PLOT FOR PRECOOLED FREEZE-DRIED BEEF AT 10% MOISTURE CONTENT (NONFAT DRY BASIS)

(5) The latent heat of vaporization for precooked freeze-dried beef at 10% moisture content (nonfat dry basis) and 38 C is 5231.6 kJ/kg.

The concept of an ideal solution has very significant considerations when applied to food products. The concept is similar to that of an ideal gas. An ideal gas is defined as having complete absence of cohesive forces between components, while an ideal solution is defined by a complete uniformity of cohesive forces. In other words, the forces interacting between the components in a solution are all the same. The fundamental law describing the behavior of an ideal solution is known as Raoult's Law and is expressed as:

$$p_A = X_A\, p_A^0 \tag{1.10}$$

where $p_A$ is the partial vapor pressure of component ($A$) above a solution which has the mole fraction ($X_A$), and $p_A^\circ$ is the vapor pressure of the pure liquid ($A$) at the same temperature. Raoult's law and variations of it can rarely be applied over an entire range of concentrations due to the lack of the ideal solution requirement. This expression, however, is extremely useful in providing indications of the magnitudes to be expected when collecting experimental data.

A typical application of Raoult's law is the development of a pressure-composition diagram of the type shown in Fig. 1.4. If the solution can be treated as ideal, the diagram will appear very similar to the one illustrated in Fig. 1.4, and the straight line will represent the dependence of total vapor pressure of the solution on the mole fraction in the liquid. The lower curved line will represent the dependence of the pressure on the composition of the vapor. The three regions shown in the background represent the three stages which must exist as a change of phase from liquid to vapor or vapor to liquid occurs.

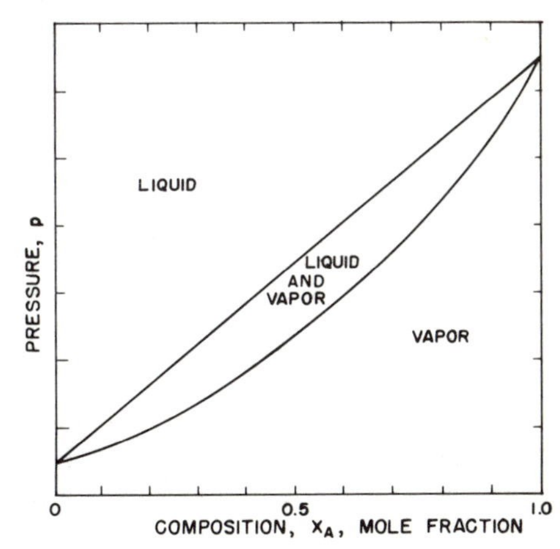


FIG. 1.4. PRESSURE-COMPOSITION DIAGRAM OF AN IDEAL SOLUTION

## 1.2 KINETICS OF REACTIONS OCCURRING IN PROCESSED FOODS

Kinetic theory is indeed fundamental to the explanation of many of the changes which occur during common processes. Kinetic theory can be used as a basis for deriving and predicting such factors as thermal conductivity of a gas, diffusion of a gas, and Brownian motion. In addition to the very fundamental factors, kinetics serves as a basis for describing the rates at which various processes occur during the processing and storage of foods.

Any reaction which is typical in nature will occur at a rate dependent upon several factors, whether the reaction is the conversion of sucrose to glucose and fructose or the rate at which some component of a food product is reduced in concentration by heat. The rate of the reaction is indicated by a rate constant ($k$), and can be described by the following equation:

$$-\frac{dC}{dt} = kC^m \tag{1.11}$$

where $C$ represents the component concentration at any time ($t$), and $m$ represents the order of the reaction.

The changes that occur in a system during a chemical reaction can be described in the following way. For the components of the system (called reactants) to take part in the reaction, their energy level must be increased by some finite amount before the reaction can proceed. As the reaction proceeds, the products of the reaction are formed. The energy change which must be supplied to the reactants before the reaction can occur is called the activation energy. The changes in the system and the concept of activation energy are illustrated in Fig. 1.5.

Although many reactions may be of zero order, the first-order reaction described by the following equation is common in food products:

$$-\frac{dC}{dt} = kC \tag{1.12}$$

In this particular type of reaction, the reaction rate is directly proportional to the concentration of the reacting substance ($C$). The application of a first-order reaction equation is more evident if equation (1.12) is solved and expressed in the following form:

$$\ln\frac{C}{C_0} = -kt \tag{1.13}$$

A well-known example of a first-order reaction is the process which describes bacterial death. In Fig. 1.6, a survivor curve for microorganisms exposed to 115 C is plotted versus time on semi-logarithmic coordinates.

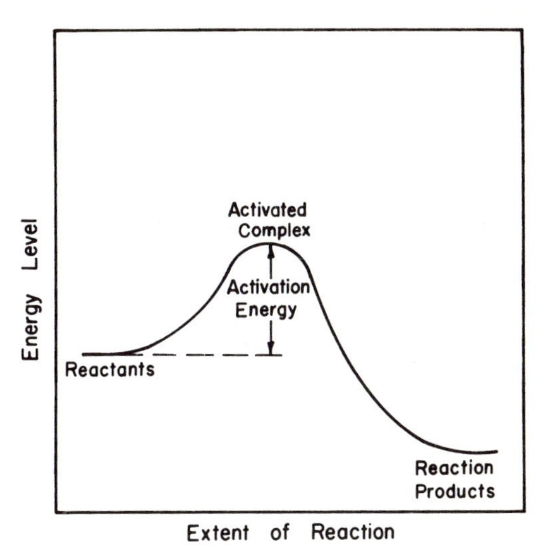


FIG. 1.5. SCHEMATIC ILLUSTRATION OF THE CONCEPT OF ACTIVATION ENERGY

The straight line obtained for experimental data indicates that the survivor curve can be described by a first-order reaction.

EXAMPLE 1.3 Evaluate the reaction rate constant ($k$) for describing death rate of bacterial spores at 115 C when given the following experimental data:

| Time (min) | Concentration |
|---|---|
| 0 | $10^6$ |
| 5 | $2.8 \times 10^5$ |
| 10 | $7.8 \times 10^4$ |
| 15 | $2.2 \times 10^4$ |
| 20 | $6.1 \times 10^3$ |
| 25 | $1.7 \times 10^3$ |

SOLUTION

Equation (1.13) can be rewritten in the following manner:

$$\log C = -\frac{kt}{2.303} + \log C_0$$

By plotting the data on semi-logarithmic coordinates, as illustrated in Fig. 1.6, the death rate constant ($k$) can be evaluated directly from the slope of the resulting curve. Since the slope is 0.111, $k$ = 0.2555/min.

From the form of the general expression of the rate equation and equation (1.12), it is obvious that an expression for a second-order rate equation would indicate that the rate of the reaction is equal to the product of the rate constant and the square of the concentration at any time:

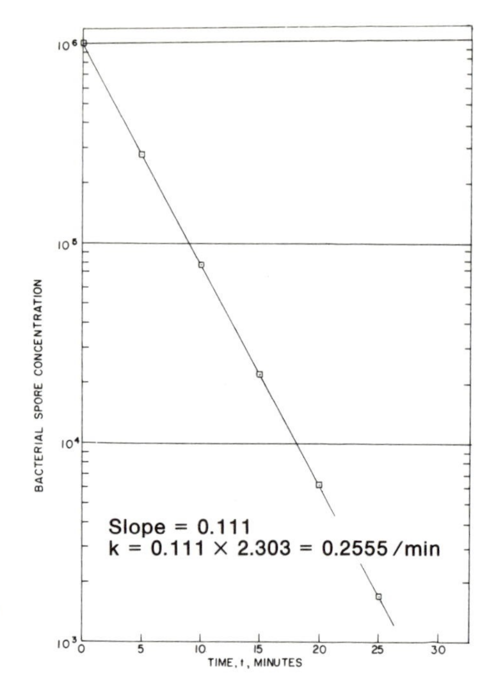


FIG. 1.6. A SURVIVOR CURVE FOR BACTERIAL SPORES AT 115C ILLUSTRATING FIRST ORDER REACTION

$$-\frac{dC}{dt} = kC^2 \tag{1.14}$$

Integration of equation (1.14) leads to the expression:

$$\frac{C - C_0}{CC_0} = -kt \tag{1.15}$$

indicating that plotting the ratio on the left-hand side of the equation versus time would allow evaluation of the rate constant ($k$).

In most cases, a second-order reaction occurs when two types of components in the system enter into the reaction. This type of situation can be described by letting the initial concentration of the two components equal (a) and (b) and by letting $x$ equal $C_0 - C$. For the case when the initial concentrations of the two components are equal ($a = b$), then equation (1.15) can be written as:

$$\frac{x}{a(a - x)} = kt \tag{1.16}$$

When the initial concentration ($a$ and $b$) are not equal, it is easier to obtain a solution by rewriting the original equation as follows:

$$\frac{dx}{dt} = k(a - x)(b - x) \tag{1.17}$$

Integration of equation (1.17) leads to the following expression describing the second-order reaction:

$$\ln \frac{b(a-x)}{a(b-x)} = (a-b)kt \tag{1.18}$$

Again it is obvious that evaluation of the rate constant ($k$) could be accomplished by plotting the ratio on the left-hand side of equation (1.18) on semi-logarithmic coordinates versus time ($t$).

EXAMPLE 1.4 Determine the rate constant which describes the loss of thiamine in pork heated at 133 C. The following data were obtained by Greenwood *et al.* (1944):

| Time (min) | Thiamine Retained (%) |
|---|---|
| 1.5 | 95 |
| 5 | 90 |
| 7 | 80 |
| 20 | 50 |

SOLUTION

(1) The experimental data are plotted as illustrated in Fig. 1.7.

(2) As indicated in Fig. 1.7, a linear relationship with slope = −0.0151 on semi-logarithmic coordinates is obtained.

(3) Based on the slope value, the rate constant will be:

$$k = 2.303 \times 0.0151 = 0.035/\text{min}.$$

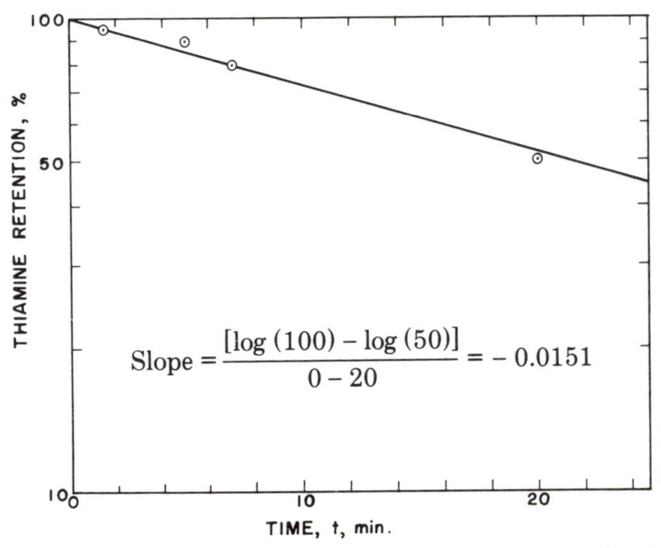


*From Greenwood et al. (1944)*

FIG. 1.7. CONCENTRATION–TIME CURVE FOR THIAMINE IN PORK HEATED AT 133 C

One of the more important factors in the study of reaction rates during food processing is the influence of temperature on the reaction. One of the widely used methods of describing this was presented by Arrhenius in 1899, when he proposed that the following expression should describe the influence of temperature on the reaction rate constant:

$$\frac{d(\ln k)}{dT} = \frac{E_a}{RT^2} \tag{1.19}$$

In equation (1.19), $T$ is absolute temperature and $E_a$ is the activation energy. Integration of equation (1.19) leads to the following equation:

$$\ln k = -\frac{E_a}{RT} + \ln B \tag{1.20}$$

where $B$ is a constant of integration and is usually referred to as a frequency factor when discussing the Arrhenius equation. Plotting the reaction rate constant ($k$) versus the inverse of absolute temperature on semi-logarithmic coordinates leads to the evaluation of the activation energy ($E_a$).

There are numerous examples in food science literature where the Arrhenius equation has been or could be applied to reaction rate data. One example is in the description of the influence of temperature on the death rate of microorganisms. Figure 1.8 illustrates this situation in which death rate constants obtained at various temperatures between 93 C and 126 C are plotted versus the inverse of the temperature to obtain a straight line, which is a relatively good description of the experimental data.

EXAMPLE 1.5 The influence of temperature on death rate of bacterial spores is illustrated by the following experimental data:

| Temperature (C) | Rate Constant (1/s) |
|---|---|
| 105 | 0.00061 |
| 107 | 0.00114 |
| 110 | 0.00222 |
| 113 | 0.00412 |
| 116 | 0.00758 |

Determine the activation energy involved in describing this reaction.

SOLUTION

From equation (1.20), it is obvious that by plotting the rate constant ($k$) versus the inverse of absolute temperature on semi-logarithmic coordinates, the slope of the resulting curve will represent $E_a/R$. As illustrated in Fig. 1.8, a linear relationship is obtained for the situation present and results in:

$$\text{Slope} = \frac{2.303\,(\log .007 - \log .0007)}{2.573 \times 10^{-3} - 2.647 \times 10^{-3}} = \frac{2.303}{-7.4 \times 10^{-5}}$$

$$= -3.11 \times 10^4 = \frac{E_a}{R}$$

and

$$E_a = -3.11 \times 10^4 \times 8.314 = -2.588 \times 10^5 \text{ kJ/kg mole}$$

From the intercept in Fig. 1.8, the frequency factor ($B$) can be computed as:

$$\ln B = \ln(0.015) + \frac{2.588 \times 10^5}{8.314}(2.55 \times 10^{-3}) = 75.18$$

$$B = 4.46 \times 10^{32} \text{ /s}$$

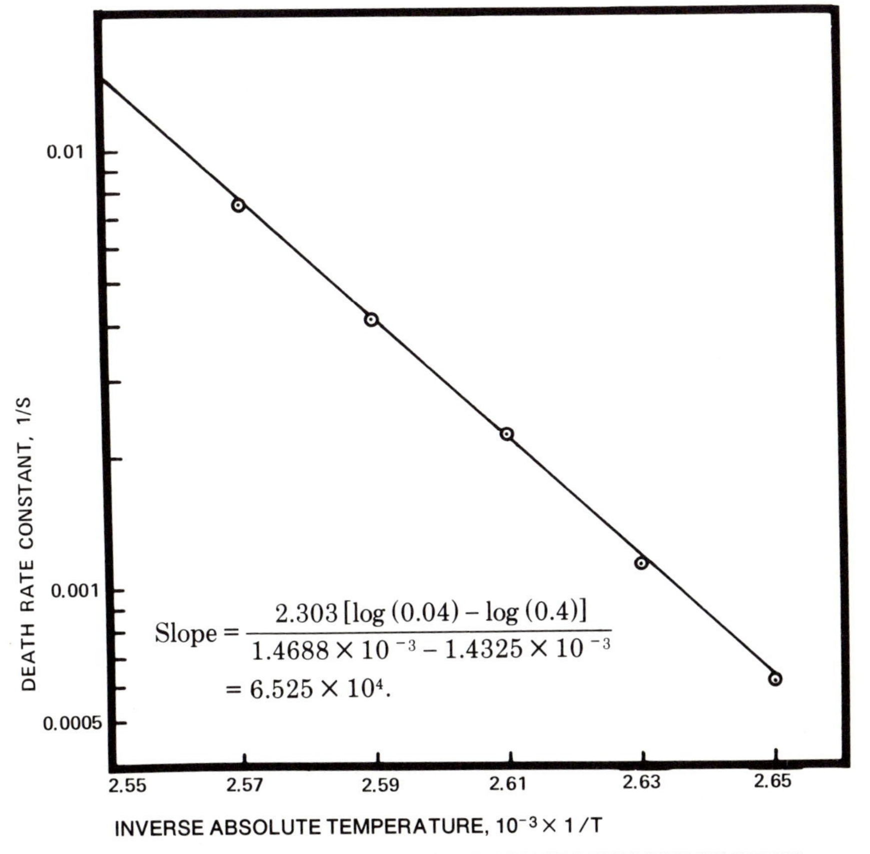


FIG. 1.8. ARRHENIUS PLOT ILLUSTRATING INFLUENCE OF TEMPERATURE ON DEATH RATE OF BACTERIAL SPORES

Another area where the Arrhenius plot could have considerable application is the investigation of the influence of temperature on bacterial growth rates and the corresponding influence on product degradation during storage. In this application, bacterial growth rate constants would be plotted versus the inverse of absolute temperature on semi-logarithmic coordinates to obtain the appropriate constants describing what should be straight-line curves related to the activation energy for these particular reactions. Examples of applications in this area are very limited at the present time.

## 1.3 FUNDAMENTALS OF MASS TRANSFER IN FOOD PROCESSING

Many operations in the food-processing plant involve mass transfer of one form or another. Many of these operations will be discussed in considerable detail in later chapters. Such operations as dehydration, distillation, liquid extraction and leaching are of considerable significance in the food-processing industry. The basic concepts of mass transfer will be introduced at this point so that the appropriate expressions can be utilized in later chapters.

Mass transfer may occur as a result of molecular diffusion or convection mass transfer. The basic equation describing molecular diffusion in one dimension is known as Fick's first law:

$$\frac{N_a}{A} = -D\frac{dC}{dz} \tag{1.21}$$

In equation (1.21), the ratio $(N_a/A)$ represents the mass flux of diffusing component, $C$ is the concentration of diffusing component, $z$ refers to direction of diffusion, and $D$ = mass diffusivity.

By conducting a mass balance on an element of interest, Fick's second law is obtained:

$$\frac{\partial C}{\partial t} = D\frac{\partial^2 C}{\partial^2 x} \tag{1.22}$$

where $D$ is assumed to be constant with respect to direction of diffusion.

Many applications of mass transfer involve the transfer of gaseous phases. By assuming that the perfect gas law will apply, equation (1.22) can be re-written as:

$$\frac{N_a}{A} = -\frac{D}{RT}\frac{dp}{dz} \tag{1.23}$$

where $p$ is the partial pressure of the diffusing gas in a mixture. Integration of equation (1.23) leads to the following expression of the mass flux of the diffusing gas:

$$\frac{N_a}{A} = -\frac{D}{RT}\frac{(p_2 - p_1)}{(z_2 - z_1)} \tag{1.24}$$

where $p_2$ represents the partial pressure in the plane ($z_2$), and $p_1$ represents the partial pressure in the plane ($z_1$). Equation (1.24) would then allow computation of the mass flux of a diffusing gas component in a mixture by knowing the partial pressure at two different locations, the mass diffusivity and the temperature.

EXAMPLE 1.6 A food product is being packaged in a 4.5-mil polyethylene film. Since the product is sensitive to oxidation, the rate of oxygen diffusion through the film is being calculated. Due to the reaction of the product with oxygen, the oxygen partial pressure within the package will be maintained at 2.53 kPa.

SOLUTION

(1) The diffusivity of oxygen through polyethylene at 25C is obtained from Table A.1 as $1.7 \times 10^{-5}$ m²/s.

(2) The appropriate gas constant ($R$) is obtained from Table A.13 as 8314.34 m³ Pa/kg mole K.

(3) Using equation (1.24) and an oxygen partial pressure of 21.28 kPa for the atmosphere around the package, the following computation can be made:

$$\frac{N_a}{A} = -\frac{(1.7 \times 10^{-5}\ \text{m}^2/\text{s})}{(8314.34\ \text{m}^3\ \text{Pa/kg mole K})(298\ \text{K})} \cdot \frac{(21.28\ \text{kPa} - 2.53\ \text{kPa}) \times 1000\ \text{Pa}}{(4.5\ \text{mil})(2.54 \times 10^{-5}\ \text{m/mil})\ 1\ \text{kPa}}$$

$$\frac{N_a}{A} = 1.126 \times 10^{-4}\ \text{kg mole/m}^2\ \text{s}$$

A common example of the diffusion of a gaseous component is the diffusion of water vapor from an air-water interface. The situation which exists when air flows over a water surface is illustrated in Fig. 1.9. The diffusion of water vapor occurs through the aerodynamic boundary layer which exists and the thickness for mass transfer ($\delta$). The resistance of this boundary layer thickness to molecular diffusion must be the same as that offered to total diffusion by the laminar sublayer, buffer layer and the turbulent zone. Since this effective thickness cannot be measured, equation (1.24) is usually rewritten as:

$$\frac{N_a}{A} = k_m (p_2 - p_1) \tag{1.25}$$

where $k_m$ is called the mass transfer coefficient and must be measured experimentally. Similar expressions describing diffusion in the liquid phase can be developed. These expressions are the same as those developed for gaseous components except for the replacement of the partial pressure by a concentration of the diffusing component.

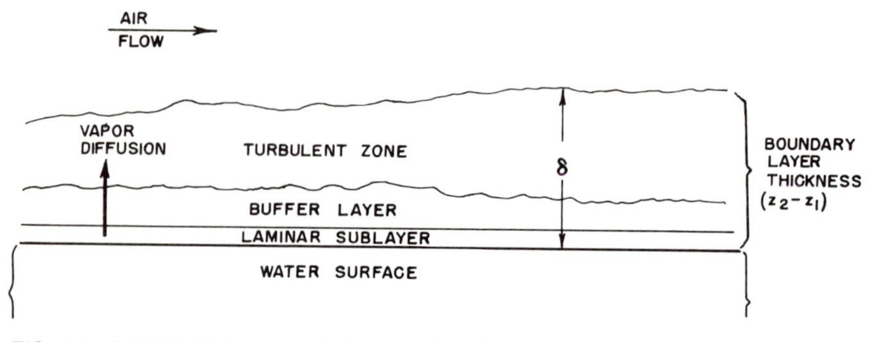


FIG. 1.9. SCHEMATIC ILLUSTRATION OF DIFFUSION OF VAPOR FROM A WATER SURFACE THROUGH AN AERODYNAMIC BOUNDARY LAYER

Since mass transfer coefficients must be evaluated experimentally, the attempts of most research are to develop empirical relationships which relate the mass transfer coefficient to the properties of the fluid and the flow characteristics. By dimensional analysis the following expression is obtained:

$$\frac{k_m L}{D} = \varphi\left(\frac{\rho u L}{\mu}\right) \Psi \left(\frac{\mu}{\rho D}\right) \tag{1.26}$$

where $L$ equals a characteristic dimemension of the system, $\rho$ = fluid density, $u$ = free-stream velocity of the fluid and $\mu$ = fluid viscosity. The three dimensionless groups present in equation (1.26) have specific names when referred to in mass transfer literature. The group on the left-hand side of equation (1.26) and variations of it are referred to as the Sherwood number. The first group on the right-hand side of equation (1.26) is the Reynolds number and the second group on the right-hand side of equation (1.26) is called the Schmidt number. Mass transfer literature contains numerous situations in which dimensionless expressions of the type given in equation (1.26) are utilized in developing empirical relationships between mass transfer coefficients and the properties of fluid and flow characteristics involved.

EXAMPLE 1.7 Estimate the rate at which moisture will be transferred from a free water surface to air flowing over the surface at a velocity of 1.5 m/s. The surface is 0.3 m wide and 0.5 m long in the direction of air flow. The average air temperature is 27 C and the mass diffusivity is 2.71 x $10^{-5}$ $m^2/s$.

SOLUTION

(1) Assuming an exact analogy between heat transfer and mass transfer, the following expression will be used:

$$N_{Sh} = 0.664\ N_{Re}^{0.5}\ N_{Sc}^{0.33}$$

(2) Computation of *Reynolds number:*

$$N_{Re} = \frac{\rho u L}{\mu} = \text{with} : \rho = 1.14 \text{ kg/m}^3 \text{ (from Table A.3)}$$

$$\mu = 1.85 \times 10^{-5} \text{ Pa s (from Table A.3)}$$

$$N_{Re} = \frac{(1.14)(1.5)(0.5)}{(1.85 \times 10^{-5})} = 46{,}216$$

(3) Computation of *Schmidt number*:

$$N_{Sc} = \frac{\mu}{\rho D} = \frac{1.85 \times 10^{-5}}{1.15 \times 2.71 \times 10^{-5}} = 0.598$$

(4) Then:

$$N_{Sh} = 0.664\ (46{,}216)^{0.5}\ (0.598)^{0.33} = 120$$

(5) Computation of *mass transfer coefficient* (for $R$ = 8314.34 $m^3$Pa/kg mole K)

$$k_m = 120 \frac{D}{LRT} = \frac{120 \times 2.71 \times 10^{-5}}{0.5 \times 8314.34 \times 300}$$

$$= 2.61 \times 10^{-9} \frac{\text{kg mole}}{\text{Pa m}^2\text{s}}$$

where the gas constant ($R$) and absolute temperature ($T$) are introduced to assure appropriate units for the mass transfer coefficient ($k_m$).

## PROBLEMS

1.1. *Enthalpy change during heating:* Blackberry syrup is at an initial temperature of 5C. Calculate the change in enthalpy if the syrup is heated to 35C.

1.2. *Shelf-life of fresh blueberries:* Fresh blueberries stored at 10C were evaluated for market quality. At 0, 4, 8, 12 and 16 days, the acceptable quality was measured to be 100, 60, 38, 25, and 17 percent. Assuming the quality degradation can be modeled by first-order kinetics, calculate the first-order rate constant.

1.3. *Kinetics of oxygen depletion in liquid food:* Oxidative reactions occurring during storage result in depletion of dissolved oxygen. The first-order rate constant for kinetics of oxygen depletion was measured to be 0.142 per hour. The initial dissolved oxygen concentration in the liquid food is 8.7 mg/l. Calculate the dissolved oxygen concentration after 24 hours.

1.4. *Activation energy and kinetics of vitamin degradation:* The activation energy for riboflavin degradation in whole milk stored in blow-molded polythylene (BMP) bottles under light was found to be 49,695 kJ/kg mole. The initial concentration of riboflavin in milk is 2 mg/l. If the first order rate constant at 2 C is $16 \times 10^{-4}$

per hour, calculate the riboflavin concentration in milk stored at 10°C for 72 hours in BMP bottles.

## COMPREHENSIVE PROBLEM I

Prediction of Food Product Quality During Storage

### Objectives:

1. To become acquainted with the application of rate reaction and mass transfer equations in food packaging.
2. To gain appreciation for the use of the digital computer in food stability problems.
3. To evaluate the factors which influence food product stability during storage.

### Background:

There has been increasing interest recently in the use of computers to predict the changes which occur in packaged foods during storage as indicated by research reported by Karel *et al.* (1971), Mizrahi *et al.* (1970), Quast and Karel (1972) and Simon *et al.* (1971). In all reported situations, there is a need for considerable amount of experimental information before the computer simulation becomes feasible.

The approach being proposed for use as a laboratory exercise will be a simplified version of previously published evidence. The approach will depend on experimentally determined relationships in order to provide realistic results.

The reaction occurring during storage of many food products is "browning." This chemical reaction is of particular significance in dehydrated foods and the rate of browning is highly dependent on moisture content. The extent of browning can be detected using colorimetry with the results expressed in Klett units. The storage stability of the product to be investigated depends on limiting the extent of browning and therefore on the integrity of the moisture barrier provided by the package.

There are three basic equations required in the simulation of storage stability. The first describes the relationship between reaction rate and the primary environmental parameter of concern. In the case of browning, Mizrahi *et al.* (1970) have shown that the following equation describes the influence of moisture content on browning of dehydrated cabbage:

$$K = k_1 \left[ 1 + \sin \left( -\frac{\pi}{2} + \frac{M\pi}{18} \right) \right]^m \quad (1)$$

where:

$K$ = rate of browning, Klett units/day
$k_1$ = constant = 10
$M$ = moisture content, g water/100 g solids
$m$ = constant = 1.025.
18 = moisture content for maximum browning rate.

The second equation describes moisture diffusion rate through the package film. Although this process would normally be described in the form of equations (1.22) or (1.23), the assumption that the product equilibrates instantaneous to the vapor content inside the package leads to the following:

$$\frac{dM}{dt} = k_m (a_w^o - a_w^i) \tag{2}$$

where: $a_w^o$ = water activity outside package
$a_w^i$ = water activity inside package
$k_m$ = moisture transfer coefficient for film.

Water activity ($a_w$) is equilibrium relative humidity divided by 100 and describes the equilibrium characteristics of air as well as the food product. The moisture transfer coefficient ($k_m$) can be defined as:

$$k_m = 100 \left[ \frac{k_w p A}{W d} \right] \tag{3}$$

where: $k_w$ = permeability of water vapor through film
$A$ = package surface area
$p$ = water vapor pressure
$W$ = product mass or weight
$d$ = package film thickness.

The third basic equation is the relationship between product moisture content (M) and water activity ($a_w$). Although there are several equations for this purpose, Mizrahi *et al.* (1970) used the following empirical equation for dehydrated cabbage:

$$a_w = \frac{k_2 + M}{k_3 + M} \tag{4}$$

where:

$k_2$ = constant = −1.2
$k_3$ = constant = 6

Equations (1), (2) and (4) can be used to simulate product storage stability and predict the extent of product browning with time.

## Procedures:

A. Develop a computer program which will predict product moisture content and extent of browning with time.
   1. Use the following values for the moisture coefficient in equation (2).
      a. Polyethylene film (2.5 mil); $k_m = 1.37$
      b. Scotchpak 48 film; $k_m = 0.245$.
      Note: These values of $k$ were determined experimentally by Mizrahi *et al.* (1970) for the same size package.

B. Determine the influence of water activity (or relative humidity) outside the package on extent of browning with time.
   1. Evaluate at least four levels; $a_w = 0.1, 0.3, 0.6, 0.95$.

C. Evaluate the influence of package film and storage relative humidity on product shelf-life.
   1. Assume 200 Klett units of browning makes the product unacceptable.
   2. Develop plots of product shelf-life versus packaging film and storage relative humidity.

D. Discuss the following:
   1. Significance of storage relative humidity.
   2. Significance of packaging film type; what are opportunities for improving product shelf-life?
   3. Improvements in the approach to prediction of food stability.

## NOMENCLATURE

$A$ = area through which diffusion or heat transfer occurs, $m^2$.
$a$ = concentration of one component in multicomponent reaction, kg component/kg.
$a_w$ = water activity
$B$ = constant in integrated form of Arrhenius equation.
$b$ = concentration of one component in multicomponent reaction, kg component/kg.
$C$ = concentration of primary component in first-order reaction, kg component/kg.
$c$ = specific heat, kJ/kgC.
$D$ = diffusion coefficient, $m^2/s$.
$d$ = package film thickness, m.
$E$ = internal energy in a system.
$E_a$ = activation energy, kJ/kg.
$G$ = free energy in a system.
$H$ = enthalpy.
$k$ = rate constant in kinetic reaction, 1/hr.
$k_m$ = mass transfer coefficient, kg-mole/s $m^2$ Pa.

$K$ = reaction rate, change in quality per unit time.
$L$ = characteristic dimension in a system, m.
$m$ = order of kinetic reaction.
$n$ = number of moles in system.
$p$ = total pressure of system (unless given subscript), Pa.
$q$ = heat or energy transfer rate, W.
$R$ = gas constant, kJ/kg K.
$S$ = entropy of system.
$T$ = absolute temperature, K.
$t$ = time, sec.
$u$ = velocity, m/s.
$V$ = volume of system, $m^3$.
$\overline{V}$ = molar volume.
$W$ = product mass or weight, kg.
$X$ = mole fraction.
$x$ = concentration change = $C_o - C$.
$\lambda$ = latent heat for phase change, kJ/kg-mole.
$z$ = direction normal to a surface.
$\eta$ = chemical potential.
$\rho$ = density, $kg/m^3$.
$\mu$ = viscosity, Pa s.

## Subscripts

$A$ = specific component in system.
$a$ = plane normal to direction of diffusion.
$i$ = specific component of system.
0 = time = zero.
$v$ = water vapor.
1,2 = locations in system.

## BIBLIOGRAPHY

BRENNAN, J.G., BUTTERS, J.R., COWELL, N.D., and LILLY, A.E.V. 1976. Food Engineering Operations. 2nd Ed. Elsevier Publishing Co., New York.

CHARM, S.E. 1978. The Fundamentals of Food Engineering, 3rd. Ed. AVI Publishing Co., Westport, Conn.

DIETRICH, W.C., BOGGS, M.M. NUTTING, M.D., and WEINSTEIN, N.E. 1960. Time-temperature tolerance of frozen foods. XXIII. Quality changes in frozen spinach. Food Technol. *14,* 552.

EARLE, R.L. 1966. Unit Operations in Food Processing. Pergamon Press, Elmsford, N.Y.

FARRALL, A.W. 1976. Food Engineering Systems. Vol. 1 - Operations. AVI Publishing Co., Westport, Conn.

FARRALL, A.W. 1979. Food Engineering Systems. Vol. 2 - Utilities. AVI Publishing Co., Westport, Conn.

GREENWOOD, D.A., KRAYBILL, H.R., FEASTER, J.R., and JACKSON, J.M. 1944. Vitamin retention in processing meat. Ind. Eng. Chem. *36,* 922.

HARPER, J.C. 1976. Elements of Food Engineering. AVI Publishing Co., Westport, Conn.

JOSLYN, M.A., and HEID, J.L. 1963. Food Processing Operations, Vol. 1, 2, and 3. AVI Publishing Co., Westport, Conn.

KAREL, M.S., MIZRAHI, and T.P. LABUZA. 1971. Computer prediction of food storage. Modern Packaging. *44* (8) 54–58.

LONCIN, M. and R.L. MERSON. 1979. Food Engineering; Principles and Selected Applications. Academic Press, Inc. New York.

MIZRAHI, S., T.P. LABUZA and M. KAREL. 1970. Computer aided prediction of extent of browning in dehydrated cabbage. J. Food. Sci. *35*: 799–803.

MOORE, W.J. 1972. Physical Chemistry, 4th Edition. Prentice-Hall, Englewood Cliffs, N.J.

OTHMER, D.F. 1940. Correlating vapor pressure and latent heat data. Ind. Engr. Chem. *32,* 841.

QUAST, D.G. and M. KAREL. 1972. Computer simulation of storage life of foods undergoing spoilage by two interacting mechanisms. J. Food Sci. *37*: 679–683.

SIMON, I.B., T.P. LABUZA and M. KAREL. 1971. Computer-aided predictions of food storage stability: Oxidative deterioration of a shrimp product. J. Food Sci. *36*:280.

SLADE, F.H. 1967. Food Processing Plant, Vol. 1. CRC Press, Cleveland.

SLADE, F.H. 1971. Food Processing Plant, Vol. 2. CRC Press, Cleveland.

TOLEDO, R.T. 1980. Fundamentals of Food Process Engineering. AVI Publishing Co., Westport, Conn.

TREYBAL, R.E. 1968. Mass Transfer Operations, 2nd Edition. McGraw-Hill Book Co., New York.

# 2

# Rheology of Processed Foods

The term rheology has been defined as follows: "a science devoted to the study of deformation and flow." Rheology encompasses the area of fluid flow, which is important to many segments of the processed food industry. The advantage of using the term rheology to describe this phenomenon is that many of the concepts associated with fluid flow can be utilized in the description of: (a) suspension flow, (b) the flow of granular products or powders, and (c) solid food products, which are of considerable importance when discussing food texture. It should not be assumed that the theories developed for other materials will apply directly and ideally to food products. Several factors tend to make the circumstances more complex. Factors such as temperature, humidity and chemical or microbiological reactions which occur in food products may influence the rheological properties of the products considerably. These complexities lead to the more frequent use of experimentally measured parameters and the development of concepts which are not utilized in other fields. The basic theories and concepts that have been developed for non-food materials must be analyzed, however, so that they can be applied in the development of the optimum approaches to be utilized in food-processing situations.

## 2.1. INTRODUCTION TO STRESS-STRAIN BEHAVIOR IN MATERIALS

Ideally, a material may behave in any of three ways: elastic, plastic, or viscous. It follows that the three fundamental rheological parameters are elasticity, plasticity, and viscosity. Since food products only rarely exhibit any of these behaviors, the concepts serve only as a basis for development of more complex theory.

The ideal elastic behavior exists when the stress ($\tau$) on a body is directly proportional to the strain ($\gamma$). The expression of this relationship is known as Hooke's Law, and is given as follows:

$$\tau = E\,\gamma \tag{2.1}$$

where $E$ is the modulus of elasticity or Young's modulus. Equation (2.1) will only apply when the body being considered is under tensile or compression stress. The same type of relationship will apply when the body is under shear stress or hydrostatic pressure; the coefficients involved become the shear modulus and bulk modulus, respectively. Although several investigations have been conducted in an attempt to apply this type of relationship to food products, an ideal elastic food product does not exist.

The ideal plastic behavior is described best by visualizing a block of a material on a flat surface. Application of force to this block does not result in movement until some yield stress is attained. After reaching this yield stress, flow or movement of the material continues indefinitely.

Unlike elastic and plastic behavior, many fluid foods can be described by ideal viscous behavior. Figure 2.1 illustrates the response of an ideal viscous fluid when a force is applied on a plane at a distance ($dy$) above the lower surface. If the upper surface is moving at a velocity ($u + du$) and the lower surface is moving at velocity ($u$), the response of an ideal viscous fluid between the two surfaces would be as follows:

$$\tau = -\mu \frac{du}{dy} \tag{2.2}$$

In equation (2.2), $\tau$ is the shear stress or force per unit area, $\mu$ is the coefficient of viscosity, and $du/dy$ is the velocity gradient which exists between the two surfaces and is equivalent to the rate of shear. The behavior of a fluid described by equation (2.2) is known as Newtonian behavior. Water and similar fluids are known as Newtonian fluids.

Many food products behave somewhat like a combination of elastic and viscous materials. These types of materials are characterized by having relationships between stress and strain which are more complex than the ideal elastic or ideal viscous behavior. In addition, the stress-strain relationship may depend on rate of strain, which introduces time-dependency and characterized viscoelastic behavior. If the ratio of stress to strain is a function of time alone, the material is said to have a linear viscoelastic response. If, however, the ratio of stress to strain is a function of stress as well as time, the behavior is said to be nonlinear viscoelastic. Most evidence indicates that many food products are characterized by the latter behavior. Due to the lack of knowledge related to nonlinear viscoelasticity, the theory of linear viscoelasticity

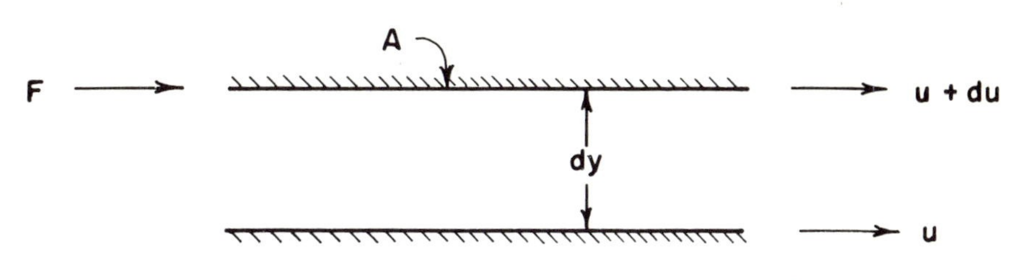


FIG. 2.1. SCHEMATIC ILLUSTRATION OF CONDITIONS DESCRIBING AN IDEAL VISCOUS BEHAVIOR

has been utilized most frequently in the development of parameters for food products. The application of these theories for solid food products will be discussed later in this chapter.

## 2.2. PROPERTIES OF FLUID FOODS

The design of transport systems for fluid foods is dependent on description of the flow characteristics of the product in the transport tube being utilized. Since these flow characteristics are dependent on the properties of the fluid, it is necessary to become thoroughly acquainted with methods utilized in measurement of these properties. Due to the variability among fluid foods and in the conditions which they exist, the methods to be presented will not be adequate for all situations. These methods and the parameters used to describe fluid properties are the best available, and should lead to acceptable results in most design situations.

### 2.2.1 Rheological Models

The strict definition of viscosity is the resistance to flow indicated by the coefficient of viscosity as given in equation (2.2). In the food industry, however, the term is utilized quite broadly as a single parameter to describe the consistency of the product under any given condition. This approach leads to considerable confusion because of the non-Newtonian behavior of many fluid food products.

The shear stress-rate of shear relationships for a variety of fluids considered to be non-Newtonian are illustrated in Fig. 2.2. The Bingham-plastic behavior is one in which a finite yield stress ($\tau_y$) is required before a viscous response is obtained. Such a response is described by the following equation:

$$\tau = m\left(-\frac{du}{dy}\right) + \tau_y \tag{2.3}$$

which represents a two-parameter model required to describe the material. The parameter ($m$) in this model is referred to as plastic viscosity.

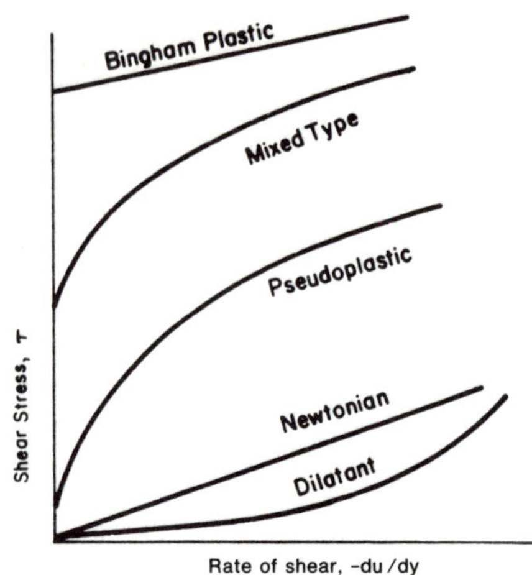


FIG. 2.2. COMPARISON OF THE SHEAR STRESS-RATE OF SHEAR RELATIONSHIPS FOR NON-NEWTONIAN FLUIDS AND NEWTONIAN FLUIDS.

Two-parameter models of the following type:

$$\tau = m \left(-\frac{du}{dy}\right)^n \tag{2.4}$$

where $m$ = consistency coefficient and $n$ = flow behavior index, can be used to describe pseudoplastic and dilatant fluids. As indicated in Fig. 2.2, a response from a dilatant fluid is a curve which is concave upward. Pseudoplastic fluids, which exhibit a concave downward curve on the shear stress-rate of shear axes, are probably the most common of the non-Newtonian fluids. This response implies a shear thinning as will occur with products which have a lower shear stress as shear rate increases.

A general power-law equation given as follows:

$$\tau = m \left(-\frac{du}{dy}\right)^n + \tau_y \tag{2.5}$$

could be utilized to describe fluids which would be called quasi-plastic or mixed types. In this case, the fluids would have either pseudoplastic or dilatant response after an initial yield point is reached. For these types of fluids, three parameters are required to describe the flow characteristics of the fluid.

EXAMPLE 2.1 Experimental results with a coaxial cylinder viscometer used for banana purée at 340 K were as follows:

| *Shear Rate* ($10^{-3} \times 1/s$) | *Shear Stress* ($10^{-4} \times Pa$) |
|---|---|
| 1 | 1.06 |
| 1.5 | 1.22 |

| *Shear Rate* ($10^{-3} \times 1/s$) | *Shear Stress* ($10^{-4} \times$ Pa) |
|---|---|
| 2 | 1.37 |
| 3 | 1.62 |
| 4 | 1.8 |
| 5 | 2.01 |
| 6 | 2.1 |
| 7 | 2.21 |

Determine the rheological parameters required to describe the product.

SOLUTION

(1) By plotting the results on regular coordinate graph paper, it is observed that the shear stress is zero at zero shear rate.

(2) By plotting the logarithm of shear stress versus the logarithm of shear rate (Fig. 2.3), the following form of equation (2.4) is described:

$$\log \tau = \log m + n \log (-du/dy)$$

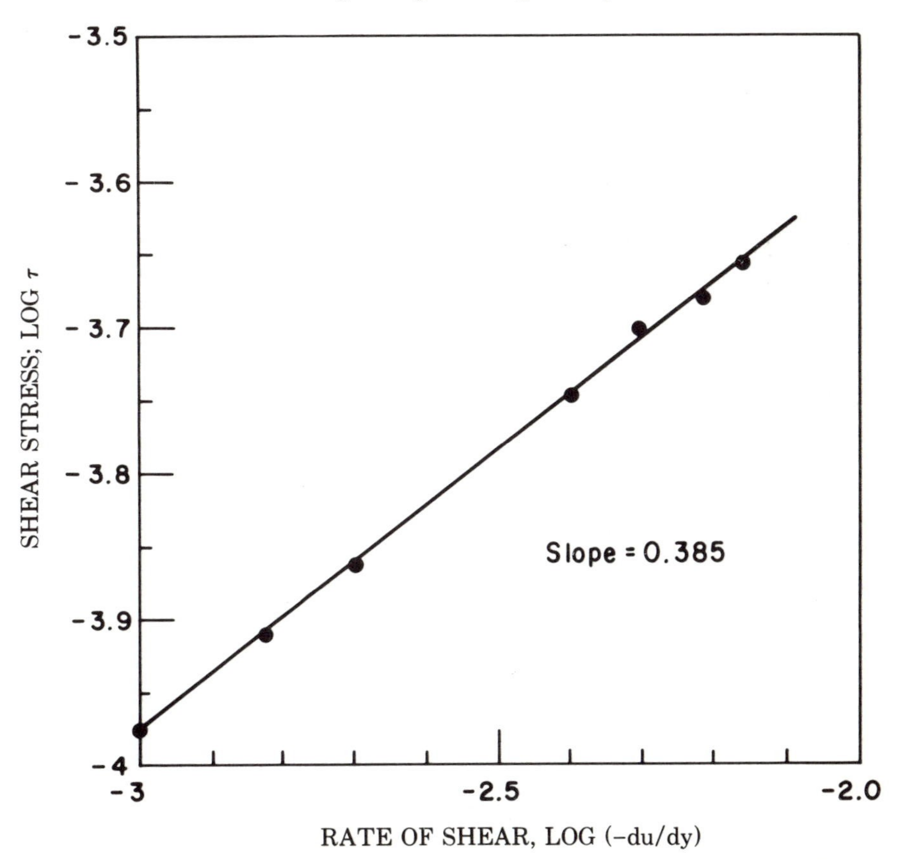


FIG. 2.3. EVALUATION OF RHEOLOGICAL PARAMETERS FOR BANANA PUREE

(3) From Fig. 2.3, the slope = 0.385 and by using an intercept value of −3.975, then

$$-3.975 = \log m + 0.385(-3)$$

$$\log m = -2.82$$

$$m = 1.514 \times 10^{-3} \text{ Pa s}^n$$

$$\text{and } n = 0.385$$

(4) These rheological properties ($m$ and $n$) describe the flow characteristics of banana purée and can be used for computation of pumping requirements.

In most cases in the food industry, an attempt is made to measure a Newtonian viscosity regardless of whether the characteristics of the fluid are known. This results in a measurement of an apparent viscosity which represents the viscosity of a Newtonian liquid exhibiting the same resistance to flow at the chosen rate of shear. In the most common case of a pseudoplastic fluid, the apparent viscosity decreases with increasing shear rate.

If the shear stress becomes a function of time, somewhat different terms are used to describe the behavior of the fluid. Although the shear stress will rarely increase with time, this type of material is called rheopectic. A behavior which may occur among food products is a decrease in shear stress with time; this type of material is called thixotropic.

Three other models have been proposed for utilization in describing non-Newtonian behavior of materials, namely, (a) the Eyring model, (b) the Ellis model, and (c) the Reiner-Philippoff model. Although these models have been applied successfully to other materials, their utilization for description of fluid food properties is not common.

One particular environmental parameter which influences viscosity or any rheological parameter considerably is temperature. When considering food products, there has been very little attempt to explain the influence of temperature on the rheological parameters on a theoretical basis. One of the more common proposals in the general area of rheology is the utilization of the Arrhenius equation as given in equation (1.19). In this particular application, a rheological parameter would be utilized as the rate constant in the determination of activation energy for viscous flow. By obtaining data at a rather limited number of temperatures (as low as 3) the influence of temperature over a range of temperatures can be predicted.

EXAMPLE 2.2 The consistency coefficients ($m$) for apricot purée vary with temperature in the following manner:

| *Temperature* (K) | *Consistency Coefficient* (Pa s$^n$) |
|---|---|
| 277 | 13.0 |
| 298 | 9.0 |
| 327 | 4.5 |
| 333 | 3.8 |

Develop a relationship between temperature and the rheological parameter using the Arrhenius-type equation.

SOLUTION

(1) The Arrhenius equation (1.19) can be written in the following form:

$$\ln m = \ln B_A - \frac{E_a}{R_g T_A}$$

(2) By plotting the data on semi-logarithmic coordinates as illustrated in Fig. 2.4, the slope of the curve is $E_a/R_g$. (See Fig. 2.4.)

(3) Using the plot in Fig. 2.4:

$$\text{slope} = \frac{\log(13) - \log(4)}{(3.6 \times 10^{-3}) - (3 \times 10^{-3})} = -853.14$$

$$\text{and} \frac{E_a}{R_g} = (2.303)(-853.14) = -1.965 \times 10^3 \text{ K}$$

(4) Using the above form of the Arrhenius equation and an intercept value:

$$\ln(13) = \ln B_A + (1.965 \times 10^3)(3.6 \times 10^{-3})$$

$$\ln B_A = -4.509$$

$$B_A = 0.011 \text{ Pa s}^n$$

(5) The influence of temperature on the consistency coefficient ($m$) for apricot purée becomes:

$$\ln m = 1.965 \times 10^3/T_A - 4.509$$

(6) The above expression can be used to compute consistency coefficients at any temperature within the range of the experimental data.

## 2.2.2 Measurement of Rheological Parameters

A significant number of instruments have been utilized to measure viscosity and, more generally, the rheological properties of fluid food products. These instruments fall into two classifications as follows: (a) capillary tube rheometers and (b) rotational rheometers including the coaxial-cylinder type and the cone and plate type.

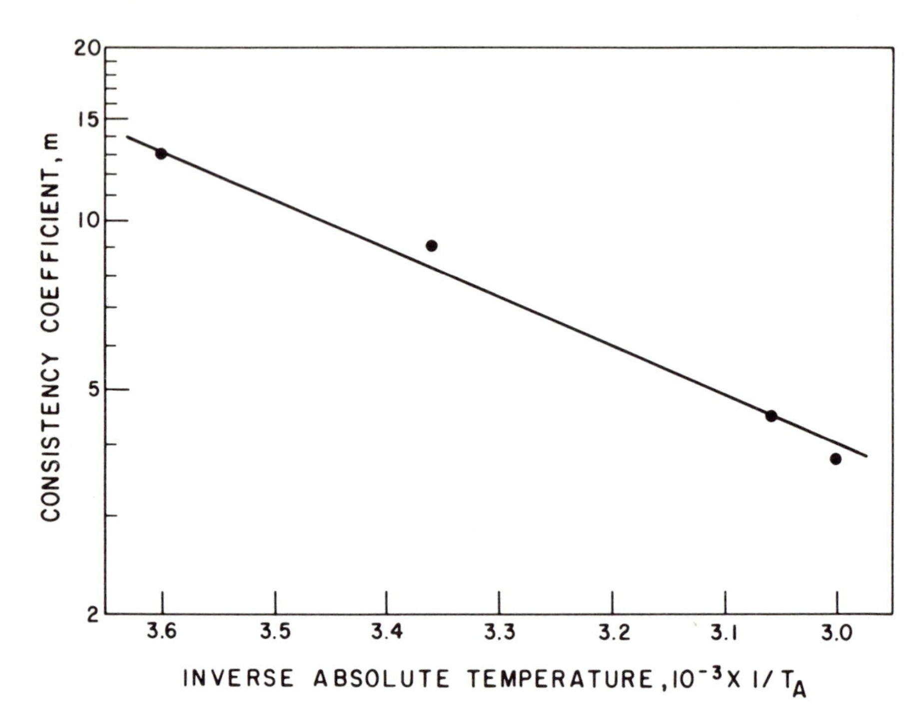


FIG. 2.4. USE OF AN ARRHENIUS PLOT TO DESCRIBE THE INFLUENCE OF TEMPERATURE ON CONSISTENCY

**2.2.2a Capillary tube rheometers.**—In general, capillary tube rheometers include a variety of instruments, all of which force the fluid through a tube of known geometry. For these types of rheometers the relationship between shear rate and shear stress is obtained from the measurement of the pressure gradient and volumetric flow rate of the fluid through the tube. The following assumptions are common to all instruments which measure rheological properties by this method: (a) flow is steady, (b) properties are independent of time, (c) flow is laminar, (d) fluid velocity has no radial or tangential components, (e) the fluid exhibits no slippage at the wall, (f) the fluid is incompressible, (g) the fluid viscosity is not influenced by pressure, and (h) the measurement is conducted under isothermal conditions.

The relationship between shear stress and pressure required to force the fluid through the capillary tube is obtained by conducting a force balance on the cross-section of the tube. Figure 2.5 illustrates the key parameters involved in conducting the force balance which leads to the following expression:

$$\tau = \frac{\Delta Pr}{2L} \tag{2.6}$$

In equation (2.6), $\Delta P$ is the applied pressure resulting in the movement of fluid through the tube, while $r$ is the variable distance from the center of the tube at which the shear stress ($\tau$) results.

If the capillary tube is used to measure the rheological properties of a fluid which can be described by the power law equation (2.4), the resulting expression is:

$$m\left(-\frac{du}{dr}\right)^n = \frac{\Delta Pr}{2L} \tag{2.7}$$

when the rate of shear for the capillary tube is $du/dr$. Integration of equation (2.7) is conducted in the following manner:

$$-\int_u^0 du = \left[\frac{\Delta P}{2Lm}\right]^{1/n} \int_r^R r^{1/n}\, dr \tag{2.8}$$

and results in the following expression for the velocity distribution in the capillary tube:

$$u(r) = \left[\frac{\Delta P}{2Lm}\right]^{1/n} \left[\frac{n}{n+1}\right]\left[R^{(n+1)/n} - r^{(n+1)/n}\right] \tag{2.9}$$

The volumetric flow rate is obtained by integration of the velocity distribution in the following manner:

$$Q = \int_0^R u(r)\, 2\pi r\, dr = \pi \left[\frac{\Delta P}{2mL}\right]^{1/n} \left[\frac{n}{3n+1}\right] R^{(3n+1)/n} \tag{2.10}$$

Utilizing data obtained from a capillary tube rheometer and equation (2.10), the rheological parameters ($m$ and $n$) can be evaluated. This is accomplished by plotting the logarithm of the volumetric flow rate ($Q$) versus the logarithm of ($\Delta P/2L$). The parameter ($n$) will be the slope of the straight line obtained from the log-log coordinates while the parameter ($m$) can be evaluated from the intercept.

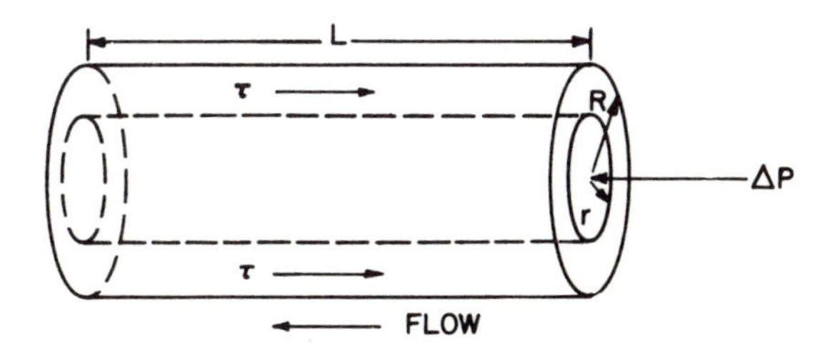


FIG. 2.5. FORCE BALANCE AND FLUID VELOCITY PROFILE LEADING TO MEASUREMENT OF RHEOLOGICAL PROPERTIES IN A CAPILLARY TUBE

EXAMPLE 2.3 A tube viscometer with 0.267 cm diameter and 0.91 m length was used to obtain the following data for applesauce (data from Saravacos (1968)).

| $\Delta P$ ($10^5 \times$ Pa) | $Q$ ($10^{-4} \times m^3/s$) |
|---|---|
| 1.30 | 0.91 |
| 1.45 | 2.50 |
| 2.56 | 2.10 |
| 1.99 | 3.20 |
| 2.13 | 5.20 |
| 2.41 | 8.50 |
| 2.70 | 12.49 |

SOLUTION

(1) Equation (2.10) can be rewritten as:

$$\log\left(\frac{\Delta P}{2L}\right) = \left[\log m - n \log \pi - n \log\left(\frac{n}{3n+1}\right)\right.$$

$$\left.(3n+1)\log R\right] + n \log Q$$

(2) By plotting log ($\Delta P/2L$) versus log $Q$, the value of the flow behavior index ($n$) can be determined from the slope and the consistency coefficient ($m$) can be computed from the equation.

(3) Based on the data provided:

| $\Delta P/2L$ ($\times 10^5$) | log ($\Delta P/2L$) | log $Q$ |
|---|---|---|
| 0.715 | 4.85 | −4.04 |
| 0.795 | 4.90 | −3.82 |
| 0.855 | 4.93 | −3.68 |
| 1.095 | 5.04 | −3.50 |
| 1.17 | 5.07 | −3.29 |
| 1.33 | 5.12 | −3.07 |
| 1.49 | 5.17 | −2.90 |

These data are plotted in Fig. 2.6. Since the slope of the curve is 0.28; $n$ = 0.28.

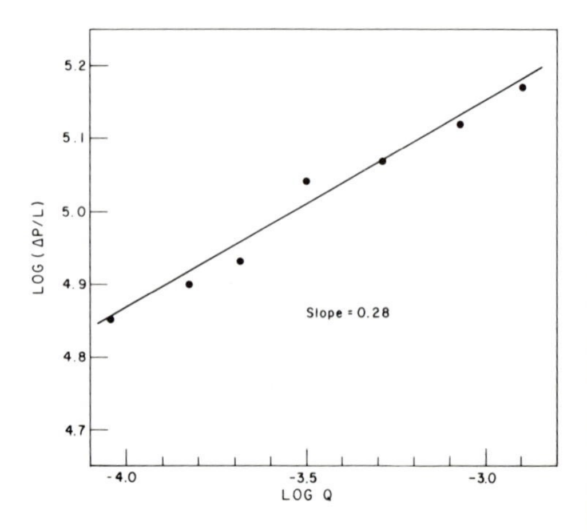


FIG. 2.6. EVALUATION OF RHEOLOGICAL PARAMETERS FROM TUBE VISCOMETER DATA

(4) The consistency coefficient ($m$) can be evaluated as follows (at $\log Q = -3.0$)

$$5.15 = \log m - (0.28)\log \pi - (0.28)\log\left[\frac{0.28}{3(0.28)+1}\right]$$

$$-[3(0.28)+1]\log\left(\frac{0.269}{2\times 100}\right) + (0.28)(-3.0)$$

$$\log m = 5.15 + 0.139 - 0.229 - 5.29 + 0.84 = 0.61$$

$$m = 4.074 \text{ Pa s}^n$$

Since equation (2.10) was obtained from the power-law equation, it will apply to pseudoplastic and dilatant fluids. In addition, it is a more general equation than would be used to describe Newtonian flow. If $n = 1$, equation (2.10) reduces to:

$$Q = \frac{\pi R^4 \Delta P}{8Lm} = \frac{\pi R^4 \Delta P}{8L\mu} \quad (2.11)$$

which would be utilized to evaluate viscosity if the capillary tube rheometer were being used for a Newtonian fluid. In this situation the consistency coefficient ($m$) becomes identical to viscosity ($\mu$).

Similar equations can be obtained for fluids with other characteristics. For example, the relationship between volumetric flow rate and pressure for a Bingham plastic fluid described by equation (2.3) would be:

$$Q = \frac{\pi R^4 \Delta P}{8mL}\left[1 - \frac{4}{3}(\tau_y/\tau_w) + \frac{1}{3}(\tau_y/\tau_w)^4\right] \quad (2.12)$$

The same type of expression would be obtained from the general power-law, equation (2.5), but it is very complex and will not be presented here. This equation or set of expressions is given by Charm (1978).

The following derivation is based on research of Rabinowitsch (1929) and Mooney (1931) and was presented in detail by Skelland (1967). The derivation requires the following assumptions: (a) the flow is steady and laminar, (b) fluid flow is independent of time and (c) no slip occurs at the wall. By considering the fluid flowing in an annulus between $r$ and $r + dr$, the following expression for volumetric flow rate results:

$$dQ = u 2\pi\, r dr \quad (2.13)$$

where $u$ is the fluid velocity at $r$. By recognizing that $2r\, dr = d(r^2)$ and with integration by parts, the following equation is obtained:

$$Q = \pi\left[ur^2 - \int r^2\, du\right]_0^{R^2} \quad (2.14)$$

By considering a general function between shear stress and rate of shear as follows:

$$du = -f(\tau)\, dr \tag{2.15}$$

and the following expressions:

$$dr = \frac{R}{\tau_w}\, d\tau \tag{2.16}$$

and:

$$\tau_w = -\frac{R}{r}\,\tau \tag{2.17}$$

appropriate substitutions can be made into equation (2.14). Since assumption (*c*) requires that the $ur^2$ in equation (2.14) will be zero, substitution of equations (2.15), (2.16) and (2.17) into equation (2.14) leads to the following:

$$Q = \pi \int_0^{\tau_w} \frac{\tau^2\, R^2}{\tau_w^2}\, f(\tau)\, \frac{R}{\tau_w}\, d\tau \tag{2.18}$$

If both sides of equation (2.18) are multiplied by $\tau_w^3$ and a differentiation with respect to $\tau_w$ is conducted, the following equation is obtained:

$$\frac{3Q}{\pi R^3} + \frac{R\Delta P}{2L}\,\frac{d(Q/\pi R^3)}{d(R\Delta P/2L)} = f(\tau_w) = \left(-\frac{du}{dr}\right)_w \tag{2.19}$$

where $\tau_w$ has been replaced by $R\Delta P/2L$. Equation (2.19) is the Rabinowitsch-Mooney equation, which expresses the rate of shear at the tube wall in terms of flow rate, pressure gradient and tube geometry. This equation is sufficiently general to allow design of the fluid transport system from experimental flow rate and pressure-drop data.

**2.2.2b Rotational rheometers.**—Rotational rheometers subject the sample to a nearly uniform rate of shear while measuring the shear stress. Physically, this is accomplished by measuring the torque required to turn the inside cylinder in Fig. 2.7 at a given number of revolutions per unit time. One of the obvious advantages of this particular type of rheometer is that it allows continuous measurement of the shear stress-rate of shear relationship and will allow analysis of a time-dependent behavior in the sample.

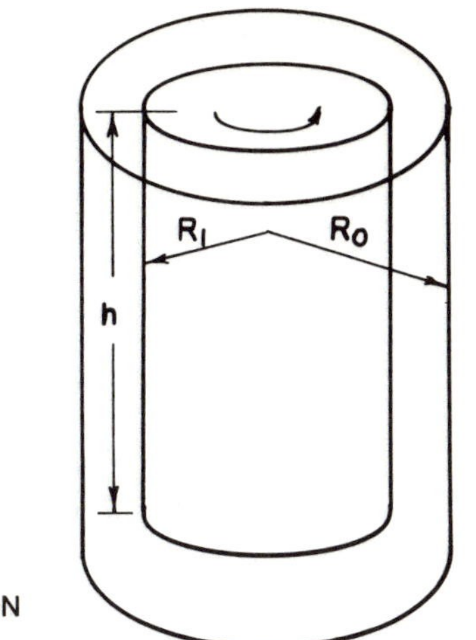


FIG. 2.7. SCHEMATIC ILLUSTRATION OF A ROTATIONAL RHEOMETER

One popular type of rotational rheometer is the coaxial-cylinder rheometer. The assumptions listed for the capillary tube rheometer also apply to this particular type. One of the main factors that must be taken into account when describing the response of a fluid in any type of rotational rheometer is the expression for the rate of shear. As indicated in Fig. 2.8 the linear velocity at some radius ($r$) from the center axis is $r\omega$, where $\omega$ is the angular velocity of the center cylinder. At the radius of $r + dr$ the linear velocity changes from $u$ to $u + du$ and the following expression results:

$$u + du = (r + dr)\,(\omega + d\omega) \tag{2.20}$$

By neglecting the second-order terms, the following expression for rate of shear is obtained:

$$\frac{du}{dr} = \omega + r\,\frac{d\omega}{dr} \tag{2.21}$$

Since $\omega$ is the angular velocity of the center cylinder, the second term of equation (2.21) is the only one which contributes to shear in the product and the following expression is obtained:

$$\frac{du}{dr} = r\,\frac{d\omega}{dr} \tag{2.22}$$

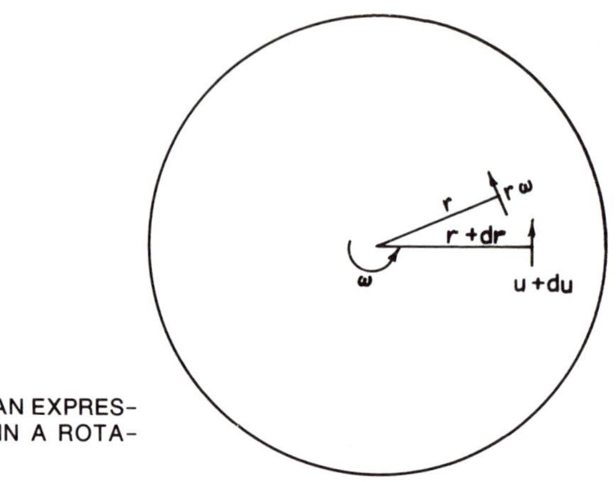


FIG. 2.8. DEVELOPMENT OF AN EXPRESSION FOR RATE OF SHEAR IN A ROTATIONAL RHEOMETER

By taking the derivative of $\omega$ in the following manner:

$$\frac{d\omega}{dr} = \frac{d}{dr}\left(\frac{u}{r}\right) = -\frac{u}{r^2} + \frac{1}{r}\frac{du}{dr} \tag{2.23}$$

giving the more common expression for rate of shear in a rotational rheometer results:

$$\gamma = \frac{du}{dr} - \frac{u}{r} \tag{2.24}$$

For a Newtonian fluid, the relationship between shear stress and rate of shear in this type of rheometer is:

$$\tau = \mu\left(\frac{u}{r} - \frac{du}{dr}\right) \tag{2.25}$$

The relationship between torque and angular velocity in a coaxial cylinder rheometer is obtained by first determining a relationship from the forces acting on the rotating cylinder. This force balance results in the following equation describing the relationship between torque and shear stress:

$$\Omega = 2\pi\, rh\tau r \tag{2.26}$$

Utilizing equation (2.26) and the following expression for a power-law fluid:

$$\tau = m\left(-r\frac{d\omega}{dr}\right)^n = m\left(\frac{u}{r} - \frac{du}{dr}\right)^n \tag{2.27}$$

an expression for a pseudoplastic or dilatant type fluid can be developed. By equating equations (2.26) and (2.27) and integrating between the radius of the inside cylinder and the radius of the outside cylinder, the following expression is obtained:

$$-\int_0^{\omega_i} d\omega = \left(\frac{\Omega}{2\pi\, hm}\right)^{1/n} \int_{R_0}^{R_i} \frac{dr}{r^{(n+2)/n}} \tag{2.28}$$

where $\omega_i$ is the angular velocity at which the inner cylinder is turning and the angular velocity is zero at the outer cylinder, which is stationary. Equation (2.28) can be integrated to give the following expression:

$$\omega_i = 2\pi N' = \frac{n}{2}\left(\frac{\Omega}{2\pi hm}\right)^{1/n}\left[\frac{1}{R_i^{2/n}} - \frac{1}{R_0^{2/n}}\right] \tag{2.29}$$

where $N'$ is the number of revolutions per unit time at which the inner cylinder is turning. Again, the rheological parameters ($m$ and $n$) can be evaluated from experimental data by plotting log $N'$ versus the log $\Omega$. The slope of the plot can be used to compute the flow behavior index ($n$) and the consistency coefficient ($m$) is determined from equation (2.29) after selection of an appropriate intercept value.

The expression for a Newtonian fluid can be obtained directly from equation (2.29) by letting $n = 1$. This results in Margules equation for Newtonian viscosity as follows:

$$\mu = \frac{\Omega}{4\pi h\omega_i}\left[\frac{1}{R_i^2} - \frac{1}{R_0^2}\right] \tag{2.30}$$

The second useful expression obtained directly from equation (2.29) is for the case of a rheometer with a single cylinder. For this situation, $R_0$ is equal to infinity and the following equation results:

$$\omega_i = 2\pi N' = \frac{n}{2}\left(\frac{\Omega}{2\pi hm}\right)^{1/n}\left[\frac{1}{R_i^{2/n}}\right] \tag{2.31}$$

When the fluid being measured has a Bingham plastic behavior, the following expression can be derived:

$$\omega_i = 2\pi N' = \left(\frac{1}{m}\right)\frac{\Omega}{4\pi\, h}\left[\frac{1}{R_i^2} - \frac{1}{R_0^2}\right] - \left(\frac{\tau_y}{m}\right)\ln\left(\frac{R_0}{R_i}\right) \tag{2.32}$$

For the case of the narrow gap coaxial-cylinder rheometer, equation (2.26) can be used directly to measure shear stress with $r$ being the radius of the cylinder. The rate of shear becomes the linear velocity of the rotating cylinder divided by the gap width. This conclusion is drawn directly from equation (2.24) when realizing that the *du/dr* term will be significantly greater than the *u/R*.

EXAMPLE 2.4 The following viscosity measurements were obtained for molasses at 274 K using a single cylinder rheometer with the cylinder having a 0.1143 m length and 0.159 m diameter.

| $N'$ *(RPM)* | *Viscosity* (Pa s) |
|---|---|
| 2.5 | 16.6 |
| 5.0 | 16.0 |
| 10.0 | 15.5 |
| 20.0 | 15.4 |
| 50.0 | 14.6 |
| 100.0 | 14.2 |

Determine the rheological parameters needed to describe the molasses.

SOLUTION

(1) Since the "viscosity" values decrease with increasing angular velocity or shear rate, the molasses must be non-Newtonian.

(2) An expression to describe the relationship between apparent viscosity and the rheological parameters ($m$ and $n$) can be obtained by comparing equations (2.31) and (2.30) with $R_0 \rightarrow \infty$. Then:

$$\mu_A = \left(\frac{1}{n}\right)^n (4\pi N')^{n-1} m$$

or:

$$\log \mu_A = n \log (1/n) + \log m + (n - 1) \log (4\pi N')$$

(3) The logarithms of the RPM and apparent viscosity values are as follows; where N′ values have been expressed as revolutions per second:

| $\log (4\pi N')$ | $\log (\mu_A)$ |
|---|---|
| −0.281 | 1.220 |
| 0.020 | 1.204 |
| 0.321 | 1.190 |
| 0.622 | 1.188 |
| 1.020 | 1.164 |
| 1.321 | 1.152 |

The values are plotted on Fig. 2.9.

(4) The slope of the curve in Fig. 2.9 is − 0.0417 and according to the above equation, the slope = $n-1$ = − 0.0417 and $n$ = 0.9583.

(5) Using the value of $\log (\mu_A)$ = 1.208 at $\log (4\pi N')$ = 0, then:

$$1.208 = (0.9583) \log (1/0.9583) + \log m$$

or:

$$\log m = 1.208 - 0.0177 = 1.1903$$

$$m = 15.5 \text{ Pa s}^n$$

(6) It is interesting to note that the computed consistency coefficient corresponds in magnitude to the apparent viscosity at 10 RPM.

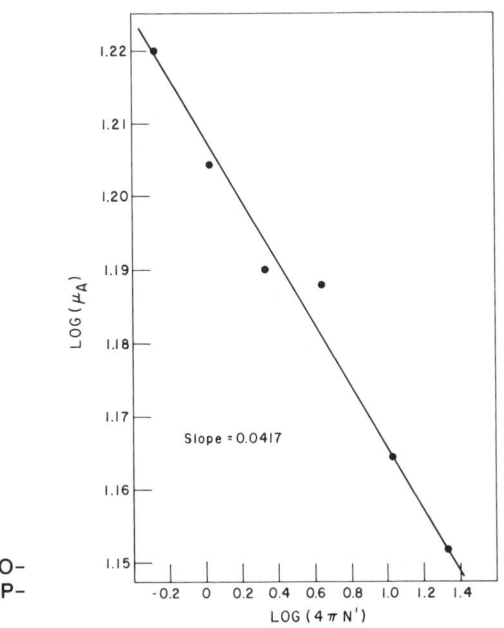


FIG. 2.9. EVALUATION OF RHEOLOGICAL PARAMETERS FROM APPARENT VISCOSITY VALUES

A second type of rotational rheometer which involves direct measurement of the shear stress-rate of shear relationship is the cone and plate rheometer. As illustrated in Fig. 2.10, this rheometer consists of a flat plate and a rotating cone forming a very small angle with the plate. Since the cone and plate actually touch at the cone point, the fluid fills the narrow gap formed by the two. The angle between the cone and the plate is usually between 2.5 and 3 degrees. The expression for shear stress in this particular case is:

$$\tau = \frac{3\Omega}{2\pi R^3} \tag{2.33}$$

while rate of shear is:

$$\gamma = \frac{\omega}{\tan \psi} \tag{2.34}$$

This type of rheometer could have considerable use in the measurement of non-Newtonian food products but its use up to the present time has been limited.

**2.2.2c Experimental Values.**—Investigations on the rheological properties of fluid foods indicate that the properties vary considerably depending on the type of product considered. Examples of fluid foods which can be considered Newtonian are corn syrup, fruit juices, and honey. Products which have been found to be pseudoplastic include fruit and vegetable purées; these have been investigated in considerable detail by Harper (1960), Saravacos (1968) and Watson (1968). The results, along with those obtained by Charm (1960), are compared in Table A.8. It is interesting to note that these results allow the comparison of three different instruments on the same products. The instruments include the coaxial-cylinder, single-cylinder, and capillary-tube rheometers. All three types of instruments have been used to measure the properties of apricot purée; in general the results are in agreement, although three different investigators were involved. All three investigations agree that this particular product is pseudoplastic and, as indicated in Table A.8., the flow behavior index ($n$) varies from 0.29 to 0.42, while the consistency

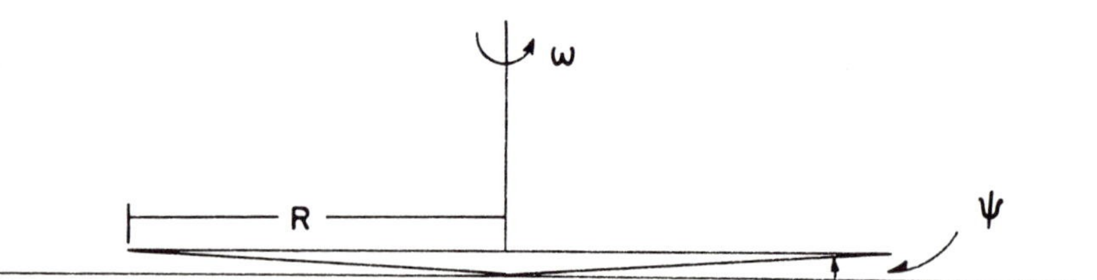


FIG. 2.10. SCHEMATIC ILLUSTRATION OF CONE AND PLATE RHEOMETER

coefficient ($m$) varies from 5.4 to 7.2 Pa s$^n$. In addition, these results were obtained from products with slightly different total solids content and at a slightly different temperature in one case. Table A.8. indicates that excellent agreement was obtained for pear puree using coaxial-cylinder and the capillary-tube rheometers. The agreement on peach purée is less acceptable. Saravacos (1968) reported that applesauce exhibited a thixotropic behavior when measurements were obtained using a capillary-tube rheometer. Results obtained by Watson (1968) did not indicate this behavior, even though he used instrumentation which would allow its evaluation.

Products which may be considered semi-solid have been investigated in considerable detail. Butter, for example, has been studied extensively, and the results have been used to determine the relationship of the rheological properties to product texture and pumping characteristics. Investigations which have utilized the power-law equations, however, are very limited due to the lack of experimental data in most cases. The more general equation (2.5) is required to describe these particular products. Data presented for butter by Hanck *et al.* (1966) and by Sterling and Wuhrmann (1960) for cocoa butter provide the information necessary to evaluate the rheological parameters of these products. In most cases, capillary-tube rheometers have been used and the data are presented in the form of pressure drop versus volumetric flow rate of the product. Scherr and Witnauer (1967) used a capillary-tube rheometer to measure the rheological parameters of lard, which they found could be described by the 2-parameter power-law equation and evaluated as:

$$n = 0.36$$

and

$$m = 169.6 \text{ Pa s}^n$$

Probably the main weakness in the data accumulated on rheological properties of fluid foods is the rather predominant use of apparent viscosity as a parameter. Since the products are usually pseudoplastic or quasi-plastic (mixed type), the value measured varies with the rate of shear and the values obtained can only be compared if the rates of shear were identical. The power-law equations allow computation of parameters which can be compared regardless of the type of instrument or the rates of shear used. The expressions do not account for the influence of temperature and at least one additional equation must be used to describe the influence of temperature on the rheological parameters.

### 2.2.3 Friction

One of the primary factors to be included in the application of rheological data to the computation of fluid flow problems in food-processing operations is friction. Friction influences the flow properties of a food product in several ways, including the flow of one layer of product over another, the flow of the product over a wall surface, and the flow of the product through some change in the transport system. In most applications, friction is defined or explained as a force which opposes the flow of a fluid in the system of concern. This leads to the following expression for a unit mass of fluid:

$$F = A\,(\mathrm{KE})\,f \tag{2.35}$$

in which the force is defined as the product of area ($A$), kinetic energy (KE) and a friction factor ($f$). In this case, the characteristic area is the area on which the friction is occurring. For example, the friction or force opposing flow occurring on a wall would have an area equal to the wall surface area. The friction factor ($f$) is a dimensionless quantity and has been used extensively in the development of charts which allow the computation of a friction factor as a function of Reynolds number. Probably the best known of these charts was developed by Moody (1944) for Newtonian fluids; it indicates the importance of surface roughness when the flow in the tube is turbulent (Fig. 2.11).

From equation (2.35), the friction factor can be defined in the following way:

$$f = \frac{F/A}{\rho \mathrm{KE}} = \frac{\tau_w}{\rho \mathrm{KE}} \tag{2.36}$$

where the fluid density is introduced to assure that the friction factor ($f$) is dimensionless. The friction factor expression for laminar flow of a Newtonian fluid in a tube can be derived in the following way. The expression for shear stress in a tube given by equation (2.6) leads to the following expression for shear stress at the wall in the same situation:

$$\tau_w = \frac{\Delta P R}{2L} \tag{2.37}$$

By inspection of equations (2.6) and (2.37), the obvious relationship between shear stress ($\tau$) and shear stress at the wall ($\tau_w$) is as follows:

$$\tau = \tau_w \frac{r}{R} \tag{2.38}$$

Utilizing equation (2.38) and equation (2.2), integration leads to the relationship between mean velocity in the tube and shear stress at the wall given by the following expression:

$$\tau_w = \frac{4\mu u}{R} \tag{2.39}$$

By combining equations (2.39) and equation (2.36), the well-known relationship between friction factor and Reynolds number is obtained.

$$f = \frac{16\mu}{\rho \bar{u} D} = \frac{16}{N_{Re}} \tag{2.40}$$

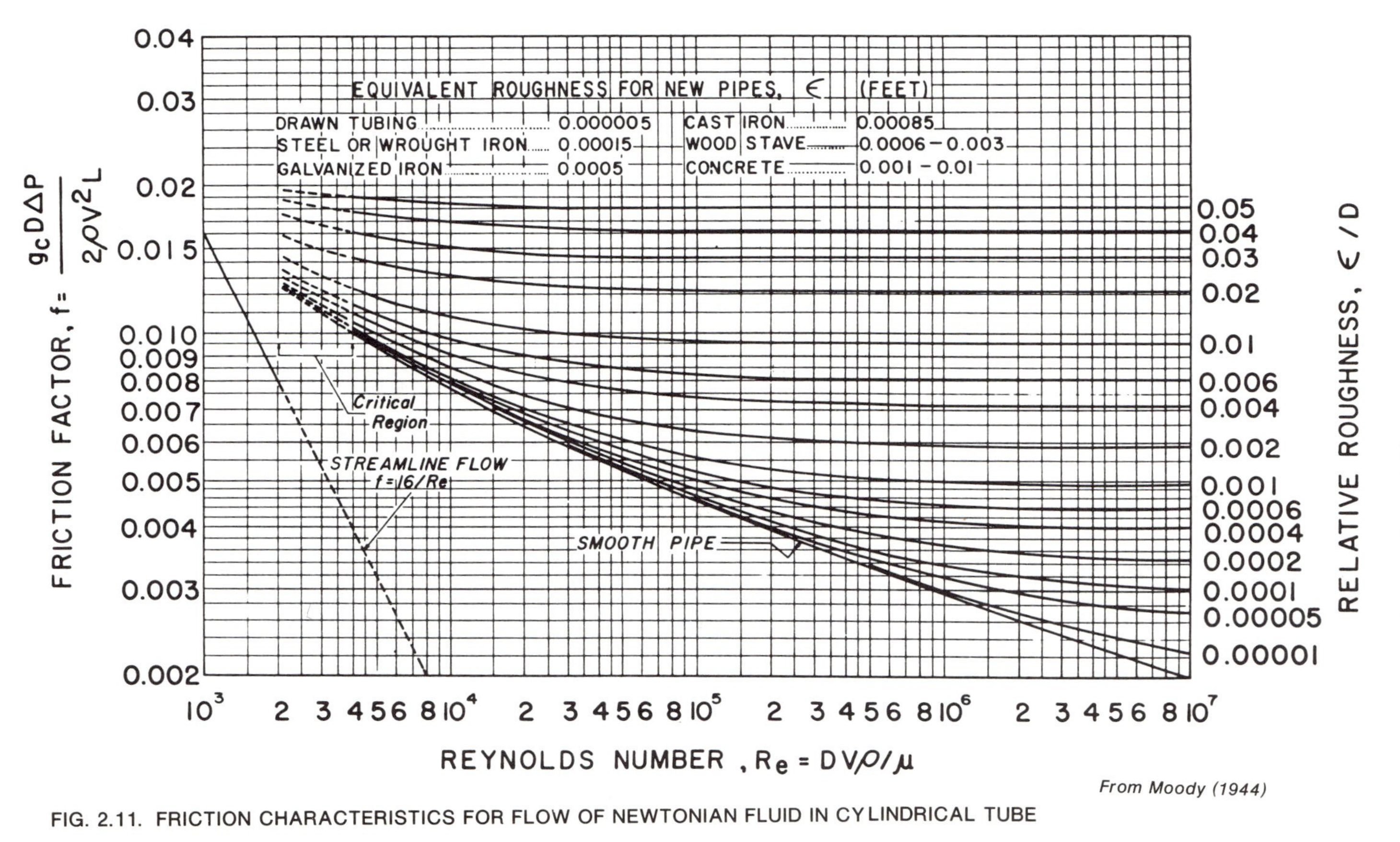


FIG. 2.11. FRICTION CHARACTERISTICS FOR FLOW OF NEWTONIAN FLUID IN CYLINDRICAL TUBE

As shown on the Moody chart (Fig. 2.11), this relationship applies for Reynolds numbers up to about 2100.

The relationship given in equation (2.40) has rather limited application when considering the types of flow problems existing in food-processing operations. Expressions of the type given as equation (2.40) will be developed for non-Newtonian fluids in the following sections. In addition, the application of these expressions to specific problems involved in pumping food products will be presented.

### 2.2.4 Laminar Flow

Laminar or streamline flow exists in a fluid system when the fluid particles are moving in a relatively straight or streamline fashion. It is defined more specifically for particular types of flow systems. For example, flow in tubes is usually laminar when the Reynolds number is less than 2100. This number will vary, however, depending on the flow system being considered.

In order to develop expressions for the friction factor of the type given by equation (2.40), which will be most useful in computing the information needed to describe flow of fluid food products, expressions must be developed for non-Newtonian fluids. The development of such expressions can be initiated with equation (2.18), which relates volumetric flow rate to shear stress for laminar flow in a cylindrical tube. Equation (2.18) can be expressed in terms of the mean fluid velocity in the tube in the following manner:

$$\bar{u} = \frac{R}{\tau_w^3} \int_0^{\tau_w} \tau^2 f(\tau)\, d\tau \tag{2.41}$$

The mean velocity can be computed from equation (2.41) for any type of fluid for which the shear stress-rate of shear relationship can be described. Utilizing equation (2.41) and equation (2.4) the mean velocity of a power-law fluid in a cylindrical tube can be determined from the following expression:

$$\bar{u} = \frac{R}{\tau_w^3} \int_0^{\tau_w} \tau^2 \left(\frac{\tau}{m}\right)^{1/n} d\tau \tag{2.42}$$

Integration leads to:

$$\bar{u} = \frac{R}{m^{1/n}} \left(\frac{n}{3n+1}\right)^{1/n} \tau_w \tag{2.43}$$

or by solving for the shear stress at the wall, the following expression is obtained:

$$\tau_w = m \left(\frac{\bar{u}}{R}\right)^n \left(\frac{3n+1}{n}\right)^n \tag{2.44}$$

Equation (2.44) can be utilized in equation (2.36) to obtain an expression for the friction factor for non-Newtonian fluid; more specifically, a power-law fluid:

$$f = \frac{2m}{\rho \bar{u}^2} \left(\frac{\bar{u}}{R}\right)^n \left(\frac{3n+1}{n}\right)^n = \frac{16m \left(\frac{3n+1}{n}\right)^n 2^{n-3}}{\rho \bar{u}^{2-n} D^n} \tag{2.45}$$

Equation (2.45) can be used directly to compute the friction factor for flow of power-law fluids in cylindrical tubes, as long as the Reynolds number to be defined is less than 2100. Based on equation (2.45), a different Reynolds number must be utilized in these situations. This Reynolds number, better known as the generalized Reynolds number, is defined as follows:

$$N_{GRe} = \frac{\rho \bar{u}^{2-n} D^n}{2^{n-3} m \left[\frac{3n+1}{n}\right]^n} \tag{2.46}$$

which must be less than 2100 for laminar flow conditions to exist. It is evident that the generalized Reynolds number will vary with the same parameters as the normal Reynolds number, but in addition the value will vary with the flow behavior index ($n$).

Another factor of concern when describing the flow of fluid foods is the kinetic energy of the flowing product. Utilizing the cylindrical tube, the kinetic energy of a flowing fluid is described by the following expression:

$$\mathrm{KE} = \frac{1}{\bar{u}\, \pi R^2 \rho} \int_0^R \frac{u^2}{2} \rho^2\, \pi r u dr \tag{2.47}$$

Equation (2.47) can be simplified to give:

$$\mathrm{KE} = \frac{1}{\bar{u} R^2} \int_0^R u^3 r dr \tag{2.48}$$

which can be utilized in the integration of the velocity distribution for a power-law fluid as given by equation (2.9). This integration leads to the following expression for kinetic energy for a power-law fluid:

$$KE = \frac{\bar{u}^2}{\alpha} \quad (2.49)$$

where the constant ($\alpha$) is equal to:

$$\alpha = \frac{(4n + 2)(5n + 3)}{3(3n + 1)^2} \quad (2.50)$$

Equation (2.49) is of considerable value when computing the losses which occur at the entrance regions of tubes, during expansion regions, and in fittings. The constant $\alpha$ equals 2 during turbulent flow, but varies according to equation (2.50) during laminar flow conditions.

EXAMPLE 2.5 Evaluate the generalized Reynolds number for apricot purée (19% T.S. at 25 C) flowing through a 2.54 cm diameter pipe at a mean velocity of 0.6 m/s.

SOLUTION

(1) Based on product properties given in Table A.8.,

$$m = 20 \text{ Pa s}^n \text{ and } n = 0.3.$$

(2) Since the density of the product would be expected to be slightly greater than water, a product density of 1040 kg/m³ is assumed.

(3) Using equation (2.46):

$$N_{GRe} = \frac{(1040 \text{ kg/m}^3)(0.6 \text{ m/s})^{1.7} \left(\frac{2.54}{100} m\right)^{0.3}}{(2^{-2.7})(20 \text{ Pa s}^n)\left(\frac{0.9 + 1}{0.3}\right)^{0.3}}$$

$$N_{GRe} = 27.1$$

(4) The computation indicates that the flow characteristics are laminar ($N_{GRe} << 2100$).

## 2.2.5 Turbulent Flow

When conditions become appropriate, the flow of a fluid in any type of system will become turbulent. Turbulent flow of a fluid exists when the fluid particles are no longer flowing in a stream-line fashion but are moving with a more random type of motion. The mean velocity in a cylindrical tube is much nearer the maximum velocity than under conditions of laminar flow. Probably the predominant parameter in determining whether flow is laminar or turbulent is the fluid velocity.

A close examination of fluid flow in a cylindrical tube reveals that many flow regions exist even when turbulent flow is fully developed. As illustrated in Fig. 2.12, turbulent flow is confined to a core in the central portion of the tube. This is the region of maximum velocity and represents the rather flat portion of the velocity profile. By moving closer to the tube wall, a transition or buffer zone is encountered. Flow in this particular region is highly unstable and at any given time it may be either turbulent or somewhat similar to laminar flow. The region directly adjacent to the tube wall is called the laminar sublayer. In this region, the flow is influenced by the shear characteristics of the fluid at the tube wall. Since the fluid velocity must be zero at the tube wall, a region directly adjacent to the wall will have velocities between zero and those velocities which exist in the turbulent core and buffer zone. These velocities must be low enough so that laminar flow can exist, and therefore the laminar sublayer is developed.

Developments which illustrate and provide working relationships and charts for non-Newtonian fluids in turbulent flow are very limited. The following development given by Skelland (1967) leads to a working relationship and chart for turbulent flow of a power-law fluid in a cylindrical smooth tube. This should improve considerably the limitations involved in using charts which have been developed for Newtonian fluids in the computation of friction factors for fluid food products.

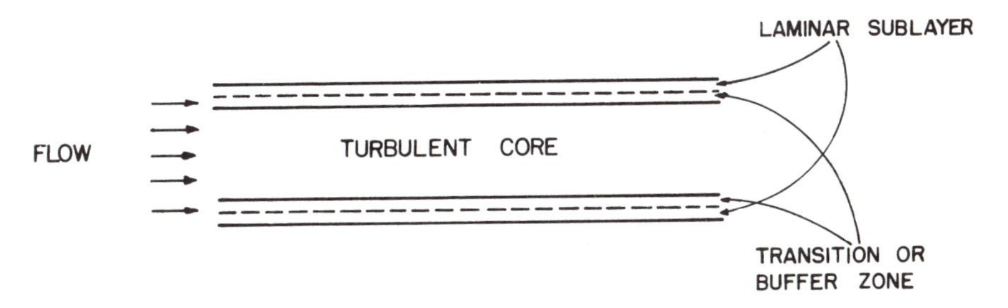


FIG. 2.12. TURBULENT FLOW REGIONS IN A CYLINDRICAL TUBE

The velocity distribution for turbulent flow of a power-law fluid in a smooth cylindrical tube may be a function of several variables, as indicated by the following equation:

$$u = f(N_{Re}, \rho, \tau_w, m, n, r) \qquad (2.51)$$

Utilizing dimensional analysis, Skelland (1967) has shown that the following four dimensionless groups can be derived from equation (2.51).

$$\frac{u}{u*} = f(Z, \xi, n) \tag{2.52}$$

where

$$Z = \frac{R^n \ \rho(u*)^{2-n}}{m} \tag{2.53}$$

$$u* = \sqrt{\frac{\tau_w}{\rho}} \tag{2.54}$$

$$\xi = \frac{r}{R} \tag{2.55}$$

Equation (2.52) is a functional expression of the velocity distribution in the tube, while equations (2.53) and (2.54) and (2.55) define the parameters in equation (2.52). Utilizing the assumption that the difference between the maximum velocity and the velocity in any other part of the velocity distribution is independent of the consistency coefficient ($m$), Skelland (1967) was able to develop an additional dimensionless group defined as follows:

$$P_n = \frac{u_m - \bar{u}}{u*} \tag{2.56}$$

The assumptions made in deriving equation (2.56) should be acceptable as long as the laminar sublayer is relatively thin, which is the case for Newtonian and pseudoplastic fluids. The assumptions become less acceptable for dilatant materials when the flow behavior index ($n$) becomes considerably larger than 1. From the definition of the friction factor given by equation (2.36) the following expression can be written:

$$\bar{u}\sqrt{\frac{\rho}{\tau_w}} = \frac{\bar{u}}{u*} = \sqrt{\frac{2}{f}} \tag{2.57}$$

which, when combined with equation (2.13) and equation (2.52) at the tube centerline results in the following:

$$\sqrt{\frac{2}{f}} = f(Z, n) - P_n \tag{2.58}$$

In the next step of the development, Skelland (1967) utilizes an integration procedure to develop the following expression:

$$\frac{u_m}{u^*} = f(Z, n) = A_n \ln Z + B_n \tag{2.59}$$

where $A_n$ and $B_n$ are constants of integration. Equations (2.58) and (2.59) can be combined to give the following expression for the friction factor:

$$\sqrt{\frac{2}{f}} = A_n \ln Z + B_n - P_n \tag{2.60}$$

Dodge and Metzner (1959) inserted the value of $Z$ from equation (2.53) into equation (2.60) to derive the following expression

$$\sqrt{\frac{1}{f}} = 1.628\, A_n \log [N_{GRe}(f)^{1-n/2}] + C_n \tag{2.61}$$

where:

$$C_n = -0.4901\, A_n (1 + n/2) + \frac{B_n - P_n}{\sqrt{2}} \tag{2.62}$$

Experimental evaluation of the constants $A_n$ and $B_n$ by Dodge and Metzner (1959) leads to the following expression:

$$\sqrt{\frac{1}{f}} = \frac{4}{n^{0.75}} \log [N_{GRe}(f)^{1-n/2}] - \frac{0.4}{n^{1.2}} \tag{2.63}$$

Utilizing equation (2.63) as a correlation between the friction factor, the generalized Reynolds number and the flow behavior index ($n$), Dodge and Metzner (1959) developed the chart shown in Fig. 2.13. As is evident the friction factor ($f$) can be determined from this chart if knowledge of the generalized Reynolds number and the flow behavior index ($n$) is available. It must be emphasized, however, that this chart can be used only for smooth pipes or tubes and the roughness of the tube surface cannot be taken into account.

EXAMPLE 2.6 Determine the friction for pumping applesauce through a 5 cm diameter pipe with a mean velocity of 3 m/s at 24 C.

SOLUTION

(1) Based on data from Table A.8., the property values for applesauce at 24 C will be:

$m$ = 0.66 Pa $s^n$ and $n$ = 0.408.

(2) Since the density of applesauce is greater than water, a value of 1100 $kg/m^3$ will be used.

(3) Using equation (2.46), the generalized Reynolds number will be computed:

$$N_{GRe} = \frac{(1100\ kg/m^3)(3\ m/s)^{1.592}(0.05\ m)^{0.408}}{(2^{-2.592})(0.66\ Pa\ s^n)\left[\frac{1.224 + 1}{0.408}\right]^{0.408}}$$

and:

$$N_{GRe} = 8519$$

(4) Using Fig. 2.13, the friction factor can be determined from $N_{GRe} = 8519$ and $n = 0.408$

$$f = 0.0045$$

The information describing the influence of tube surface roughness on the flow characteristics of non-Newtonian fluids is insufficient to be utilized. For conditions such as these, Metzner (1961) recommends the utilization of the existing relationships for Newtonian fluids in rough tubes as given in the Moody chart (Fig. 2.11). It would appear that this procedure would be most acceptable when the influence of surface roughness on the flow characteristics is more significant than the influence of the fluid being non-Newtonian. If, however, the influence of the non-Newtonian behavior is most significant, Fig. 2.13 should be utilized so that the influence of the flow behavior index ($n$) can be taken into account.

### 2.2.6 Considerations in Pumping Fluid Foods

Probably one of the more widely used expressions in the computation of pumping requirements for fluids is the mass flow rate equation given as follows:

$$w = \rho A \bar{u} \tag{2.64}$$

Equation (2.64) establishes a direct relationship between the fluid velocity ($\bar{u}$) and the flow rate ($w$) of the fluid mass. In addition, it establishes the influence of cross-sectional area of the transport tube ($A$) on the mean fluid velocity.

The expression which represents the entire basis for fluid-flow computations and is used extensively in determining the pumping requirements for a fluid system is the mechanical energy balance equation given as follows:

$$Z_1' + \frac{P_1}{\rho} + (\mathrm{KE})_1 + W = Z_2' + \frac{P_2}{\rho} + (\mathrm{KE})_2 + E_f \tag{2.65}$$

Equation (2.65) accounts for all factors which influence the flow of the fluid in a given system. In addition, the equation provides the necessary combination of parameters to allow computation of the work which must be accomplished in order to provide the desired flow. Evaluation of all parameters except the work ($W$) leads to the computation of the pump size required to achieve the desired fluid transport.

One of the common factors which must be accounted for in computing requirements for pumping fluid foods is the influence of the entrance

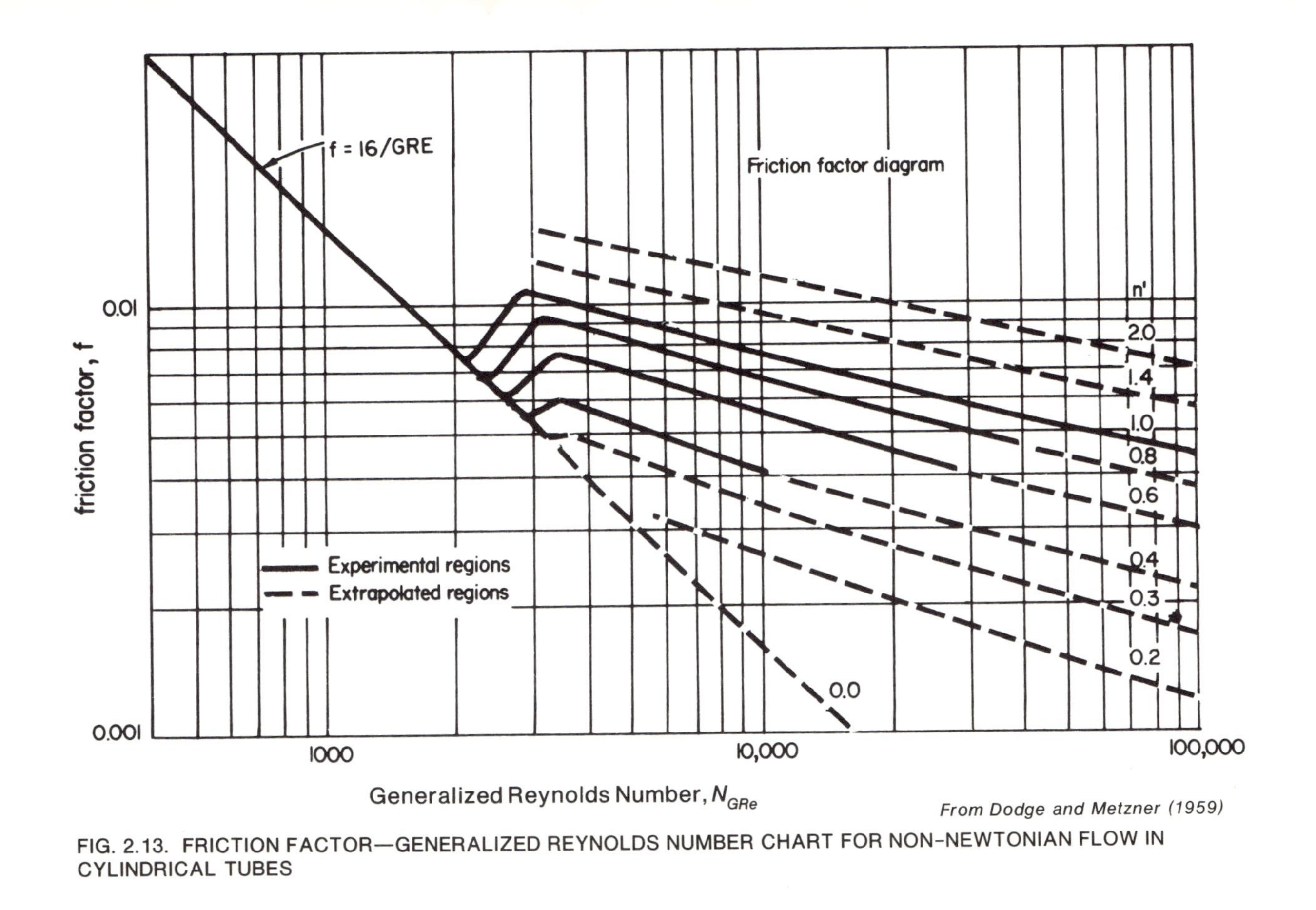


FIG. 2.13. FRICTION FACTOR—GENERALIZED REYNOLDS NUMBER CHART FOR NON-NEWTONIAN FLOW IN CYLINDRICAL TUBES

region to the cylindrical tube. Skelland (1967) provides an expression which is most useful for computing the influence of this particular factor.

$$\frac{P_1 - P_2}{\rho \bar{u}^2/2} = \frac{32(L/R)}{N_{GRe}} + C_a \tag{2.66}$$

In the application of equation (2.66), the pressure ($P_1$) is at some point upstream from the entrance to the cylindrical tube, as illustrated in Fig. 2.14. The pressure ($P_2$) is at some point downstream from the entrance to the cylindrical tube and at a sufficient distance for the boundary layer of fluid to be fully developed, as illustrated in Fig. 2.14. The distance ($L$) is the distance from the entrance of the cylindrical tube to location 2. The constant ($C_a$) has been evaluated experimentally by Weltmann and Keller (1957) and is normally assigned a value of 2.2 for pseudoplastic fluids. The kinetic energy terms in equation (2.65) are evaluated by utilizing equation (2.49), which does account for the non-Newtonian behavior of the fluid.

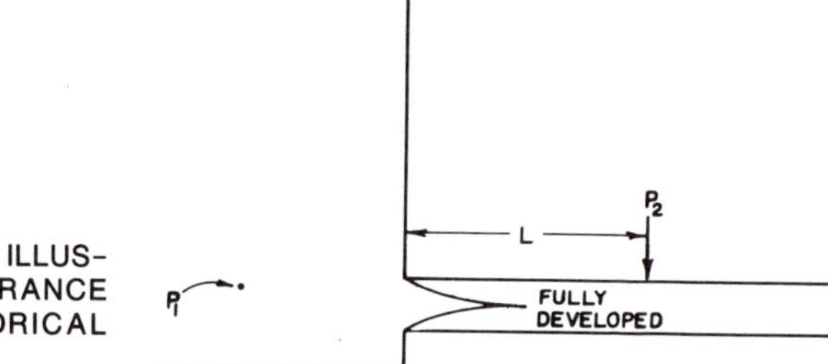


FIG. 2.14. SCHEMATIC ILLUSTRATION OF THE ENTRANCE REGION TO A CYLINDRICAL TUBE

The friction loss term in equation (2.65) is a summation of several contributing factors. The first and one of the more important factors is the loss due to friction in the cylindrical tube itself. The standard procedure for determining this loss is to first compute the generalized Reynolds number in the case of a non-Newtonian fluid and then consult a chart of the type given in Fig. 2.13 to determine the friction factor. After finding the friction factor, the friction loss can be computed from the following well-known expression:

$$\frac{\Delta P}{\rho} = f \frac{\bar{u}^2 L}{R} \tag{2.67}$$

Other common types of contributions to friction losses in a fluid transport system are contractions or expansions in the cylindrical tubes used for the transport. These friction losses can be computed from rather standard expressions such as the following one for the contraction in the tube:

$$\frac{\Delta P_c}{\rho} = K_f \frac{\bar{u}^2}{\alpha} \tag{2.68}$$

where:

$$K_f = 0.4\,(1.25 - D_2^2/D_1^2) \quad \text{at } D_2^2/D_1^2 < 0.715 \tag{2.69}$$

$$K_f = 0.75\,(1 - D_2^2/D_1^2) \quad \text{at } D_2^2/D_1^2 > 0.715 \tag{2.70}$$

In equation (2.68), the parameter $\alpha$ accounts for the non-Newtonian behavior of the fluid. In the case of an expansion or increase in the cylindrical tube diameter, the energy loss due to friction can be computed from the following equation:

$$\frac{\Delta P_c}{\rho} = \frac{u_1^2}{\alpha}\left[1 - \left(\frac{A_1}{A_2}\right)\right]^2 \tag{2.71}$$

where parameters with subscript (1) are at locations upstream from the expansion in tube diameter.

Accounting for the friction loss in other types of fittings such as valves and elbows or tees may be accomplished in one of two ways. The first procedure is to utilize experimental data obtained on various types of fittings expressed as equivalent lengths of the tube being utilized. Such information is given in Table 2.2. The second approach is to utilize an expression of the following form:

$$\frac{\Delta P_f}{\rho} = \frac{K_f\,\bar{u}^2}{\alpha} \tag{2.72}$$

in which the constant ($K_f$) is determined experimentally and would be available in a table such as Table 2.2. It must be emphasized that in both approaches, that of using equivalent length and that of using constant factors in equation (2.72), the available information is for Newtonian fluids. Most available information indicates that the error involved is small. In the case of utilizing equivalent lengths for standard fittings, the length values are added to the length of tubing in the computation conducted using equation (2.67). When using equation (2.72), the friction losses for each standard fitting are summed separately to obtain the total value for friction losses.

After computing all parameters in equation (2.65) except the work ($W$), the computation of the latter parameter can be accomplished and leads directly to the sizing of the pump required to perform the desired fluid transport.

TABLE 2.2.

FRICTION LOSSES FOR STANDARD FITTINGS

| Fitting | Friction Constant, $K_f^1$ | Equivalent Length, Pipe Diameters[2] |
|---|---|---|
| Elbow, 90° standard | 0.74 | 32 |
| Elbow, 90° medium sweep | 0.5 | 26 |
| Elbow, 90° long sweep | 0.25 | 20 |
| Elbow, 90° square | 1.5 | 60 |
| Tee, used as elbow, entering run | 1.5 | 60 |
| Gate valve, open | 0.13 | 7 |
| Globe valve, open | 6.0 | 300 |
| Angle valve, open | 3.0 | 170 |

[1]Earle (1966).
[2]Perry and Chilton (1973).

EXAMPLE 2.7 The system illustrated in Fig. 2.15 is being used to pump applesauce at 25°C through a 5.08 cm smooth pipe from storage tank A to storage tank B at a rate of 50 kg/min. Compute the size of pump required to accomplish the product transport.

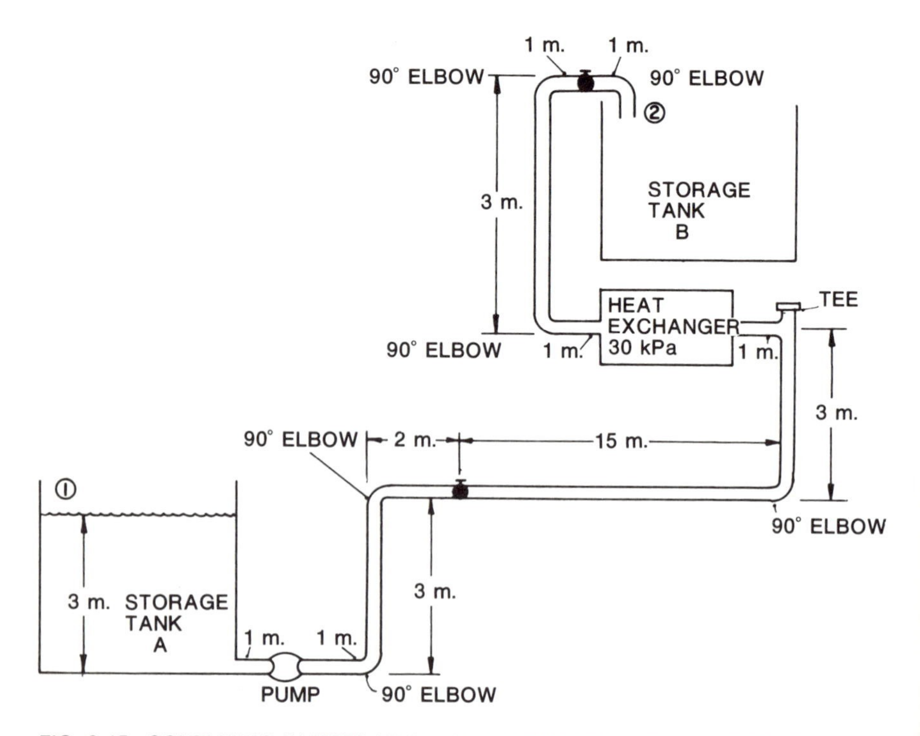


FIG. 2.15. SCHEMATIC ILLUSTRATION OF SYSTEM USED IN EXAMPLE 2.7

SOLUTION

(1) Based on data in Table A.8., the properties of applesauce include $m = 22$ Pa s$^n$ and $n = 0.4$.

(2) The solution requires evaluation of all portions of equation (2.65). By assigning reference points (1) and (2) as illustrated in Fig. 2.15, equation (2.65) can be written in the following form:

$$3 + 0 + 0 + W = 9 + 0 + (KE)_2 + E_f$$

or:

$$W = 6 + (KE)_2 + E_f$$

(3) In order to evaluate $(KE)_2$ and $E_f$, the mean velocity of product flow must be computed:

$$\bar{u} = \frac{w}{\rho A} = \frac{50\ \text{kg/min}}{(1100\ \text{kg/m}^3)(\pi)(\frac{5.08}{200}m)^2} = 22.4\ \text{m/min}$$

$$\bar{u} = 0.37\ \text{m/s}$$

(4) The kinetic energy $(KE)_2$ can be evaluated using equation (2.49):

$$(KE)_2 = \frac{(0.37\ \text{m/s})^2}{\alpha}$$

where:

$$\alpha = \frac{[4(0.4) + 2][5(0.4) + 3]}{3[3(0.4) + 1]^2} = 1.24$$

so:

$$(KE)_2 = \frac{(0.37)^2}{1.24} = 0.11\ \text{J/kg}$$

(5) The energy loss due to friction will be caused by several factors including pipes, fittings and equipment components. The generalized Reynolds number is a key parameter involved in determining friction:

$$N_{GRe} = \frac{(1100\ \text{kg/m}^3)(\frac{5.08}{100}\ \text{m})^{0.4}(0.37\ \text{m/s})^{1.6}}{(2)^{-2.66}(22\ \text{Pa s}^n)\left[\frac{3(0.4) + 1}{0.4}\right]^{0.4}}$$

$$N_{GRe} = 9.9$$

(6) Since the generalized Reynolds number is much less than 2100, the friction can be computed from equation (2.40):

$$f = \frac{16}{N_{GRe}} = \frac{16}{9.9} = 1.62$$

(7) The energy loss due to friction in the pipes can be computed using equation (2.67):

$$E_f = \frac{\Delta P}{\rho} = 1.62\left[\frac{(0.37)^2(32)}{5.08/200}\right] = 279.4\ \text{J/kg}$$

(8) The energy loss due to friction at the entrance to the pipe from Tank A can be evaluated from equations (2.69) and (2.70) (with

$D_2^2/D_1^2 = 5.08^2/\infty = 0 < 0.715$):

$$K_f = 0.4\,(1.25 - 0) = 0.5$$

and:

$$E_f = \frac{\Delta P_c}{\rho} = 0.5\left[\frac{0.37^2}{1.24}\right] = 0.055\ \mathrm{J/kg}$$

(9) The energy loss due to friction in 6 long-sweep elbows ($K_f$ = 0.25) (see Table 2.2):

$$E_f = \frac{\Delta P}{\rho} = 0.25\left[\frac{0.37^2}{1.24}\right](6) = 0.166\ \mathrm{J/kg}$$

(10) The energy loss due to friction in the heat exchanger is obtained by:

$$E_f = \frac{\Delta P}{\rho} = \frac{30\ \mathrm{kPa}}{1100\ \mathrm{kg/m^3}} = 27.3\ \mathrm{J/kg}$$

(11) The energy loss for friction in two angle values ($K_f$ = 3):

$$E_f = \frac{\Delta P}{\rho} = 3\left[\frac{0.37^2}{1.24}\right](2) = 0.66\ \mathrm{J/kg}$$

(12) The energy loss from friction due to the tee ($K_f$ = 1.5) will be:

$$E_f = \frac{\Delta P}{\rho} = 1.5\left[\frac{0.37^2}{1.24}\right] = 0.166\ \mathrm{J/kg}$$

(13) The total energy loss due to friction in the system becomes:

$$E_f = 279.4 + 0.055 + 0.166 + 27.3 + 0.66 + 0.166$$

$$E_f = 307.75\ \mathrm{J/kg}$$

(14) By incorporating all information into the mechanical energy balance:

$$W = 6 + 0.11 + 307.75 = 313.86\ \mathrm{J/kg}$$

(15) The power requirement can be determined by:

$$\text{Power} = 313.86\ \mathrm{J/kg} \times 50\ \mathrm{kg/min} = 15693\ \mathrm{J/min}$$

$$= 261.55\ \mathrm{W}$$

$$\textit{Power} = 0.26\ \mathrm{kW}\ (\text{at } 100\%\ \text{efficiency})$$

(16) Note that this problem could be solved by computation of fluid head for each contribution to work accomplished by the pump and by expressing energy loss due to friction in fittings in terms of equivalent lengths of pipe.

## 2.3. PROPERTIES OF SUSPENSIONS AND CONCENTRATED PRODUCTS

Many food products are in the form of suspensions or concentrated products at some point during processing in preparation for consumer use. In some cases the raw food product which enters the food-processing plant will be in one of these forms and must be handled and transported to the first processing operation. In some situations, food components are mixed together to form suspensions, while in other cases high-moisture liquid products are reduced in moisture content to give more highly concentrated products. Finally, there are some products which are delivered to the consumer as a suspension or in a concentrated state. Since these products are a special situation of the non-Newtonian case, similar consideration must be given to investigating and studying the factors involved in fluid transport.

Suspensions and concentrated products include a variety of different particle-liquid and particle-air mixtures. In the following sections we will be considering only those which seem to have the most application in the food industry. A product slurry is a particle-liquid mixture which is normally considered dilute in that the ratio of particles to liquid is very low. This mixture can be transported in laminar flow with insignificant settling. The particle-liquid mixture which is considered a suspension is one in which the ratio of particle to liquid is much higher, and forms a product that cannot be easily transported in laminar flow without deposition of the particles. In other words, the transport must be conducted under turbulent conditions in order to maintain the particles in suspension. The third type of product is the particle-air mixture, which will be discussed primarily in its relationship to pneumatic transport.

### 2.3.1 Viscosity

One of the basic properties of concern in flow problems, as illustrated in the previous section, is viscosity or some similar rheological property. As might be expected, the measurement of such properties for any kind of suspension has been limited by the complexity involved in describing the physical system. Several expressions have been proposed to be used for prediction of the product properties under these conditions. In many cases, the expressions still lack experimental verification and support.

For slurries, the proposed expressions attempt to relate the slurry viscosity to the viscosity of the liquid fraction and the concentration of the particles per unit volume. Probably the most widely used expression is one proposed by Einstein (1906, 1911):

$$\mu = \mu_L \, (1 + 2.5X_c) \tag{2.73}$$

in which the slurry viscosity is related almost directly to the particle concentration. Equation (2.73) was derived utilizing several assumptions as follows: (a) particles are spherical in shape, (b) particles are large in comparison to the liquid molecules, (c) particles are uncharged, (d) no slip occurs between particles and liquid, and (e) turbulence is not present. Eirich *et al.* (1936) and Ford (1960) were able to confirm equation (2.73) for concentrations of 5 and 10% when the particles were spherical in shape. Guth and Simha (1936) and Manley and Mason (1952) developed similar expressions for somewhat higher particle concentrations. The Guth and Simha equation is:

$$\mu = \mu_L (1 + 2.5X_c + 14.1X_c^2) \tag{2.74}$$

while the Manley and Mason equation is:

$$\mu = \mu_L (1 + 2.5X_c + 10.05X_c^2) \tag{2.75}$$

As is evident, these expressions [equations (2.74) and (2.75)] are identical except for the coefficient on the higher-order term. Most investigators of these expressions feel that the use of additional terms in the expressions will allow the prediction of viscosities for even higher concentrations. The Guth and Simha equation is considered valid up to concentrations of 20% by volume.

EXAMPLE 2.8 Compare equations (2.73), (2.74) and (2.75) when predicting the viscosity of a 15% by volume slurry and a 30% by volume suspension. The viscosity of the liquid suspending medium is $1.3 \times 10^{-3}$ Pa s.

SOLUTION

(1) For 15% suspension:
equation (2.73) gives:

$$\mu = 1.3 \times 10^{-3} [1 + 2.5(0.15)] = 1.788 \times 10^{-3} \text{ Pa s.}$$

equation (2.74) gives:

$$\mu = 1.3 \times 10^{-3} [1 + 2.5(0.15) + 14.1(0.15)^2] = 2.2 \times 10^{-3} \text{ Pa s.}$$

equation (2.75) gives:

$$\mu = 1.3 \times 10^{-3} [1 + 2.5(0.15) + 10.05(0.15)^2]$$

$$= 2.081 \times 10^{-3} \text{ Pa s.}$$

(2) For 30% suspension:
equation (2.73) gives:

$$\mu = 1.3 \times 10^{-3} [1 + 2.5(0.3)] = 2.275 \times 10^{-3} \text{ Pa s.}$$

equation (2.74) gives:

$$\mu = 1.3 \times 10^{-3} [1 + 2.5(0.3) + 14.1(0.3)^2] = 3.925 \times 10^{-3} \text{ Pa s.}$$

equation (2.75) gives:

$$\mu = 1.3 \times 10^{-3} [1 + 2.5(0.3) + 10.05(0.3)^2] = 3.451 \times 10^{-3} \text{ Pa s.}$$

(3) The computations indicate that differences in viscosity predicted by the equation increase as the suspension concentration increases. There is no assurance about the accuracy of values from any of the equations when concentrations are above 20% by volume. Equations (2.74) and (2.75) should be more acceptable at concentrations between 10 and 20%.

No attempts have been made to relate the viscosities or rheological properties of suspensions to their particle concentration. As indicated previously, these concentrations would be considerably higher than in slurries and differences would be related to the extent that turbulent conditions would be needed during flow to keep the particles in suspension.

Most particle-in-air suspensions have sufficiently low concentrations of particles to total volume that the air viscosity is not influenced significantly by the presence of the particles. At least no attempts have been made to investigate this area to date. It would appear that an expression of the form proposed by Einstein [equation (2.73)] could be used for this purpose.

Measurements of rheological properties of slurries and suspensions might be accomplished by evaluation of at least two parameters, as described by equation (2.4). This expression and its application in flow problems have been discussed in detail in the previous section. If equation (2.4) cannot be used, the Rabinowitsch-Mooney equation (2.19) should be applied.

## 2.3.2 Flow in Pipes

Although most types of suspensions and concentrated products would be considered non-Newtonian in that the rheological properties can be described by the power-law expression, some variations are introduced because of the manner in which the particle solids are contained in the fluid.

**2.3.2a Suspension Transport.**—The flow properties of slurry-like materials have been investigated by Charm and Kurland (1962). In this investigation the relationship between pressure drop and flow rate for a suspension during laminar flow was developed as follows:

$$\left[\frac{\Delta P}{\bar{u}^2 \rho}\left(\frac{D^n \bar{u}^{2-n} \rho}{m}\right)^{0.634n+0.364}\right] + \log \frac{D}{L} = 1.207 \qquad (2.76)$$

As is evident, the properties of the material were evaluated utilizing the two-parameter power-law equation (2.4). Probably one of the more significant factors found by the investigators is that a thin layer of suspending fluid exists near the wall of the tube which is transporting the slurry. The observation that friction loss for transporting a slurry may be less than for transporting a homogeneous fluid is attributed to the existence of this layer of suspension fluid near the tube wall. Charm and Kurland (1962) have suggested the following equation for use in evaluation of the thickness ($\delta$) of the gap or layer of suspension fluid existing at the wall of a tube which is transporting a slurry-like material:

$$\log \left[ Q - R^{1/n+2} \left( \frac{\Delta P}{2mL} \right)^{1/n} \delta \right] = \frac{1}{n} \log \frac{\Delta P}{L} + \log K \tag{2.77}$$

where $K$ is a constant to be determined experimentally.

The transport of suspensions during horizontal or vertical flow is discussed in detail by Orr (1966). Since turbulent motion is a very critical factor in keeping the particles in suspension during transport in horizontal tubes, the suspension velocity must be evaluated and determined in such a way as to provide this turbulent motion. Lowenstein (1959) used existing experimental data to develop the following expression for describing the minimum velocity for maintaining a suspension:

$$\frac{\rho_L \bar{u}_s^2}{\Delta \rho g X} = K_s \left( \frac{D \bar{u}_s \rho_s}{\mu_L} \right)^{0.775} \tag{2.78}$$

where $K_s$ is a constant that varies depending on the type of particle suspension desired. Values of $K_s$ vary from a value of 0.0251 for minimum velocity to maintain a suspension to a value of 0.0741 for a velocity required to maintain a suspension with no concentration gradient for particles across the velocity profile. These values for the constant ($K_s$) allow the use of equation (2.78) with units in the SI system.

The energy loss due to friction during transport of a suspension through a vertical tube was investigated by Newitt *et al.* (1961). Based on their investigation and available data, the following expression was developed:

$$\frac{H_s - H_L}{H_L} = 0.0037 \left[ \frac{X_c D}{x} \right] \left[ \frac{gD}{\bar{u}_s} \right]^{0.5} \left[ \frac{\rho_s}{\rho_L} \right]^2 \tag{2.79}$$

In equation (2.79) the parameters $H_s$ and $H_L$ represent the relationship between head loss of the suspension and head loss of the liquid carrying the suspension. The parameter ($X_c$) is the volumetric concentration of particles being delivered through the vertical pipe. The equation (2.79) applied as long as a Reynolds number defined as:

$$\mathrm{Re}_s = \frac{x u_T \rho_s}{\rho_L} \tag{2.80}$$

is greater than 2000. A layer of clear suspending fluid near the tube wall has been observed during transport of suspensions through vertical tubes in a manner similar to horizontal tubes.

**EXAMPLE 2.9** A suspension of 500-micron particles is being transported through a 10-cm diameter horizontal tube using 15 C water as a conveying liquid. If the volumetric concentration is 0.3 and the suspension is being conveyed at a velocity that maintains a uniform concentration gradient in the pipe, compute the energy loss due to friction per unit length of pipe. The density of the particle solids in suspension is 1525 kg/m$^3$.

SOLUTION

(1) The velocity of the suspension required to maintain a uniform particle concentration can be computed using equation (2.78) with $K_s$ = 0.0741. The following property values are needed:

$\mu_L = 1.14 \times 10^{-3}$ Pa s; from Table A.4 for water at 15 C.

$\rho_L = 999.1$ kg/m$^3$; from Table A.4 for water at 15 C.

$\rho_c = 1525$ kg/m$^3$; given.

$\rho_s = 1525\,(0.3) + 999.1\,(0.7) = 1156.87$ kg/m$^3$.

$x = 500 \times 10^{-6}$ m; given.

$D = 0.1$ m; given.

(2) Using equation (2.78):

$$\frac{(999.1\text{ kg/m}^3)(\bar{u}_s)^2}{(1525-999.1\text{ kg/m}^3)(9.815\text{ m/s}^2)(500\times10^{-6}\text{ m})}$$

$$= (0.0741)\left[\frac{(0.1\text{ m})(\bar{u}_s)(1156.87\text{ kg/m}^3)}{1.14\times10^{-3}\text{ Pa s}}\right]^{0.775}$$

$$387.12\,\bar{u}_s^2 = 0.0741\,(7584.8)\,\bar{u}_s^{0.775}$$

$$\bar{u}_s^{1.225} = 1.45$$

$$\bar{u}_s = 1.36\text{ m/s}$$

(3) After computation of the required suspension velocity, the energy loss due to friction can be determined using equation (2.79); the following parameters are used:

$X_c = 0.3$ $\qquad$ $\bar{u}_s = 1.36$ m/s

$D = 0.1$ m $\qquad$ $\rho_s = 1156.87$ kg/m$^3$

$x = 500 \times 10^{-6}$ m $\qquad$ $\rho_L = 999.1$ kg/m$^3$

and:

$$\frac{H_s - H_L}{H_L} = 0.0037\left[\frac{(0.3)(0.1)}{5\times10^{-4}}\right]\left[\frac{(9.815)(0.1)}{1.36}\right]^{0.5}\left[\frac{1156.87}{999.1}\right]^2$$

$$= (0.0037)\,(60)\,(0.8495)\,(1.34)$$

$$\frac{H_s - H_L}{H_L} = 0.253$$

(4) The results of the computation indicate that the energy loss due to friction for the suspension is 25% higher than the value for transport of pure water.

**2.3.2b Pneumatic Transport.—**The pneumatic transport of dry food products is a commonly used procedure in many processing plants. Most expressions available for computing the pressure-drop requirements for transporting suspensions of dry products in air are based on expressions which assume that the total pressure loss is the sum of the two components which are due to the friction between the gas and pipe wall as well as the friction due to particle and particle-to-wall contact. The total pressure drop due to friction was described by an expression presented by Orr (1966) as follows:

$$\Delta P_T = \Delta P_g + \frac{f_c u_c^2 \rho_c L}{2\ D} \qquad (2.81)$$

where:

$$f_c = \frac{3\rho_g C_D D}{2x\rho_c} \left[\frac{u_g - u_c}{u_c}\right]^2 \qquad (2.82)$$

and the drag coefficient ($C_D$) has values between 0.4 and 0.44.

An equation describing the relationship between air velocity and the conveying velocity of a particle was presented by Zenz and Othmer (1960):

$$\bar{u}_c = \bar{u}_g \left[1 - 0.0639 x^{0.3}\ \rho_c^{\ 0.5}\right] \qquad (2.83)$$

where the coefficient (0.0639), the particle diameter ($x$), and particle density ($\rho_c$) has been modified for use with SI units.

According to Orr (1966), equation (2.81) may have limitations due to the procedures used to compute the contribution of the gas phase flow to the pressure drop due to friction. An improvement suggested by Orr (1966) involved the use of a mixed friction factor ($f_m$) and the following equation:

$$\Delta P_T = f_m L u_g^2 \rho_g \left[\frac{1 + \left(\dfrac{u_c^2 \rho_{cg}}{u_g^2 \rho_g}\right)^a}{2gD}\right] \qquad (2.84)$$

where $\rho_{cg}$ represents the density of the air-particle mixture. Equation (2.84) would be used for horizontal flow, but could be adapted to vertical conveying by adding a term ($\rho_c L$) to the right side of the equation. This term accounts for energy required to maintain the solid particles in suspension. The constant ($a$) is a function of particle size and density along with the type of flow maintained. Mehta *et al.* (1957) found that $a = 1$ for 97 micron diameter glass beads and $a = 0.3$ for 36 micron diameter particles. The constant ($a$) should be measured for each specific type of flow situation and for each type of particle being transported.

EXAMPLE 2.10 Nonfat dry milk a particle density of 1450 kg/m$^3$ and mean particle diameter of 50 micron is being conveyed with 27 C air through a 10 cm diameter pipe with a length of 15 m. Determine the pressure drop due to friction for conveying the dry product through the pipe when the air velocity is 6 m/s.

SOLUTION

(1) The conveying velocity of the particle can be computed using equation (2.83):

$$\bar{u}_c = (6\ \text{m/s})\,[1 - (0.0639)(50 \times 10^{-6}\ \text{m})^{0.3}(1450\ \text{kg/m}^3)^{0.5}]$$

$$\bar{u}_c = (6)(1 - 0.1247)$$

$$\bar{u}_c = 5.25\ \text{m/s}$$

(2) The friction factor ($f_c$) can be computed using equation (2.82):

$$f_c = \frac{3\,(1.138\ \text{kg/m}^3)\,(0.4)\,(10/100\ \text{m})}{2\,(50 \times 10^{-6}\ \text{m})\,(1450\ \text{kg/m}^3)}\left[\frac{6\ \text{m/s} - 5.25\ \text{m/s}}{5.25\ \text{m/s}}\right]^2$$

where:

$$\rho_g = 1.138\ \text{kg/m}^3 \text{ is from Appendix Table A.3}$$

$$C_D = 0.4$$

and:

$$f_c = 0.01922$$

(3) The pressure drop due to friction for conveying the product is computed from equation (2.81):

$$\Delta P_T = \Delta P_g + \frac{(0.01922)\,(5.25\ \text{m/s})^2\,(1450\ \text{kg/m}^3)\,(15\ \text{m})}{2(10/100\ \text{m})}$$

$$\Delta P_T = \Delta P_g + 57.6 \times 10^4\ \text{Pa} = \Delta P_g + 57.6\ \text{kPa}$$

(4) The results indicate that the pressure drop due to friction during product conveying would be 57.6 kPa greater than the pressure drop due to friction for flow of air in the pipe at the same velocity.

## 2.4. PROPERTIES OF GRANULAR FOODS AND POWDERS

Dry food products make up a considerable portion of the total number of food products available to the consumer. Like fluid food products, they are handled in various ways in different parts of the processing plant. The design of the handling system for dry products requires knowledge of the properties of the products being handled. The following sections will review the basic relationships required in the prediction and computation of the properties of granular food products.

### 2.4.1 Density

Density is one of the basic properties of any material, but in the case of granular food products various types of density have been defined. Bulk density is defined by the following expression:

$$\rho_B = \frac{M}{V} \tag{2.85}$$

Although equation (2.84) appears to be a relatively simple expression, two types of bulk density have been designated for dry food products. Loose bulk density is measured after placing the product in the constant-volume container without vibration. Packed bulk density is measured after the sample placed in the constant-volume container has been vibrated until the volume seems constant. As might be expected, the bulk density value will be dependent on the particle-size characteristics and any factors which influence these characteristics.

The apparent particle density of a dry food product is a measure of the amount of air which may be trapped within the individual particles. This parameter can be measured by utilizing a picnometer in which a solvent of known density replaces air within the particles and allows computation of the apparent particle density.

Two additional properties of granular products which relate to density are void and porosity. The void can be defined as the ratio of the volume of space between particles to the entire volume, and can be computed by the following expression:

$$v = \rho_B\left(\frac{1}{\rho_B} - \frac{1}{\rho_p}\right) = 1 - \frac{\rho_B}{\rho_p} \tag{2.86}$$

Equation (2.85) reveals that void ($v$) can be computed after measurement of bulk density ($\rho_B$) and apparent particle density ($\rho_P$). Porosity ($\psi$) can be defined as the ratio of the air volume within the particles to the total particle volume and can be computed from the following expression:

$$\Psi = 1 - \frac{\rho_B}{\rho_t} \tag{2.87}$$

where $\rho_t$ is the true density of the air free solids making up the product. Normally this density can be computed from knowledge of the density of the various chemical components which make up the product solids.

### 2.4.2 Particle Size and Size Distributions

A very important property of granular foods or powders is the particle size and size distribution. Although some notation of the mean particle diameter may be useful in describing the product, the complete picture is not apparent until the size distribution is described, also. One of the important factors to consider when discussing the mean diameter of a particle size distribution is the type of mean diameter being utilized. Mugele and Evans (1951) developed a generalized expression which can be used to define all types of mean diameters. This expression is as follows:

$$x_{q\,p}^{q-p} = \frac{\displaystyle\int_{x_o}^{x_u} x^q \frac{d\,\Sigma\,N}{dx}\,dx}{\displaystyle\int_{x_o}^{x_u} x^p \frac{d\,\Sigma\,N}{dx}\,dx} = \frac{\Sigma(x^q N)}{\Sigma(x^p N)} \tag{2.88}$$

where the symbols and values of the parameters $p$ and $q$ are presented in Table 2.3. For example, the arithmetic or number mean diameter would be defined as:

$$x_L = \frac{\Sigma x N}{N} \tag{2.89}$$

which is obtained from equation (2.87) when $q = 1$ and $p = 0$. Another commonly used notation is the volume-surface diameter, usually called the Sauter mean diameter, defined as follows:

$$x_{vs} = \frac{\Sigma\, x^3\, N}{\Sigma\, x^2\, N} \tag{2.90}$$

and describes the ratio of volume to surface area that is equal to the ratio for the entire distribution of particles.

TABLE 2.3

VARIOUS NOTATIONS FOR MEAN PARTICLE SIZE

| Symbol | Name of Mean Diameter | p | q | Order |
|---|---|---|---|---|
| $x_L$ | Linear (arithmetic) | 0 | 2 | 1 |
| $x_s$ | Surface | 0 | 2 | 2 |
| $x_v$ | Volume | 0 | 3 | 3 |
| $x_m$ | Mass | 0 | 3 | 3 |
| $x_{sd}$ | Surface-diameter | 1 | 2 | 3 |
| $x_{vd}$ | Volume-diameter | 1 | 3 | 4 |
| $x_{vs}$ | Volume-surface | 2 | 3 | 5 |
| $x_{ms}$ | Mass-surface | 3 | 4 | 7 |

Source: from Mugele and Evans (1951).

EXAMPLE 2.11 Compare the arithmetic, surface and volume-surface mean diameter for particles in a dry food product with the following distribution of sizes:

| Number | Size (micron) |
|---|---|
| 1 | 40 |
| 4 | 30 |
| 25 | 20 |
| 20 | 15 |
| 10 | 10 |

SOLUTION

(1) Utilizing equation (2.89) for arithmetic mean:

$$x_L = \frac{40 + 30(4) + 20(25) + 15(20) + 10(10) + 5(4)}{1 + 4 + 25 + 20 + 10 + 4}$$

$$x_L = \frac{40 + 120 + 500 + 300 + 100 + 20}{64} = \frac{1080}{64}$$

$$= 16.9 \text{ microns}$$

(2) For the mean surface diameter, equation (2.88) reduces to ($p = 1; q = 2$):

$$x_s = \frac{\Sigma\, x^2 N}{\Sigma\, xN}$$

$$= \frac{(40)^2 + (30)^2\, 4 + (20)^2\, 25 + (15)^2\, 20 + (10)^2\, 10 + (5)^2\, 4}{40(1) + 30(4) + 20(25) + 15(20) + 10(10) + 5(4)}$$

$$x_s = \frac{20800}{1080} = 19.26 \text{ microns}$$

(3) For the volume-surface diameter, equation (2.90) can be utilized:

$$x_{vs} = \frac{(40)^3 + (30)^3\, 4 + (20)^3\, 25 + (15)^3\, 20 + (10)^3\, 10 + (5)^3\, 4}{(40)^2 + (30)^2\, 4 + (20)^2\, 25 + (15)^2\, 20 + (10)^2\, 10 + (5)^2\, 4}$$

$$x_{vs} = \frac{450{,}000}{20800} = 21.6 \text{ microns}$$

(4) This comparison indicates that attention must be given to the type of mean diameter used in a computation. Previous examples (2.9) and (2.10) have illustrated the importance of particle size in computation of power requirements for hydraulic and pneumatic transport.

The particle size distribution of most granular or dry food products is not described by the normal distribution. The distribution used to describe many dry food products is the following log-normal density function:

$$f(x) = \frac{1}{\ln \sigma_g \sqrt{2\pi}} \exp\left[-\frac{1}{2}\left(\frac{\ln x - \ln \bar{x}_g}{\ln \sigma_g}\right)^2\right] \qquad (2.91)$$

with the arithmetic- and log-geometric mean defined in the following way:

$$\ln \bar{x}_g = \frac{\Sigma\,(N \ln x)}{\Sigma\, N} \qquad (2.92)$$

and the geometric standard deviation ($\sigma_g$) defined as:

$$\ln \sigma_g = \left[ \frac{\Sigma[N \ln x - \ln x_g)^2]}{N} \right]^{1/2} \tag{2.93}$$

Both the geometric mean and geometric standard deviation are usually determined by using log-normal probability paper as illustrated in the Fig. 2.16. After plotting on these coordinates, the geometric standard deviation can be estimated by dividing the particle size corresponding to 15.879 cumulative percent by the particle size corresponding to 50 cumulative percent. The particle size corresponding to 50 cumulative percent is the geometric mean of the distribution.

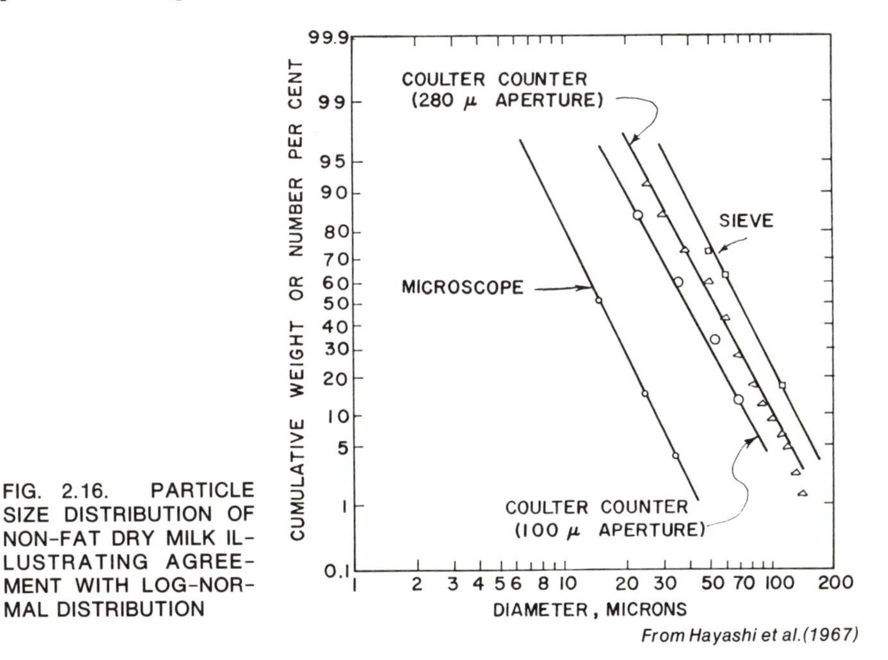


FIG. 2.16. PARTICLE SIZE DISTRIBUTION OF NON-FAT DRY MILK ILLUSTRATING AGREEMENT WITH LOG-NORMAL DISTRIBUTION

EXAMPLE 2.12 Particle size distribution data for nonfat dry milk are presented in Fig. 2.16 on log-normal probability coordinates. Determine the mean and geometric standard deviation for data obtained with a 100-micron aperture of the Coulter Counter.

SOLUTION

(1) The mean size is obtained directly form Fig. 2.16 by observing the size corresponding to the 50% cumulative value. For the conditions specified, the mean size is 40 microns.

(2) For the data in Fig. 2.16, the geometric standard deviation will be (with 65 micron corresponding to 15.879%).

$$\sigma_g = \frac{65}{40} = 1.625$$

(3) The geometric standard deviation describes the distribution of particle sizes in the dry product.

### 2.4.3 Flow of Food Powders

The manner in which granular foods or powders may flow into or out of containers is of concern in processing plants. In addition to the previously described density and particle size parameters, there are specific parameters which describe the flow properties of these types of food products. Two common parameters used for this purpose are the angle of repose and the angle of slide. Both of these parameters lack theoretical basis but do serve as a means of comparing different food powders. The angle of repose is defined in the following way:

$$\tan \beta = \frac{2\pi H'}{S} \tag{2.94}$$

where $H'$ is the height of a mound of a powder which forms as it flows from a container directly above a horizontal surface, and $S$ is the circumference of the mound at its base. The angle of slide is a parameter determined by placing the powder on a horizontal plate and the angle of the plate is changed until the powder slides from the plate. The angle from the horizontal which is required for the powder to lose its position on the plate is the angle of slide. It is obvious that this angle will be a function of the type of surface on which the powder is placed. The type of material used in the test, therefore, should be the same as that used in the actual physical situation.

A somewhat more basic and fundamental property of powder flow is the angle of internal friction ($\phi$), defined as follows:

$$\tan \phi = \frac{\tau}{\sigma_n} \tag{2.95}$$

As indicated by equation (2.95) this flow parameter for a powder is a function of the shear stress ($\tau$) and the normal stress ($\sigma_n$) and therefore is directly related to the force required to move one layer of the granular material over another. One method of determining this parameter is by direct shear measurement leading to plots of data, as illustrated in Fig. 2.17. The internal angle of friction is measured directly from this type of plot and will be a function of the properties such as bulk density, particle size and size distribution and moisture content.

One application in which the flow of the dry food is of particular concern is the flow of the product from a container through an opening. This is a complex type of problem to describe because of the many interacting forces which influence the flow characteristics. One expression which can be utilized to describe the flow rate of powder through an orifice in the bottom of a storage container is:

$$w = C_c\,(\pi/4)\,\rho_B\left[\frac{gD^5}{2\tan\beta}\right]^{0.5} \tag{2.96}$$

In equation (2.96) the mass flow rate of powder ($w$) is a function of the orifice diameter ($D$) and the angle of repose ($\beta$). The discharge coefficient ($C_c$) usually varies between 0.5 and 0.7. Orr (1966) points out that the equation (2.96) is applicable only if the ratio of particle diameter ($x$) to orifice ($D$) is less than 0.1. In addition, the expression may be somewhat in error when the powder density is high.

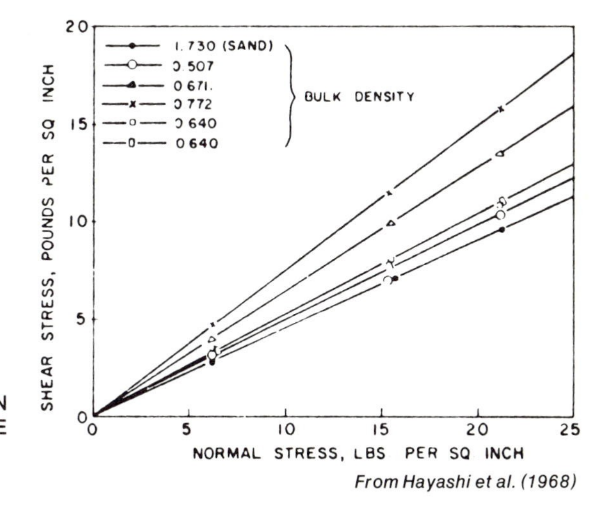


FIG. 2.17. PLOT USED IN COMPUTATION OF ANGLE OF INTERNAL FRICTION

*From Hayashi et al. (1968)*

EXAMPLE 2.13 A dry food product with bulk density of 560 kg/m$^3$ is stored in a large storage container and is removed by gravity through a 0.075 m diameter opening in the bottom of the container. The test for angle of repose gave a mound of product with 10-cm diameter and 7.5 cm height. Compute the rate at which the product will be released from the container.

SOLUTION

(1) Use the equation (2.96) to compute powder flow rate from a container requires knowledge of the angle of repose for the product. Using equation (2.94) provides:

$$\tan\beta = \frac{2\pi(7.5)}{\pi 10} = \frac{15}{10} = 1.5$$

$$\beta = 56.3^\circ$$

(2) Utilizing equation (2.96) with a discharge coefficient of 0.6:

$$w = 0.6\ \frac{\pi}{4}\ (560\,\text{kg/m}^3)\left[\frac{(9.815\,\text{m/s}^2)(0.075\text{m})^5}{2\tan(56.3)}\right]^{0.5}$$

$$w = 0.7353\ \text{kg/s} = 44.12\ \text{kg/min}.$$

One of the major problems in the flow of powders form a container through an opening in the bottom of the container is the arching or bridging of the material above the opening, which inhibits flow of the product from the container. This is another complex problem to analyze but Richmond and Gardner (1962) have shown that it is a function of the product cohesiveness, the angle of internal friction, and the bulk density of the product. The expression derived to describe the minimum distance between the two walls of the opening ($D_L$) which will not result in bridging is as follows:

$$D_L = \frac{C_b}{\rho_B}[1 + \sin\phi] \tag{2.97}$$

in which the natural cohesiveness parameter ($C_b$) is determined during the test for the coefficient of internal friction. The value would be determined from a plot similar to Fig. 2.17 in which the curve of shear stress versus normal stress does not intersect at zero. The natural cohesiveness parameter would then be the intercept of the curve on the shear stress axis.

## 2.5. PROPERTIES OF SOLID FOODS

As indicated earlier in this chapter, most foods will not respond as either ideal elastic, ideal plastic or ideal viscous materials. Most solid foods approach or come closer to the ideal elastic material than to any other; hence it appears that the elastic theory is the most useful to describe the behavior of these types of food products. Since these types of materials may exhibit a stress-strain relationship which is dependent on rate of strain, a different type of mathematical description and physical description is possible and necessary. The material which follows is not intended to be complete; for more complete information on viscoelasticity, references by Ferry (1961) and the rheology series by Eirich (1958) are suggested. For applications of this material to agricultural products, including food products, the reference by Mohsenin (1979) is suggested.

### 2.5.1 Viscoelastic Models

The two basic viscoelastic models utilize the elements of the ideal viscous behavior. The Maxwell model is a combination of the elastic and viscous elements in series, as illustrated in Fig. 2.18. The response of a material obeying the behavior of this model is described by the following differential equation:

$$\frac{d\sigma}{dt} + \frac{E}{\mu}\sigma = E\frac{d\epsilon}{dt} \tag{2.98}$$

where $E$ is the modulus of the elastic element and $\mu$ is the viscosity coefficient describing the response of the viscous element in the model. Equation (2.98) can be solved to obtain the following expression for stress as a function of time when strain is constant: ($d\epsilon/dt = 0$)

$$\sigma(t) = \sigma_0 \exp\left(-\frac{E}{\mu}t\right) \tag{2.99}$$

where

$$\sigma_o = E\epsilon$$

This response is illustrated in Fig. 2.19.

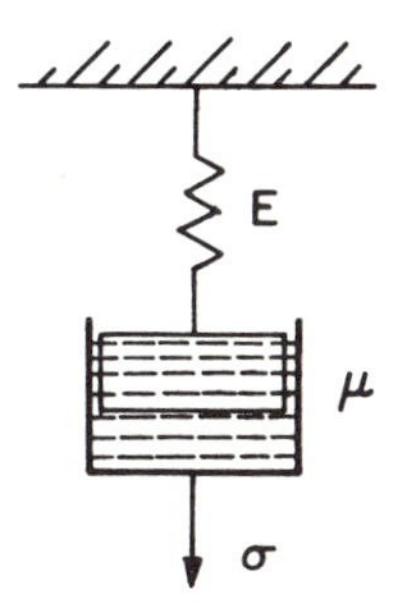


FIG. 2.18. REPRESENTATION OF MAXWELL MODEL

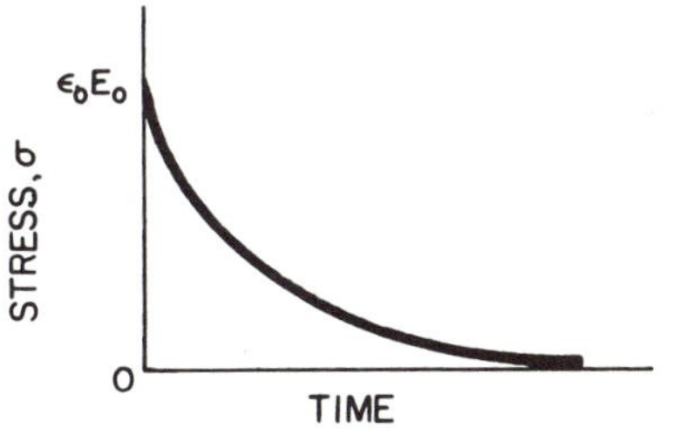


FIG. 2.19. RESPONSE OF MAXWELL MATERIAL TO STEP-CHANGE IN STRESS

EXAMPLE 2.14 A relaxation test conducted on a sample of rehydrated freeze-dried beef gave the following stress-time relationship

| Time (sec) | Stress (kPa) |
|---|---|
| 0 | 108.3 |
| 10 | 103.4 |
| 20 | 102.1 |
| 30 | 96.5 |
| 40 | 93.1 |
| 50 | 89.6 |
| 60 | 89.6 |
| 70 | 84.1 |
| 80 | 82.7 |
| 90 | 79.3 |
| 100 | 76.5 |

Compute the mechanical properties which will describe this product when the strain ($\epsilon$) is 0.15.

SOLUTION

(1) Utilizing the stress at $t = 0$ as:

$$\sigma_o = E\epsilon$$

then:

$$E = \frac{108.3}{0.15} = 722 \text{ kPa}$$

(2) By plotting the log of stress versus time as illustrated in Fig. 2.20, it is evident from equation (2.99) that the slope of the curve can be used to compute $E/\mu$. From Fig. 2.20, the slope value becomes $1.51 \times 10^{-3}$ which leads to the following computation:

$$\frac{E}{\mu} = (2.303)(1.51 \times 10^{-3}) = 3.48 \times 10^{-3}$$

or

$$\mu = \frac{722}{3.48 \times 10^{-3}} = 2.075 \times 10^{5} \text{ kPa s}$$

(3) The results of this example illustrate that the product structure can be described by a viscoelastic model.

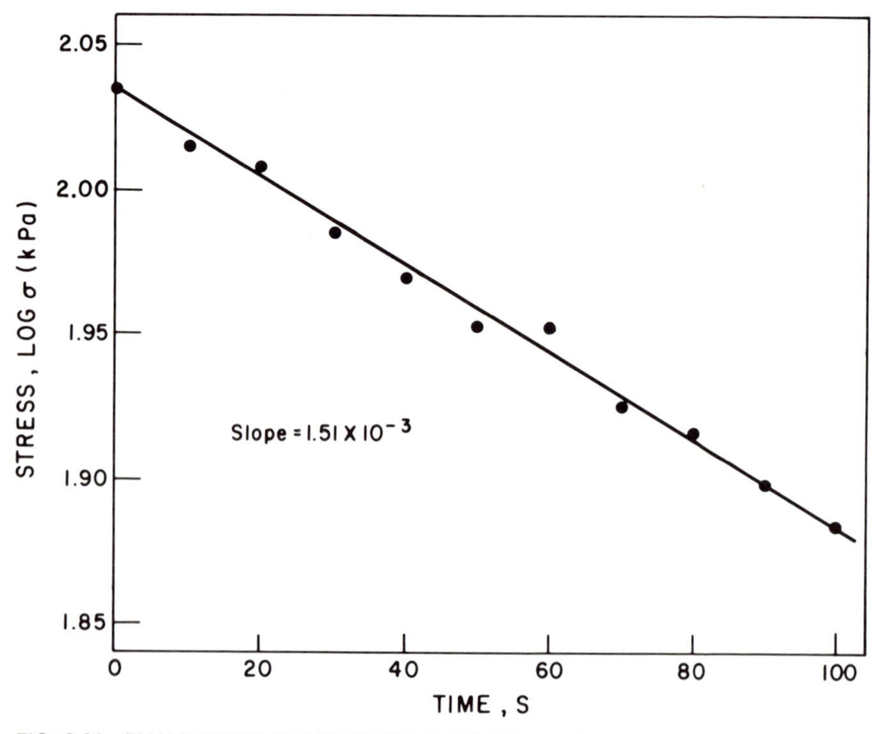


FIG. 2.20. EVALUATION OF MECHANICAL PROPERTIES WHICH DESCRIBE A SOLID FOOD PRODUCT

When the ideal elastic and ideal viscous elements are combined in parallel, the behavior is described by the Kelvin model as illustrated in Fig. 2.21. The response of this particular model is described by the following differential equation:

$$\frac{d^2\epsilon}{dt^2} + \frac{E}{\mu}\frac{d\epsilon}{dt} = \frac{1}{\mu}\frac{d\sigma}{dt} \quad \text{or} \quad \frac{d\epsilon}{dt} + \frac{E}{\mu}\epsilon = \frac{\sigma}{\mu} \tag{2.100}$$

When equation (2.100) is solved, the following expression for strain as a function of time is obtained:

$$\epsilon(t) = \frac{\sigma_0}{E}\left[1 - \exp\left(\frac{Et}{\mu}\right)\right] \tag{2.101}$$

where

$$\sigma_o = \sigma = \text{constant and } \epsilon_\infty = \frac{\sigma_o}{E}$$

and the response is illustrated in Fig. 2.22. Both the Maxwell and Kelvin models have been used extensively in obtaining the elastic and viscous properties of many solid food products.

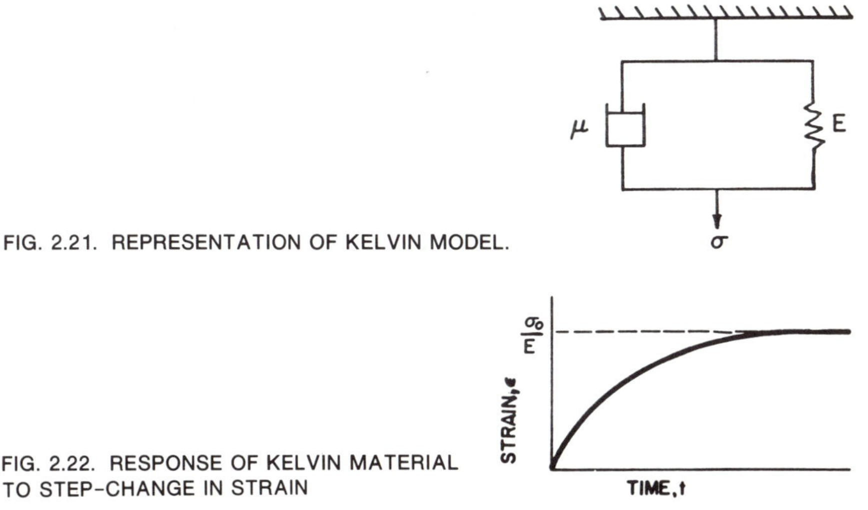


FIG. 2.21. REPRESENTATION OF KELVIN MODEL.

FIG. 2.22. RESPONSE OF KELVIN MATERIAL TO STEP-CHANGE IN STRAIN

### 2.5.2 Measurement of Viscoelastic Parameters

There are several standard types of tests which can be utilized to determine the viscoelastic properties of solid food products. These tests will be classified and discussed, in general, in the following paragraphs.

Many of the properties required to describe a viscoelastic material can be measured by a test which induces stress-strain behavior. In most cases, a compression apparatus which provides a constant rate of loading during the test is used to obtain a force-deformation relationship. By varying the rate of loading, the response which can be predicted by the Maxwell model is obtained. This leads to evaluation of the parameters in the Maxwell model.

The second type of test that can be utilized to describe the viscoelastic behavior of a food product is the creep test. In this type of test, the load or stress is applied as a step change and is held constant while measuring the deformation or strain as a function of time. The results obtained are in the form of independent parameters which depend on whether the creep is applied as a compression, shear or bulk.

When a given deformation or strain is applied to the product and a stress required to hold the deformation constant is measured as a function of time, the test is called stress relaxation. Again, a number of time-dependent parameters are obtained depending on whether the strain applied is high-compression, shear or in the form of a bulk pressure. The model utilized in the stress relaxation test is a form of Maxwell model.

The fourth type of test for measurement of viscoelastic parameters is the dynamic test. In this type of test, a specimen or product is deformed by stress which varies with time in some known manner. This test overcomes two disadvantages of the previous tests:

(a) The previous tests require several time scales during which the product may change in some way.

(b) It is physically impossible to achieve instantaneous application of load or deformation as required at the beginning of the experiment in most cases. The dynamic test allows evaluation of the elastic modulus and the mechanical damping over a wide range of frequencies not available by the other tests.

### 2.5.3 Food Texture

The viscoelastic properties of a solid food should relate very closely to the texture of the product. This relationship has not been investigated in sufficient detail to be established. Traditional parameters utilized to describe texture are very empirical in nature and allow comparison of product samples only when all aspects of the measurement are identical. In most situations, attempts at measurement by instrument are intended to simulate response in the mouth during mastication of the product.

Probably the first attempt to standardize measurement techniques and the resulting parameters was by Szczesniak (1963), who suggested eight characteristics to completely describe the product texture. The parameters included hardness, cohesiveness, adhesiveness, viscosity, elasticity, brittleness, chewiness and gumminess. This proposal led to development of Texture Profile Analysis, which consisted of instrumental texture measurement and texture evaluation by a trained sensory panel. The instrument used for texture measurement utilized a cyclic force application and the response obtained was as illustrated in Fig. 2.23. Brandt *et al.* (1963) used the profile to evaluate texture parameters in the following manner:

$$\text{Hardness} = L_1 \tag{2.102}$$
$$\text{Adhesiveness} = A_3 \tag{2.103}$$
$$\text{Cohesiveness} = A_2/A_1 \tag{2.104}$$
$$\text{Elasticity} = 68.5 - B \tag{2.105}$$
$$\text{Chewiness} = L_1\,(A_2/A_1)(68.5 - B) \tag{2.106}$$
$$\text{Gumminess} = L_1\,(A_2/A_1) \tag{2.107}$$

where 68.5 is a constant obtained for the measurement of B on a completely inelastic material. The parameter terms defined should not be confused with similar engineering parameters. Modifications which reduce potential confusion have been proposed in later publications such as by Bourne (1968). The approach to describing food texture does offer considerable flexibility and leads to easier comparison of results from food texture research by different investigators.

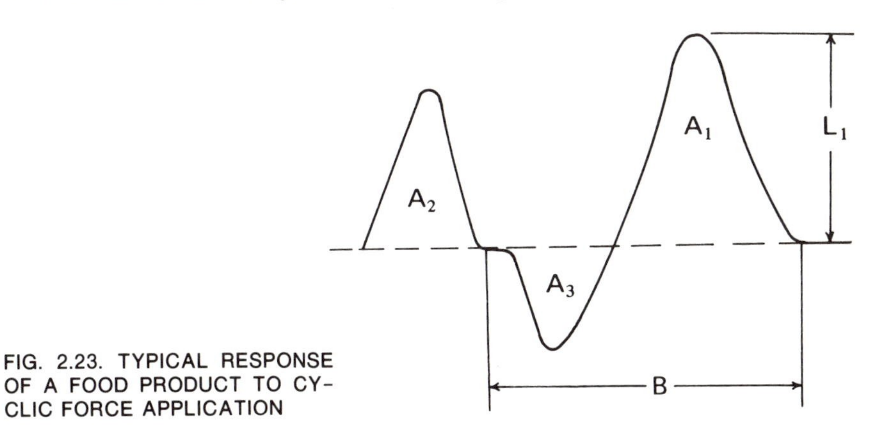


FIG. 2.23. TYPICAL RESPONSE OF A FOOD PRODUCT TO CYCLIC FORCE APPLICATION

The relationship between selected parameters from the Texture Profile Analysis and typical engineering parameters may be obvious. Reidy (1970) investigated the relationship in considerable detail by deriving mathematical models with solutions which would describe the response of a food product to cyclic loading. Using this approach, it was shown that the food texture parameters were direct functions of appropriate engineering parameters. For example, the hardness parameter correlated directly with the modulus of elasticity in the Maxwell portion of a four-element model obtained by combining Maxwell and Kelvin models in series.

## PROBLEMS

2.1. *Rheological parameter estimation.* The following shear stress-shear rate data were obtained for pear purée with 31% total solids at 82 C:

| *Shear Stress* (kPa) | *Shear Rate* (1/s) |
|---|---|
| 3.3 | 4 |
| 4.5 | 10 |
| 7.5 | 50 |
| 9.0 | 100 |
| 12.0 | 500 |
| 17.5 | 1000 |

Estimate the rheological parameters required to describe the product.

2.2. *Capillary tube rheometer.* Derive a mathematical relationship between volumetric flow rate and pressure for a food described by Bingham-plastic behavior in a capillary tube rheometer.

2.3. *Estimating rheological parameters from apparent viscosity.* The following apparent viscosity data were obtained for peach purée at 30 C using a capillary rheometer with 0.28 cm diameter and 100 cm length:

| *Apparent Viscosity* (Pa s) | *Shear Rate* (1/s) |
|---|---|
| 0.406 | 121 |
| 0.301 | 205 |
| 0.255 | 285 |
| 0.209 | 380 |
| 0.174 | 505 |
| 0.146 | 720 |
| 0.100 | 1300 |
| 0.075 | 2150 |

Determine the rheological parameters required to describe the flow characteristics of the product.

2.4. *Rotational rheometer.* Derive the mathematical relationship between revolutions per unit time and torque for a single cylinder rotational rheometer when the fluid can be described by the power law equation.

2.5. *Mechanical energy balance.* Use the equation for mechanical energy balance to derive a mathematical expression for friction factor when: (a) fluid density is given, (b) pipe is horizontal with known diameter and length, (c) pressure drop can be measured and (d) mass flow rate is measured.

2.6. *Friction factors.* Compute the influence of temperature on friction factor for flow of apricot purée at 4.4 and 25 C. The flow behavior index ($n$) is 0.37 at both temperatures while the consistency coefficient ($m$) decreases from 1.3 Pa $s^n$ at 4.4 C to 0.9 Pa $s^n$ at 25 C. The flow rate is 1 kg/s in a 0.05 m diameter pipe. A product density of 975 kg/$m^3$ can be used at both temperatures.

2.7. *Pressure drop due to friction.* Compare the pressure drop due to friction for water and banana purée flowing in a 1.25 cm inside diameter pipe at 24 C and at a flow rate of 1.25kg/s. The pipe length is 3m and the properties of the banana purée are $m = 6$ Pa $s^n$, $n = 0.454$ and density = 975 kg/$m^3$.

2.8. *Power requirements for pumping.* Estimate the power requirements for pumping (100% efficiency) apricot purée at 25 C from an open tank filled to a level of 10m above the discharge point to a second open tank at an elevation of 3m. The smooth pipeline has a 5 cm diameter and contains 4 standard 90° elbows, 2 medium sweep 90° elbows and one globe valve. The product density is 1200 kg/$m^3$ and the mass flow rate is 1.2 kg/s while 45m of pipeline are required.

2.9. *Transport of suspensions.* A suspension of 250 micron particles in water is being transported through a 6 cm diameter at a velocity of 1 m/s. The volumetric concentration of particles in water is 0.4 and the density of the particle solids is 1500 kg/$m^3$. Estimate the pressure drop due to friction for pumping the suspension.

2.10. *Pneumatic conveying.* Dry food particles with a mean diameter of 75 micron are being conveyed in 20 C air through a 15 cm diameter tube. The product is being moved through 50 m of pipe with no change in elevation. Estimate the size of pump required when an average air velocity of 5 m/s is used. The product particle density is 1400 kg/$m^3$.

2.11. *Flow of dry foods.* A dry food product with a bulk density of 650 kg/$m^3$ and an angle of repose of 55° is removed by gravity through a circular opening in the bottom of a large container. Estimate the diameter of the opening required to maintain a mass flow rate of 5 kg/min. A discharge coefficient of 0.6 can be assumed.

2.12. *Solid product properties.* A relaxation test is being conducted with a solid food product by applying an initial shear stress of 200 kPa with strain of 0.1. If the viscosity coefficient for the product is $8 \times 10^4$ kPa s, compute the shear stress 150 sec. after the strain is applied.

## COMPREHENSIVE PROBLEM II

Estimation of Rheological Parameters from Experimental Data for A Food Product

### Objectives:

1. To gain experience with procedures required to estimate rheological parameters from experimental data for a food product.
2. To illustrate the use of least squares curve fitting techniques when estimating rheological parameters.

### Procedures:

A. The pressure loss-velocity data in Fig. 2.24 were obtained by Hanck *et al.* (1966). Note that the data were obtained at four different tube lengths of 8.9, 17.8, 26.7 and 35.6 cm. and that a different pressure loss - velocity relationship is presented for each tube length.
B. Derive a volumetric flow rate - pressure drop expression similar to equation (2.10) that will allow estimation of the three rheological parameters in equation (2.5).
C. Select a least squares curve fitting computer program that will allow estimation of the three parameters from data at all four of the tube lengths.

### Discussion:

A. Discuss the expression required to describe the relationship between volumetric flow rate and pressure drop for a general power law fluid.
B. Discuss the significance of each of the three parameters estimated for the food product.
C. Discuss the curve fitting procedures used to evaluate the rheological parameters and any apparent limitations to the approach.

## NOMENCLATURE

$A$ = area, $m^2$.
$A_n$ = constant of integration used in equation (2.59).

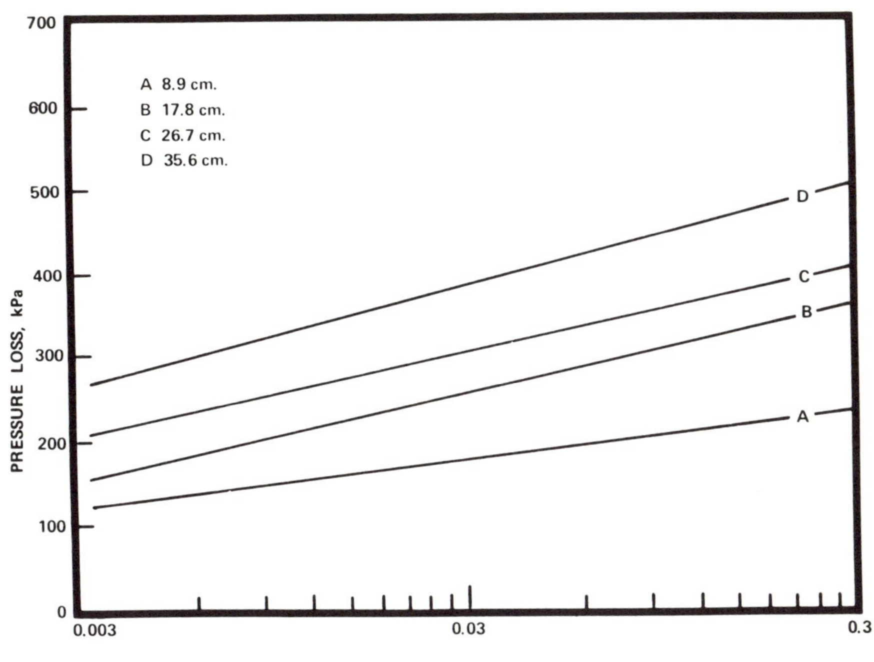


FIG. 2.24. PRESSURE LOSS VS VELOCITY IN DIFFERENT-LENGTH TUBES

$a$ = constant in equation (2.83).
$B$ = measurement defined in Fig. 2.23.
$B_A$ = Arrhenius constant.
$B_n$ = constant of integration in equation (2.59).
$C_a$ = constant used in equation (2.66).
$C_b$ = natural cohesiveness parameter in equation (2.97).
$C_c$ = discharge coefficient in equation (2.95).
$C_D$ = drag coefficient.
$C_n$ = constant defined by equation (2.62).
$D$ = tube or pipe diameter, m.
$D_L$ = distance value described by equation (2.97), m.
$\bar{d}_g$ = geometric mean size, micron.
$E$ = modulus of elasticity or Young's modulus, kPa.
$E_a$ = activation energy in Arrhenius equation, kJ/kg.
$E_f$ = energy loss due to friction, kJ/kg.
$F$ = force, N.
$f$ = friction factor.
$f(x)$ = density function describing size distribution.
$N_{GRe}$ = generalized Reynolds number.

$g$ = acceleration due to gravity, m/s$^2$.
$H_L$ = head loss for pure liquid, m.
$H_s$ = head loss for suspension, m.
$H'$ = height of powder mound in equation (2.94), m.
$h$ = height, m.
$K$ = constant in equation (2.77).
$K_f$ = coefficient defined in equations (2.69) and (2.70) or given in Table 2.2.
$K_s$ = constant in equation (2.78).
KE = kinetic energy, J/kg.
$L$ = length, m.
$L_1$ = texture parameter defined in equation (2.102).
$M$ = mass or weight, kg.
$m$ = consistency coefficient, Pa s$^n$.
$N$ = particle numbers.
$N'$ = number of revolutions per unit time.
$n$ = flow behavior index.
$P$ = pressure, Pa.
$P_n$ = dimensionless group defined by equation (2.56).
$\Delta P$ = pressure difference between two locations in a tube with flowing fluid, Pa.
$\Delta P_c$ = pressure loss due to friction in tube contraction, Pa.
$\Delta P_f$ = pressure loss due to friction in pipe fitting, Pa.
$Q$ = volumetric flow rate, m$^3$/s.
$R$ = radius of a tube or cylinder, m.
$R_g$ = gas constant, kJ/kg K.
$r$ = variable radius of a tube or pipe.
$N_{Re}$ = Reynolds number.
$S$ = circumference of powder mound [equation (2.94)], m.
$T$ = temperature, °C.
$T_A$ = absolute temperature, K.
$t$ = time, s.
$u$ = fluid velocity, m/s.
$\bar{u}$ = mean fluid velocity, m/s.
$u^*$ = friction velocity defined by equation (2.53), m/s.
$u_m$ = maximum velocity, m/s.
$V$ = volume, m$^3$.
$W$ = work parameter utilized in evaluating pump size, J/kg.
$w$ = mass flow rate defined in equation (2.64), kg/s.
$X$ = concentration expressed in decimal form.
$x$ = particle diameter, m.
$X_g$ = geometric mean diameter, m.

$y$ = coordinate perpendicular to direction of flow.
$Z$ = dimensionless group defined by equation (2.53).
$Z'$ = potential energy as used in equation (2.65), J/kg.
$\alpha$ = constant defined by equation (2.50).
$\beta$ = angle of repose.
$\gamma$ = rate of shear, l/s.
$\delta$ = thickness of fluid layer in equation (2.77), m.
$\epsilon$ = strain.
$\phi$ = angle of internal friction.
$\mu$ = coefficient of viscosity, Pa s.
$\mu_A$ = apparent viscosity, Pa s.
$v$ = particle void in a bed [equation (2.85)].
$\Omega$ = torque as defined in equation (2.26).
$\omega$ = angular velocity.
$\psi$ = angle between cone and plate of rheometer.
$\Psi$ = porosity defined by equation (2.86).
$\rho$ = density, kg/m$^3$.
$\rho_p$ = apparent particle density, kg/m$^3$.
$\rho_t$ = true particle density, kg/m$^3$.
$\sigma$ = stress, Pa.
$\sigma_g$ = geometric standard deviation.
$\sigma_n$ = normal stress, Pa.
$\sigma_o$ = stress at $t = 0$.
$\tau$ = shear stress defined in equation (2.1), Pa.
$\xi$ = dimensionless number defined by equation (2.55).

## Subscripts

$B$ = bulk.
$c$ = particle phase.
$g$ = gas phase; equation (2.81).
$i$ = inside; parameter reference.
$L$ = liquid phase.
$m$ = mixed phase or mixture.
$o$ = outside; parameter reference.
$p$ = pressure.
$s$ = solid phase or suspension.
$T$ = parameter references; terminal [equation (2.80)] or total [equation (2.81)].
$w$ = wall condition.
$y$ = parameter reference to $y$-coordinate.

## BIBLIOGRAPHY

BOURNE, M.C. 1968. Texture profile of ripening pears. J. Food Sci. *33,* 223.

BRANDT, M.A., SKINNER, E.Z., and COLEMAN, J.A. 1963. Texture profile method. J. Food Sci. *28,* 404.

CHARM, S.E. 1960. Viscometry of non-Newtonian food materials. Food Res. *25,* 351.

CHARM, S.E. 1962. Fluid consistency in food engineering applications. Advan. Food Res. *11,* 355–435.

CHARM, S.E. 1963. The determination of shear-stress-rate of shear relationships for pseudoplastic food materials using cylindrical viscometers. Ind. Eng. Chem. Process Design and level. *2,* No. 1, 62–65.

CHARM, S.E. 1978. The Fundamentals of Food Engineering, 3rd Edition. Avi Publishing Co., Westport, Conn.

CHARM, S.E., and KURLAND, G. 1962. A comparison of the tube flow behavior and shear stress-shear rate characteristics of canine blood. Am. J. Physiol. *103,* No. 3, 417.

DODGE, D.W., and METZNER, A.B. 1959. Turbulent flow of non-Newtonian systems. Am. Inst. Chem. Engrs. J. *5,* 189–204.

EARLE, R.L. 1966. Unit Operations in Food Processing. Pergamon Press, Long Island, N.Y.

EINSTEIN, A. 1906. A new method for determination of dimensions of molecules. Ann. Physik. (Leipsig) *19,* 289.

EINSTEIN, A. 1911. A correction of the article: A new method for determination of dimensions of molecules. Ann. Physik *34,* 591–592.

EIRICH, F.D. 1958. Rheology: Theory and Applications, Vol. 2. Academic Press, New York.

EIRICH, F.R., BUNZL, M., and MARGARETHA, A. 1936. Investigation of the viscosity of suspension with spherical particles. Kolloid-Z. *75,* 20.

FERRY, J.D. 1961. Viscoelastic Properties of Polymers. John Wiley & Sons, New York.

FORD, T.F. 1960. Viscosity-concentration and fluidity concentration relationships for suspensions of spherical particles in Newtonian liquids. J. Phys. Chem. *64,* 1168–1174.

FRIEDMAN, H.H., WHITNEY, J.E., and A.S. SZCZESNIAK. 1963. The texturometer—a new instrument for objective texture measurement. J. Food Sci. *28,* 390.

GUTH, E., and SIMHA, R. 1936. On the viscosity of suspensions with spherical particles. Kolloidzochr, *74,* 147.

HANCK, R.C., HALL, C.W., and HEDRICK, T.I. 1966. Pressure losses and rheological properties of flowing butter. J. Food Sci. *31,* No. 4, 534–541.

HARPER, J.C. 1960. Viscometric behavior in relation to evaporation of fruit purées. Food Technol. *14,* 557.

HARPER, J.C. 1961. Coaxial-cylinder viscometer for non-Newtonian fluids. Rev. Sci. Instr. *32*, 425.

HARPER, J.C., and EL SAHRIGI, A.F. 1965. Viscometric behavior of tomato concentrates. J. Food Sci. *30*, 470.

HARPER, J.C., and KAMEL, I. 1966. Viscosity Relationships for Liquid Products, Am. Soc. Agr. Eng. Paper 66–837. Am. Soc. Agr. Engrs., St. Joseph, Mich.

HARPER, J.C., and LEBERMANN, K.W. 1964. Rheological behavior of pear purées. Proceedings 1st Intern. Congr. Food Sci. and Technol. 719–728.

HAYASHI, H., HELDMAN, D.R., and HEDRICK, T.I. 1967. A comparison of methods for determining the particle size and size distribution of nonfat dry milk. Quart Bull. Mich. Agr. Expt. Sta. *50*, No. 1, 93-99.

HAYASHI, H., HELDMAN, D.R., and HEDRICK, T.I. 1968A. Internal friction of nonfat dry milk. Trans. Am. Soc. Agr. Engrs. *11*, No. 3, 422–425.

HAYASHI, H., HELDMAN, D.R., and HEDRICK, T.I. 1968B. Influence of spray-drying conditions on size and size distribution of nonfat dry milk particles. J. Dairy Sci. *52*, No. 1, 31–37.

LOWENSTEIN, J.G. 1959. Design so solids can't settle out. Chem. Eng. *66*, 133–135.

MANLEY, R.S.J., and MASON, S.G. 1952. Particle motions in sheared suspensions. II. Collisions of uniform spheres. Colloid Sci. *7*, 354–369.

MEHTA, N.C., SMITH, J.M., and COMMINGS, E.W. 1957. Pressure drop in air-solid flow systems. Ind. Eng. Chem. *49*, 986–992.

METZNER, A.B. 1956. Non-Newtonian technology. Advan. Chem. Eng. *1*, 113–152.

METZNER, A.B. 1958. Recent developments in the engineering aspects of rheology. Rheol. Acta Band. *1*, 205.

METZNER, A.B. 1961. *In* Handbook of Fluid Dynamics, V. L. Streeter (Editor). McGraw-Hill Book Co., New York.

METZNER, A.B., and REED, J.C. 1955. Flow of non-Newtonian fluids correlation of the laminar, transition and turbulent-flow regions. Am. Inst. Chem. Engrs. J. *1*, No. 4, 434–440.

MOHSENIN, N.N. 1978. Physical Properties of Plant and Animal Materials, Vol. 1, Part II. Gordon and Breach Science Publishers, New York.

MOODY, L.F. 1944. Friction factors for pipe flow. Trans. Am. Soc. Mech. Engrs. *66*, 671.

MOONEY, M. 1931. Explicit formulas for slip and fluidity. J. Rheol. *2*, 210.

MUGELE, K.A., and EVANS, H.D. 1951. Droplet size distributions in sprays. Ind. Eng. Chem. *43*, 1317.

NEUMANN, B.S. 1953. Powders. *In* Flow Properties of Disperse Systems, J. J. Hermanns (Editor). John Wiley & Sons, New York.

NEWITT, D.M., RICHARDSON, J.F., and GLIDDON, B.J. 1961. Hydraulic conveying of solids in vertical pipes. Trans. Inst. Chem. Engrs. (London) *39*, 93–100.

ORR, C., Jr. 1966. Particulate Technology. Macmillan Co., New York.

PERRY, J.H. 1950. Chemical Engineer's Handbook, 3rd Edition. McGraw-Hill Book Co., New York.

RABINOWITSCH, B. 1929. On the viscosity and elasticity of solutions. Z. Physik. Chem. *A145*, 1.

REIDY, G.A. 1970. Relationship Between Engineering and Texture Parameters for Low and Intermediate Moisture Foods. Ph.D. Thesis. Michigan State Univ., East Lansing. MI.

RICHMOND, O., and GARDNER, G.C. 1962. Limiting spans for arching of bulk materials in vertical channels. Chem. Eng. Sci. *17*, 1071–1078.

SARAVACOS, G.D. 1968. Tube viscometry of fruit purées and juices. Food Technol. *22,* No. 12, 1585–1588.

SCHERR, H.J., and WITNAUER, L.P. 1967. The application of a capillary extrusion rheometer to the determination of the flow characteristics of lard. J. Am. Oil Chemists Soc. *44,* 275–280.

SCOTT BLAIR, G.W. 1967. A model to describe the flow curves of concentrated suspensions of spherical particles. Rheol. Acta Band. *6,* 201–202.

SKELLAND, A.H.P. 1967. Non-Newtonian Flow and Heat Transfer. John Wiley & Sons, New York.

STERLING, C., and WUHRMANN, J.J. 1960. Rheology of cocoa butter. I. Effects of contained fat crystals on flow properties. Food Res. *25,* 460–463.

SZCZESNIAK, A.S. 1963. Objective measurement of food texture. J. Food Sci. *28,* 410.

WATSON, E.L. 1968. Rheological behavior of apricot purées and concentrates. Can. Agr. Engr. *10,* No. 1, 8–12.

WELTMANN, RUTH N., and KELLER, J.S. 1957. Some observations of the flow of linear polymer solutions. NACA Tech. Note. 3889.

WILKINSON, W.L. 1960. Non-Newtonian Fluids. Pergamon Press, Elmsford, N.Y.

WOHL, M.H. 1968. Designing for non-Newtonian fluids. Chem. Eng. *75,* No. 4, 130–136.

WOHL, M.H., 1968B. Instruments for viscometry. Chem. Eng. *75,* No. 7, 99–104.

WOHL, M.H. 1968C. Isothermal laminar flow of non-Newtonian fluids in pipes. Chem. Eng. *75,* No. 8, 143–146.

WOHL, M.H. 1968D. Dynamics of flow between parallel plates and in non-circular ducts. Chem. Eng. *75,* No. 10, 183–186.

WOHL, M.H. 1968E. Isothermal turbulent flow in pipes. Chem. Eng. *75,* No. 12, 95–100.

ZENZ, F.A., and OTHMER, D.F. 1960. Fluidization and Fluid-Particle Systems. Reinhold Publishing Corp., New York.

# 3

# Heating and Cooling Processes

Heating and cooling of food products are probably the most frequently occurring processes in the food processing plant. Only a small percentage of processed foods do not receive some type of heat process or are not cooled at some point between the time of entering the food-processing plant and arriving at the point of consumption. As heat transfer is the unit operation involved in both heating and cooling, it must receive considerable attention when designing processes for a processing plant.

Information on heat transfer may be more important when considering processes involving food products than in the design of processes for other materials. This importance is related to several factors, the most important being the heat-sensitivity of foods. Any process involving heat transfer in food products must account for the fact that the food product will tolerate only a limited amount of heat before deterioration of quality occurs. The theoretical and empirical relationships utilized in the design of heat processes assume knowledge of the thermal properties of the materials being utilized. Unfortunately this information on food products is not always readily available. The information on thermal properties that is available has been obtained by using several different techniques, and the values are not always in agreement. A rather basic understanding of heat transfer is important, therefore, in the utilization of conflicting information on thermal properties and in the design of methods which can best be utilized to measure and obtain accurate information on thermal properties of foods.

An additional factor which complicates the use of the most readily available relationships in heat transfer for food products is their non-Newtonian flow characteristics. There is an increasing amount of literature accumulating on heat transfer to non-Newtonian fluids during laminar and turbulent flow, and the application of this information to heat transfer in food products should be extremely useful.

## 3.1. MODES OF HEAT TRANSFER

Heat transfer is defined as the transmission of energy from one region to another by means of a temperature gradient which exists between the two regions. There are three recognized modes of heat transfer. Conduction is the transfer of heat within a body or from one body to another by interchange of kinetic energy between molecules without actual displacement of molecules. This mode of heat transfer describes the heat flow within a solid food product during heating or cooling.

Though convection does not conform entirely to the strict definition of heat transfer, it has been accepted as an important heat-transfer mode. In this particular mode, energy is transported by a combination of heat conduction, energy storage and mixing action. An example of convection is the transfer of heat to a product in a tubular heat exchanger where heat is transferred from the wall to the fluid by conduction, energy storage and mixing action of the fluid product.

Heat transfer due to radiation occurs when energy is transported by electromagnetic waves from a body of high temperature to one of a low temperature. The temperature difference between and the surface characteristics of the two bodies are of particular importance in this mode of heat transfer. Examples of radiation heat transfer include the net transfer of heat from the surface of a ham suspended in a cool room where the heat moves from the warm product surface to the cooler surfaces of the walls. The same type of heat transfer exists during baking of a bread loaf where the transfer of heat is from the walls of the oven at the much higher temperature to the cooler surface of the loaf of bread.

### 3.1.1 Conduction

The concept of heat conduction depends on basic properties of the solid material and rather significantly on the geometry involved. The expression utilized to describe heat conduction depends significantly on whether cartesian, cylindrical, or spherical coordinates are utilized in describing the system. The expression which describes the temperature distribution that may develop in a body due to heat conduction can be derived by utilizing a volumetric element such as illustrated in Fig. 3.1 for a cartesian coordinate system. By equating the sum of the heat flowing into the element during a given segment of time and heat generated within the element during the same segment of time to heat flowing out of the element during the same time segment plus a change in internal energy within the element during the same time period, a partial differential equation can be derived. The expression obtained describes the transient, three-dimensional heat conduction in an isotropic solid with internal heat generation, and is normally written in the following way:

$$\frac{\partial^2 T}{\partial x^2} + \frac{\partial^2 T}{\partial y^2} + \frac{\partial^2 T}{\partial z^2} + \frac{q}{k} = \frac{1}{\alpha}\frac{\partial T}{\partial t} \quad (3.1)$$

where

$$\alpha = \frac{k}{c_p \rho} \quad (3.2)$$

Equation (3.1) can be subdivided in various ways to obtain equations which can be solved for specified initial and boundary conditions. When conduction occurs at steady state and there is no internal heat generation, the well-known Laplace equation is obtained as follows:

$$\frac{\partial^2 T}{\partial x^2} + \frac{\partial^2 T}{\partial y^2} + \frac{\partial^2 T}{\partial z^2} = 0 \quad (3.3)$$

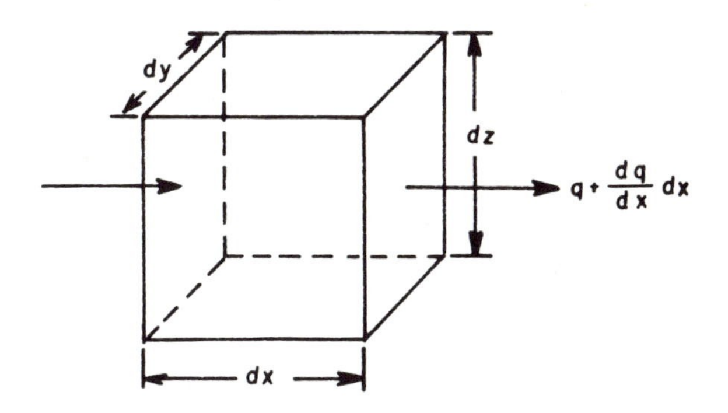


FIG. 3.1. HEAT TRANSFER IN A VOLUMETRIC ELEMENT OF SOLID

From equation (3.3), for the case of one-dimensional heat conduction, the heat flux will be directly related to the temperature gradient in the direction of heat flow. The Fourier heat conduction equation for heat flux in one dimension is:

$$q = -kA\frac{dT}{dx} \quad (3.4)$$

where the negative sign indicates that heat must flow downhill on the temperature scale, thus the second law of thermodynamics will be satisfied. Expressions similar to equations (3.1), (3.3) and (3.4) can be written for cylindrical or spherical coordinates. These expressions are given in well-known heat transfer or transport phenomena books by Carslaw and Jaeger (1959), Kreith (1965) and Bird *et al.* (1960). Use of equation

(3.4) to compute heat flux due to heat conduction through a solid material requires an integration to obtain the following expression:

$$q = \frac{A(T_2 - T_1)}{L/k} \tag{3.5}$$

where $T_2$ and $T_1$ are the surface temperatures and $L$ is the thickness of the material through which heat conduction is occurring. In many situations the material will not have uniform thermal conductivity. In these situations, a composite structure may be utilized. The following expression will apply when component materials are in series:

$$q = \frac{A\,(T_4 - T_1)}{L_1/k_1 + L_2/k_2 + L_3/k_3} \tag{3.6}$$

In equation (3.6), $L_1$, $L_2$, and $L_3$ represent the thickness of each portion of the composite structure, and $k_1$, $k_2$, and $k_3$ represent the thermal conductivity of each corresponding part of the composite structure. The $L/k$ portions of equations (3.5) and (3.6) are commonly called thermal resistances. From equation (3.5) the influence of $L/k$ on the heat flux can be examined. An increase in the thermal resistance ($L/k$) decreases heat flux. The increase in heat resistance can only be due to an increase in the thickness ($L$) of the material or a decrease in thermal conductivity ($k$). Thus, materials of low thermal conductivity ($k$) are useful as insulation since they reduce the heat flux. The insulation value of materials is commonly expressed as the 'R value' which is the same as $L/k$. This type of analysis is quite common when designing insulation for pipes conveying steam or process liquid in a food processing plant. In cylindrical coordinates the following expression can be derived from equation (3.4) for a three-layered system (Fig. 3.2):

$$q = \frac{2\pi Z(T_1 - T_4)}{\ln(r_2/r_1)/k_1 + \ln(r_3/r_2)/k_2 + \ln(r_4/r_3)/k_3} \tag{3.7}$$

EXAMPLE 3.1 A $1 \times 10^{-2}$m thick steel pipe with an internal diameter of $5 \times 10^{-2}$m is used to convey steam. The inside pipe surface temperature is 120 C. The pipe is covered with $3 \times 10^{-2}$m thick insulation. Under steady state conditions the heat loss from the pipe (with insulation) is 35 W/m. Calculate the outer insulation surface temperature. The thermal conductivity of steel is 16.3 W/m C and of the insulation is 0.038 W/m C.

SOLUTION

(1) A modified form of equation (3.7) can be used directly to compute the outer surface temperature

$$35 = \frac{2\pi(1)(120 - T)}{\ln(\frac{3.5}{2.5})/16.3 + \ln(\frac{6.5}{3.5})/0.038}$$

$$T = 29.14 \text{ C}$$

(2) The outer surface temperature of the insulation is 29.14 C.

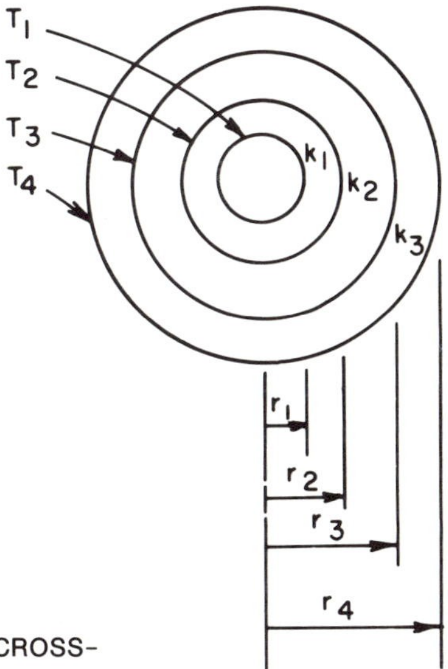


FIG. 3.2. A THREE-LAYERED CYLINDRICAL CROSS-SECTION

A typical situation which may occur in food products is that of heat generation. Such would be the case if heat of respiration were to be accounted for in heat conduction problems relating to food products. If the heat conduction were limited to one dimension at steady state, equation (3.1) would reduce to the following:

$$-k\frac{d^2T}{dx^2} = \dot{q} \tag{3.8}$$

indicating that all heat generated ($\dot{q}$) must flow through the surfaces of the one-dimensional body. A solution to equation (3.8) can be obtained by setting a surface temperature ($T_o$) for a solid material with half-thickness ($L$). The temperature distribution within the body is given by:

$$T = -\frac{\dot{q}}{2k}x^2 + \frac{\dot{q}L}{k}x + T_o \tag{3.9}$$

Equation (3.9), or a slight variation of it, indicates that the temperature distribution across the body or plate is parabolic with the maximum temperature at the distance ($L$) from the surface and the temperature decreasing to $T_o$ at the surface. An application of expressions such as equation (3.9) is in the evaluation of product temperatures which may occur due to internal heat generation under various boundary conditions.

Under many practical situations in food products, the boundary conditions result in heat flow in more than one direction within the body and require solution of a two- or three-dimensional heat conduction problem. The analytical solution of an expression of the type given as equation (3.3) is normally accomplished by a method called separation of variables. This procedure is described in detail in most heat-transfer books and results in an infinite series expression. Analytical solutions have somewhat limited value, because of the changes in the solution that result from small changes in geometry or boundary conditions. The distinct advantage is that temperature distribution can be visualized and the influence of various parameters can be evaluated.

There are several other ways of evaluating temperature distributions in two- or three-dimensional systems with heat conduction occurring. These include potential-field plotting, analogical methods, and numerical relaxation methods. All these techniques are discussed in considerable detail in heat-transfer books, including Kreith (1965). The numerical relaxation technique seems to be the most logical approach when computers are available.

### 3.1.2 Convection

Heat transfer in a fluid system will normally occur by convection. Since the transfer of heat by convection is due to the mixing action within the fluid, the rate of heat flow from a surface to the fluid will depend on the fluid flow properties as well as the thermal properties of the fluid. Figure 3.3 illustrates the type of temperature gradient that will exist when a warm wall surface is exposed to a fluid. The region near the wall within which the temperature changes from the wall temperature ($T_s$) to the free stream temperature ($T_\infty$) is defined as the thermal boundary layer in much the same way as the hydrodynamic boundary layer was defined in Chapter 2. The conditions illustrated can exist under both free convection and forced convection. The basic equation which describes heat transfer by convection is:

$$q = h_c A (T_s - T_\infty) \tag{3.10}$$

where $h_c$ is the convective heat-transfer coefficient, which will depend on the fluid flow characteristics and thermal properties of the fluid.

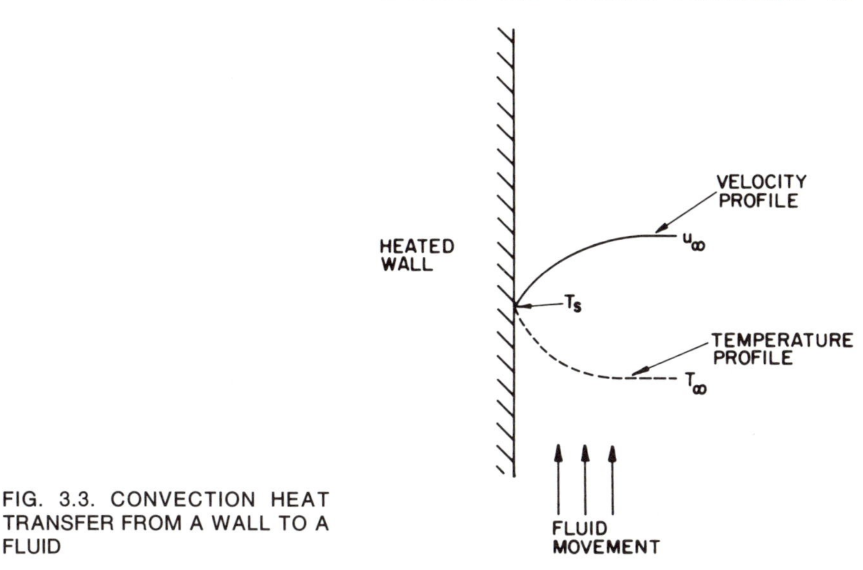


FIG. 3.3. CONVECTION HEAT TRANSFER FROM A WALL TO A FLUID

Most of the available information which leads to the evaluation of the convective heat-transfer coefficient is in the form of empirical relationships. In some cases, these empirical relationships may be derived from a theoretical basis, or the relationships may be based on dimensional analysis. The two types of convective heat transfer are forced convection and free convection.

**3.1.2a Forced convection.**—In forced convection, the fluid flow patterns around the heated surface are determined by an external force such as a pump or a fan. Dimensional analysis reveals the relationship between three dimensionless groups as shown in the following expression:

$$N_{Nu} = f\,(N_{Re}, N_{Pr}) \tag{3.11}$$

where

$$N_{Nu} = \frac{h_c l}{k_f} \tag{3.12}$$

$$N_{Pr} = \frac{c_p \mu}{k_f} \tag{3.13}$$

The dimensionless groups in equation (3.11) are the Nusselt number ($N_{Nu}$), the Reynolds number ($N_{Re}$) and the Prandtl number ($N_{Pr}$). The Nusselt number, defined in equation (3.12), is directly dependent on the convective heat-transfer coefficient ($h_c$), some characteristic dimension ($l$) and the thermal conductivity of fluid ($k_f$). The third dimensionless group is the Prandtl number which is directly dependent on thermal and physical properties of the fluid as shown in equation (3.13).

In designing heat transfer equipment it is important for an engineer to estimate the convective heat transfer coefficient. The following discussion provides information to estimate $h_c$ values for different geometries.

The classic example of the theoretical basis for the relationship given by equation (3.11) is heat transfer from a flat plate to a fluid during laminar flow. In this particular situation, the continuity equation, the momentum equation and the energy equation are combined and solved to give an exact solution of the following form:

$$N_{Nu} = 0.664\, N_{Re}^{1/2}\, N_{Pr}^{1/3} \tag{3.14}$$

Probably the most critical factor in obtaining equation (3.14) was simultaneous solution of the momentum equation and continuity equation by Blasius (1908), to obtain the velocity distribution within a laminar boundary layer on a flat plate. Equation (3.14) has been verified by approximate boundary layer analysis and by experimental methods. Attempts have been made to extend the relationship between heat and momentum transfer to turbulent flow over flat plates and other geometries. The success has been limited, however, and the attempt to find a theoretical basis for describing heat transfer during turbulent flow is continuing.

The most widely used expressions in the design of equipment for food processing are those which allow the evaluation of heat-transfer coefficients in tubular geometries, primarily tubular heat exchangers used in various processes. As might be expected, most of the available expressions represent examples of the empirical relationships available and, in addition, represent types which appear to have considerable use in the design of processing equipment for foods.

One of the obvious problems when evaluating convective heat-transfer coefficients for tubular geometries is the selection of the temperatures to be used in equation (3.10) and the temperature at which the properties of the fluid will be evaluated. The so-called free-stream temperature ($T_\infty$) in equation (3.10) will vary from the entrance to the outlet of a tube in which convective heat transfer is occurring. The actual rise in temperature is described by the expression:

$$q = w\, C_p\, \Delta T_b \tag{3.15}$$

where $\Delta T_b$ is the change in bulk fluid temperature between entrance and exit of the tube. The bulk temperature ($T_b$) is the average temperature of the fluid at the given cross-section in the tube. The convective heat-transfer coefficient will then be defined by a revised form of equation (3.10) as follows:

$$h_{cL} = \frac{q}{A(T_s - T_b)_{lm}} \tag{3.16}$$

where the convective heat-transfer coefficient is based on the log mean temperature difference between the surface and bulk temperature of the fluid. The log mean temperature difference is defined in the following manner:

$$(T_s - T_b)_{lm} = \frac{(T_s - T_{b_1}) - (T_s - T_{b_2})}{\ln\left[\dfrac{T_s - T_{b_1}}{T_s - T_{b_2}}\right]} \tag{3.17}$$

assuming the tube has a constant surface temperature in the axial direction.

Most empirical expressions derived to allow evaluation of the convective heat transfer have been derived to give the convective heat-transfer coefficient based on the log mean temperature difference as defined in equation (3.17). As would be expected, these expressions vary depending on the flow characteristics of the tube and in some cases on whether the fluid is being heated or cooled in the tube.

The following sections include information on estimating convective heat-transfer coefficient for two types of flow; namely, internal flow with fluid flowing inside a stationary conduit and external flow with fluid flowing past a stationary immersed solid.

*Laminar Flow.—Internal Flow.* Although laminar flow heat transfer has been studied extensively, true laminar flow heat transfer very rarely occurs due to presence of natural convection. Under these circumstances, empirically derived equations are most reliable. Empirical data are often correlated by Nusselt number, $(N_{Nu})_{lm}$ or $(N_{Nu})_{am}$, Graetz Number, $N_{Gz} = (N_{Re}, N_{Pr}, D/L)$, and Grashof Number, $N_{Gr}$, to account for influence of natural convection.

*Circular Tubes.* For horizontal tubes, following expressions may be used:

$$(N_{Nu})_{lm} = 3.66 + \frac{0.085\, N_{Gz}}{1 + 0.047\, N_{Gz}^{2/3}} \left(\frac{\mu_b}{\mu_w}\right)^{0.14} \quad \text{for } N_{Gz} < 100 \tag{3.18}$$

$$(N_{Nu})_{am} = 1.86\, N_{Gz}^{1/3} \left(\frac{\mu_b}{\mu_w}\right)^{0.14} \quad \text{for } N_{Gz} > 100 \tag{3.19}$$

Where *lm* and *am* refer to log mean and arithmetic mean, respectively. It can be determined through theoretical analysis that for very long circular tubes $(N_{Nu})_{lm}$ approaches a limiting value of 3.66 for constant wall temperature and 4.36 for constant heat flux, assuming that $N_{Gz} < 4.0$.

The charts developed by Pigford (1955) may be used for predicting heat transfer coefficients in laminar flow in vertical tubes.

*Non-Circular Ducts.* Equations 3.18 and 3.19 may be used for most non-circular ducts if the equivalent diameter is used as the characteristic length. The equivalent diameter is defined as 4F/wetted perimeter.

*External Flow.* Laminar flow is said to exist when flow occurs over an immersed body with a complete laminar layer over the whole body. The preceding would be true even though the flow in the mainstream is turbulent.

A general relationship to predict average heat transfer coefficient on an immersed body is given by:

$$N_{Nu} = C_r\,(N_{Re})^m\,(N_{Pr})^{1/3} \tag{3.20}$$

where $C_r$ and $m$ are given in Table 3.1. The characteristic length is used to compute both Nusselt Number and Reynolds Number. All properties are evaluated at the film temperature = $(T_s + T_\infty)/2$. The velocity term in the Reynolds Number is the undisturbed free stream velocity.

*Transition Region.*—The evaluation of the convective heat-transfer coefficient during the transition region in flow characteristics ($2100 < N_{Re} < 10{,}000$) cannot be done with simple equation. A diagrammatic relationship between heat transfer coefficient and Reynold's number is given by Perry and Chilton (1973) that may be used for obtaining estimates.

*Turbulent Flow.—Circular Tubes.* For evaluation of convective heat-transfer coefficients when flow is turbulent, expressions are based on empirical expressions proposed by Sieder and Tate (1936). The following expression is found to be acceptable when the Prandtl Number ($N_{Pr}$) is between 0.7 and 700 and the L/D ratio is greater than 60:

$$N_{Nu} = 0.023 N_{Re}^{0.8}\, N_{Pr}^{1/3}\, (\mu_b/\mu_s)^{0.14} \tag{3.21}$$

Equation (3.21) describes experimental data to within 20%. Some variations in the exponent of the viscosity ratio were found by Kays and London (1954), whose research indicated that the exponent may be 0.36 for heating and 0.2 for cooling. If the fluid is being heated at a very low viscosity during turbulent flow, the following expression proposed by Dittus and Boelter (1930) appears to be acceptable:

$$N_{Nu} = 0.023 N_{Re}^{0.8} N_{Pr}^{0.4} \tag{3.22}$$

During cooling, the exponent on the Prandtl Number ($N_{Pr}$) is changed to 0.3. In equations (3.21) and (3.22) the Reynolds Number and Prandtl Number are evaluated at bulk fluid temperatures.

TABEL 3.1. LAMINAR-FLOW HEAT TRANSFER OVER IMMERSED BODIES*

| Configuration | Characteristic length | $N_{Re}$ | $N_{Pr}$ | $C_r$ | $m$ |
|---|---|---|---|---|---|
| Flat plate parallel to flow ......... | Plate length | $10^3$ to $3 \times 10^5$ | >0.6 | 0.648 | 0.50 |
| Circular cylinder axes perpendicular to flow ....................... | Cylinder diameter | 1–4 | | 0.989 | 0.330 |
| | | 4–40 | | 0.911 | 0.385 |
| | | 40–4000 | >0.6 | 0.683 | 0.466 |
| | | $4 \times 10^3 - 4 \times 10^4$ | | 0.193 | 0.618 |
| | | $4 \times 10^4 - 2.5 \times 10^5$ | | 0.0266 | 0.805 |
| Non-circular cylinder, axis ........ | Square, short diameter | $5 \times 10^3 - 10^5$ | | 0.104 | 0.675 |
| Perpendicular to flow, characteristic | Square, long diameter | $5 \times 10^3 - 10^5$ | | 0.250 | 0.588 |
| Length perpendicular to flow ...... | Hexagon, short diameter | $5 \times 10^3 - 10^5$ | >0.6 | 0.155 | 0.638 |
| | Hexagon, long diameter | $5 \times 10^3 - 2 \times 10^4$ | | 0.162 | 0.638 |
| | | $2 \times 10^4 - 10^5$ | | 0.0391 | 0.782 |
| Sphere | Diameter | $1 - 7 \times 10^4$ | 0.6–400 | 0.6 | 0.50 |

*Adapted from Perry and Chilton (1973)

*Non-Circular Ducts.* Equation (3.21) may be used to predict heat-transfer coefficient in non-circular ducts provided equivalent diameter is used instead of characteristic length.

**3.1.2b Free Convection.**—During free or natural convection, fluid flow is not induced by external forces and the motion within the fluid is brought about by the influence of temperature on fluid density and the development of a buoyant force. Under these conditions, the fluid motion is described in terms of the Grashof number ($N_{Gr}$), which is defined in the following way:

$$N_{Gr} = \frac{\rho^2 g \beta \chi^3 \Delta T}{\mu^2} \quad (3.23)$$

where $\chi$ is the characteristic dimension of the body involved in the natural or free convection, $\beta$ is the coefficient of expansion for the fluid being heated, and $\Delta T$ is the temperature gradient between the surface and the fluid. Empirical relationships involving free convection are of the following form:

$$N_{\overline{Nu}} = K\,(N_{Gr}\,N_{Pr})^a \quad (3.24)$$

where the Nusselt number ($N_{\overline{Nu}}$) describes an average convective heat-transfer coefficient for the surface and the constants ($K$ and $a$) vary depending with the geometry and orientation of the surface. The coefficients $K$ and $a$ are given in Table 3.2 for various configurations.

One general statement which must be made concerning all the relationships presented for the evaluation of convective heat-transfer coefficients is that the expressions have been developed for Newtonian fluids. Since the flow characteristics of food products may deviate from

TABLE 3.2. *VALUE OF a AND K FOR EQUATION 3.24

| Configuration | $N_{Gr}N_{Pr}$ | K | a |
|---|---|---|---|
| Vertical Surfaces | $< 10^4$ | 1.36 | 1/5 |
| L = vertical dimension 0.9 m | $10^4 < N_{Gr}N_{Pr} < 10^9$ | 0.59 | 1/4 |
| | $> 10^9$ | 0.13 | 1/3 |
| Horizontal Cylinder | $< 10^{-5}$ | 0.49 | 0 |
| L = diameter 0.2 m | $10^{-5} < N_{Gr}N_{Pr} < 10^{-3}$ | 0.71 | 1/25 |
| | $10^{-3} < N_{Gr}N_{Pr} < 1$ | 1.09 | 1/10 |
| | $1 < N_{Gr}N_{Pr} < 10^{-4}$ | 1.09 | 1/5 |
| | $10^{-4} < N_{Gr}N_{Pr} < 10^9$ | 0.53 | 1/4 |
| | $> 10^9$ | 0.13 | 1/3 |
| Horizontal flat surface | $10^5 < N_{Gr}N_{Pr} < 2 \times 10^7$ (FU) | 0.54 | 1/4 |
| | $2 \times 10^7 < N_{Gr}N_{Pr} < 3 \times 10^{10}$ (FU) | 0.14 | 1/3 |
| | $3 \times 10^5 < N_{Gr}N_{Pr} < 3 \times 10^{10}$ (FD) | 0.27 | 1/4 |

FU = Facing upward.
FD = Facing downward.
*Adapted from Perry and Chilton (1973).

Newtonian flow considerably, the error involved in the use of expressions presented in this section may lead to significant error.

It would appear that as long as the food product has a low viscosity and approaches Newtonian behavior, these expressions should be acceptable. Some effort should be made, however, to utilized expressions which will be presented later in this chapter to describe and incorporate the non-Newtonian state existing in many food products.

### 3.1.3 Radiation

Heat transfer from a surface due to radiation is proportional to the temperature of the surface. More specifically, the energy emitted from a surface or heat flux is expressed by the following equation:

$$q/A = e\,\sigma\,T_A^4 \tag{3.25}$$

which illustrates that the heat flux is proportional to the fourth power of the surface temperature. The constant ($\sigma$) in equation (3.25) is the Stefan-Boltzmann constant, which is $5.669 \times 10^{-8}$ W/m$^2$ K$^4$. The emissivity ($e$) in equation (3.25) describes the extent to which the surface is similar to a black body. The value of emissivity for a black body would be 1.

The transfer of heat by radiation between two surfaces is dependent upon the emissivity of the radiating surface and the absorptivity of that same surface. The expression normally used to describe this type of heat transfer is as follows:

$$q_{1\text{-}2} = A\,\sigma\,[e_1\,T_{A1}^4 - \alpha_{1\text{-}2}\,T_{A2}^4\,] \tag{3.26}$$

where $e_1$ is the emissivity of the radiating surface at temperature ($T_{A1}$) and $\alpha_{1-2}$ is the absorptivity of the surface for radiation being emitted at temperature ($T_{A2}$). In most cases the heat transfer due to radiation by the surface is described in terms of a radiation heat-transfer coefficient. Such a coefficient is described in the same manner as the previous heat-transfer coefficient and can be expressed in the following way:

$$q_r = h_r\,A_1\,(T_1 - T_2) \tag{3.27}$$

The use of a radiation heat-transfer coefficient allows this type of heat transfer to be treated in the same manner as the previously discussed modes of heat transfer. This coefficient ($h_r$) will not be a constant, however, since it is a function of $T_A^3$.

### 3.1.4 Overall Heat Transfer

Under most conditions, heat transfer from one surface to a fluid or from one fluid to another will occur by more than one mode of heat transfer. To account for such situations, an overall heat-transfer coefficient is utilized, which can be described by the following expression:

$$q = UA(T_1 - T_2) \tag{3.28}$$

where $U$ is the overall heat-transfer coefficient incorporating heat transfer due to all three modes of heat transfer, if necessary. If, for example, heat transfer should occur by all three modes of heat transfer across constant surface area, the overall heat-transfer coefficient would be defined in the following manner:

$$\frac{1}{UA_{lm}} = \left(\frac{1}{h_c} + \frac{1}{h_r}\right) \frac{1}{A_{\text{outside}}} + \frac{L}{kA_{lm}} \tag{3.29}$$

In equation (3.29), each term on the right-hand side represents the resistance due to the particular mode involved and the sum $(1/U)$ represents the overall resistance to heat transfer.

## 3.2 THERMAL PROPERTIES OF FOODS

Probably one of the more important limitations in process design for food products is the lack of information on their thermal properties. The importance of these properties cannot be overemphasized, because of the influence of composition upon them and the obvious changes in composition which occur during such processes as freezing, evaporation, and dehydration. Although a rather large volume of information (primarily experimental data) can be found, the differences in the methods used to measure the thermal properties place serious limitations on the value of the available data. In addition, the amount of thermal properties data that would be required to describe all products and all possible conditions and compositions for each product would be almost infinite. A solution of this type of situation could be obtained through prediction of the thermal properties of the components within the product.

### 3.2.1 Specific Heat

Several expressions have been proposed to allow prediction of specific heat based on components of a food product. Dickerson (1969) presented the following expression which has been used considerably in higher-moisture food products:

$$c_p = 1.675 + .025 \text{ (Water Content, \%)} \tag{3.30}$$

This particular expression has been used for several meat products and appears to be consistent in the range of 26 to 100% moisture content. In addition, it has been applied to fruit juices which have moisture contents greater than 50%. A similar expression was proposed by Siebel (1892) as follows:

$$c_p = 0.837 + .034 \text{ (Water Content, \%)} \tag{3.31}$$

Although no specification is made, it would appear that this expression also is limited to higher-moisture food products.

An expression which is somewhat more dependent on specific heats of product components was given by Charm (1978) as follows:

$$c_p = 2.094 \, X_F + 1.256 \, X_S + 4.187 \, X_M \tag{3.32}$$

where the values of 2.094, 1.256 and 4.187 are specific heats of the fat, solids and water present in the product, respectively. This concept can be expanded even further to incorporate specific heats of even more basic components of the product to give the following expression:

$$c_p = 1.424 \, X_c + 1.549 \, X_p + 1.675 \, X_F + 0.837 \, X_a + 4.187 \, X_M \tag{3.33}$$

where the value of 1.675 used for specific heat of fat is for solid fat, as compared to the value of 2.094 in equation (3.32), which is for liquid fat.

EXAMPLE 3.2 A formulated food product contains the following components, water 80%, protein 2%, carbohydrate 17%, fat 0.1% and ash 0.9%. Predict the specific heat using different mathematical models presented in equations (3.30), (3.31), (3.32) and (3.33).

SOLUTION

(1) Using the model proposed by Dickerson (1969)

$$c_p = 1.675 + 0.025(80) = 3.68 \text{ kJ/kg C}$$

(2) Using Siebel's (1892) model

$$c_p = 0.837 + 0.034 \, (80) = 3.56 \text{ kJ/kg C}$$

(3) Using the expression given by Charm (1978)

$$c_p = 2.094 \times 0.001 + 1.256 \times 0.199 + 4.187 \times 0.8 = 3.60 \text{ kJ/kg C}$$

(4) Expression (3.33) may be used to give

$$c_p = 1.424 \times 0.17 + 1.549 \times 0.02 + 1.675 \times 0.001 + 0.837 \times 0.009 + 4.187 \times 0.8 = 3.63 \text{ kJ/kg C}$$

(5) The predicted values for this product are similar for all models.

The specific heats, as predicted by equations (3.30), (3.32) and (3.33), are given in Table A.9 of the Appendix. In this table, the values predicted by the expressions are compared to experimental values, and several other interesting observations can be made. Probably the most obvious is that equation (3.30) does result in considerable error for low-moisture products. Equations (3.32) and (3.33) predict values close to the experimental data in all situations.

### 3.2.2 Thermal Conductivity

Most expressions which have been proposed to predict thermal conductivity of products consider the product to be a two-phase system, and incorporate the effect of the thermal conductivity of water and of the solid components in the product. Such expressions have been used most widely in the prediction of the change in thermal conductivity of the product during a phase change, such as during freezing, when the thermal conductivity of the water portion changes significantly as it changes from liquid to solid. Riedel (1949) has proposed a completely empirical expression for fruit juices and sugar solutions as follows:

$$k = [326.575 + 1.0412T - 0.00337T^2]\,[.796 + .009346\,(\%\ \text{water})]10^{-3} \tag{3.34}$$

where the temperature ($T$) is in C. This expression is very limited in use, since it can be applied only to the indicated products.

Sweat (1974) has experimentally determined thermal conductivities of several fruits and vegetables. He provides the following regression equation to predict thermal conductivity of fruits and vegetables with water contents greater than 60%. This equation predicts thermal conductivity within ±15% of experimental values; however, it is unsatisfactory for low density foods and foods containing void spaces such as apples.

$$k = 0.148 + 0.00493\,(\%\ \text{Water}) \tag{3.35}$$

Expressions which seem to have somewhat wider applications include the so-called Maxwell equation (Maxwell 1904) as given in the following:

$$k = k_L \left[\frac{1 - (1 - a\,k_S/k_L)\,b}{1 + (a - 1)\,b}\right] \tag{3.36}$$

where:

$$a = \frac{3k_L}{2k_L + k_S}$$

$$b = \frac{X_S}{X_L + X_S}$$

which was used by Long (1955) to predict the change of thermal conductivity of cod fish during freezing. Predicted thermal conductivities were compared with experimental values and the agreement is illustrated in Fig. 3.4. As is shown, there is a considerable change in thermal conductivity as the water changes phase. Lentz (1961) used the Maxwell equation to analyze the thermal conductivities of meat and similar products. He concluded that satisfactory results were obtained when the product contained a dispersed phase of solids. There were some limitations when predicting the thermal conductivity of a fibrous material such as

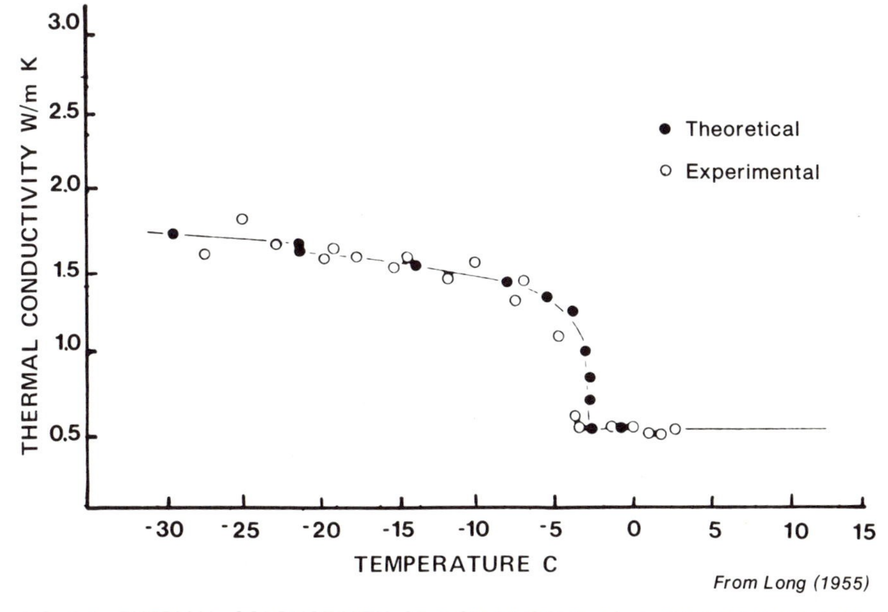


FIG. 3.4. THERMAL CONDUCTIVITY OF FISH MUSCLE AS A FUNCTION OF TEMPERATURE

beef, due to the fact that thermal conductivity parallel to the fibers was approximately 15 to 20% higher than thermal conductivity perpendicular to the fibers.

Kopelman (1966) has presented three models to investigate thermal conductivity of foods.

A. A two component-three dimensional-isotropic system. The two components may form two phases. One component is randomly dispersed in the other to form the non-continuous phase (Fig. 3.5). Examples of this model are butter (water dispersed in fat) or ice cream (air dispersed in liquid).

B. A two-component-two-dimensional-anisotropic fibrous system. The two components may form two phases. The fibers are parallel to each other and randomly distributed. The dispersed components (the fibers) are continuous in one direction and the random dispersion is two dimensional (Fig. 3.5). This model is typical of all fibrous systems such as, meat flesh, wood, or fibrous vegetables. This system is characterized with two thermal conductivities, $k_{\parallel}$, the thermal conductivity in the direction parallel to fibers; $k_{\perp}$ the thermal conductivity in the direction perpendicular to the fibers.

C. Two (or more) components-one dimensional-anisotropic layered system. In this model, the components may form more than one phase. The components are arranged in parallel layers to form some kind of a com-

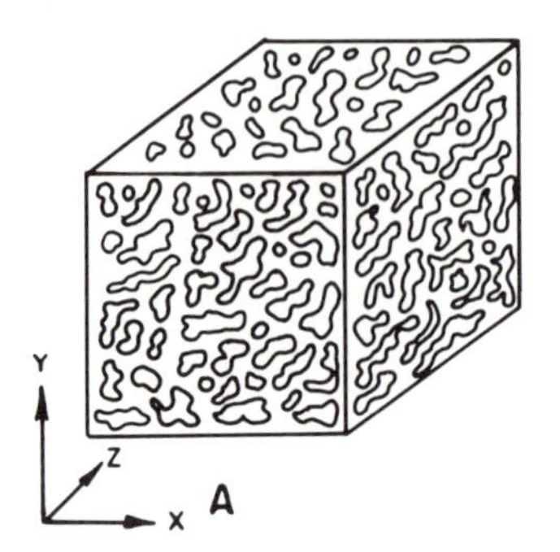


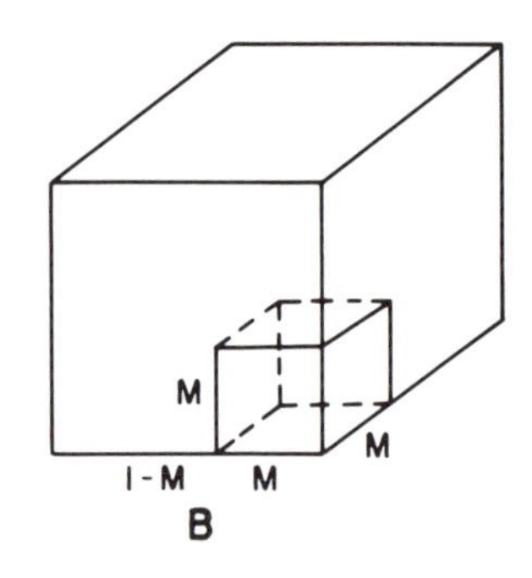


FIG. 3.5a. THE DIAGRAM OF THE TWO COMPONENT HOMOGENEOUS THREE DIMENSIONAL DISPERSION SYSTEM. A. NATURAL RANDOM STATE. B. THE REARRANGEMENT OF THE COMPONENTS

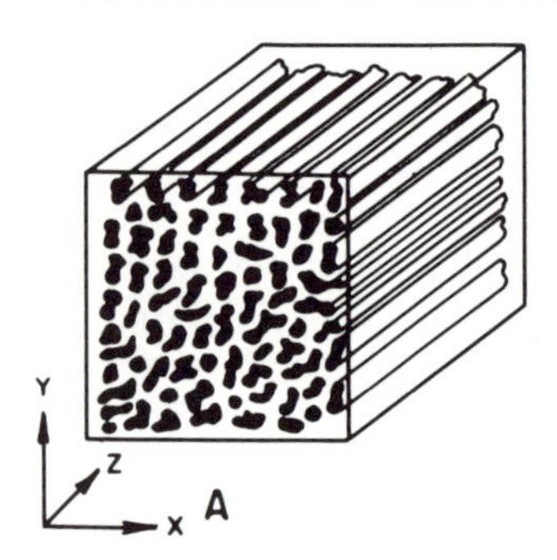


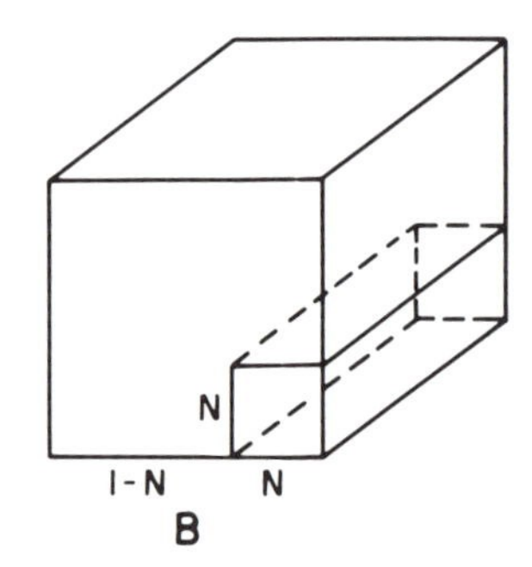


FIG. 3.5b. THE DIAGRAM OF THE TWO COMPONENT HOMOGENEOUS TWO DIMENSIONAL FIBROUS SYSTEM. A. NATURAL RANDOM STATE. B. THE REARRANGEMENT OF THE COMPONENTS

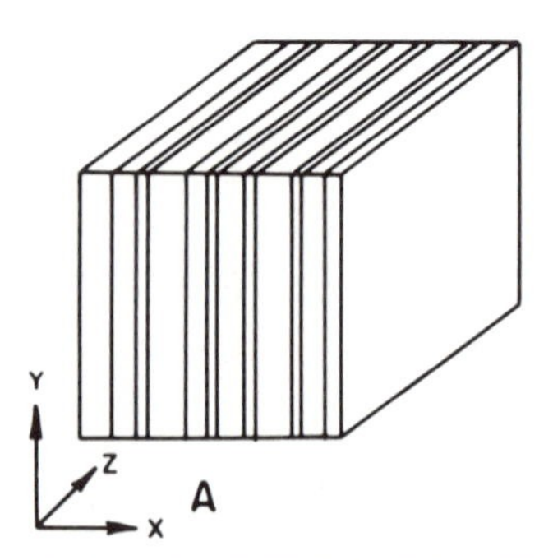


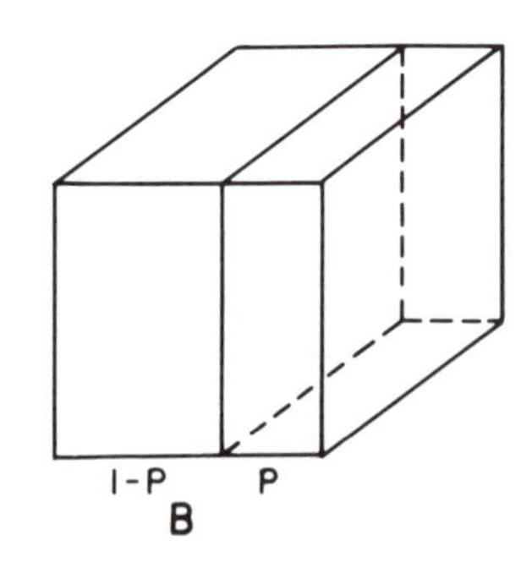


FIG. 3.5c. THE DIAGRAM OF THE TWO COMPONENT HOMOGENEOUS ONE DIMENSIONAL LAYERED SYSTEM. A. NATURAL RANDOM STATE. B. THE REARRANGEMENT OF THE COMPONENTS

*From Kopelman (1966)*

posite layer (Fig. 3.5). An example is fat layer above the flesh.

For the first isotropic model the following expression is proposed:

$$k = k_L \left[ \frac{1 - M^2}{1 - M^2 \ (1 - M)} \right] \tag{3.37}$$

where $M^3$ is the volume fraction of solids or discontinuous phase in the product, and the thermal conductivity of the continuous phase or liquid is much larger than the thermal conductivity of the discontinuous or solid phase. When the latter is not the case, the following expression will apply:

$$k = k_L \left[ \frac{1 - Q'}{1 - Q'(1 - M)} \right] \tag{3.38}$$

where:

$$Q' = M^2 \ (1 - k_s / k_L)$$

For the second model which is a two-component anisotropic system, two expressions are proposed. The first is for the case in which thermal conductivity would be measured in the direction parallel to the fibers as follows:

$$k_{\|} = k_L \ [1 - N^2 (1 - k_s / k_L)] \tag{3.39}$$

where $N^2$ is the volume fraction of solids or discontinuous phase in the fibrous product. If the thermal conductivity of interest is perpendicular to the fibers, the following expression is proposed:

$$k_{|} = k_L \quad \frac{1 - Q''}{1 - Q''(1 - N)} \tag{3.40}$$

where:

$$Q'' = \frac{N}{(1 - k_s / k_L)}$$

Another set of expressions is for the third model for a two-component product with a layered component constituting the discontinuous phase. For this situation, the thermal conductivity parallel to the layer is as follows:

$$k_{\|} = k_L \ [1 - P(1 - k_S / k_L)] \tag{3.41}$$

Where $P$ is the volume fraction solids or discontinuous phase. The following expression is proposed for prediction of thermal conductivity perpendicular to the layer in the system:

$$k_{\perp} = k_L \left[ \frac{k_S}{Pk_L + k_S(1 - P)} \right] \tag{3.42}$$

Since equations (3.37) through (3.42) as proposed by Kopelman (1966) have not been applied to food systems to a great extent, the validity of the proposed expression may be somewhat limited until more application is attempted. It would appear however, that an application should be of considerable value in systems containing fibers or layers.

EXAMPLE 3.3 Compare the Maxwell and Kopelman equations for predicting the thermal conductivity of lean beef with 74% moisture content at 7°C. A thermal conductivity of 0.26 W/m°C can be utilized for the solids fraction of beef.

SOLUTION

(1) The first factor to be resolved before using either of the two equations is the quantity representing the solids fraction. Since this quantity is dependent on volume fractions of the two phases, the density of the product must be known or assumed. Let us assume that the density of beef is 1280 kg/m³ and the density of water is 1000 kg/m³.

$$\text{The specific volume of water} = \left(\frac{74\text{ kg water}}{100\text{ kg beef}}\right) / \left(\frac{1000\text{ kg water}}{1\text{ m}^3\text{ water}}\right)$$

$$= 7.4 \times 10^{-4}\text{ m}^3\text{ water/kg beef}$$

Also, the specific volume of solids

$$= \frac{1\text{ m}^3\text{ beef}}{1280\text{ kg beef}} - 0.00074\text{ m}^3 \frac{\text{water}}{\text{kg beef}}$$

$$= 4.125 \times 10^{-5}\text{ m}^3\text{ solids/kg beef}$$

Using these values, the solids fraction ($b'$) in the Maxwell equation can be computed as follows:

$$b' = \frac{4.125 \times 10^{-5}}{7.4 \times 10^{-4} + 4.125 \times 10^{-5}} = 0.0528$$

(2) By using a thermal conductivity of .578 W/m C for water at 7C the value of the constant ($a$) in the Maxwell equation becomes

$$a = \frac{3(0.578)}{2(0.578) + 0.26} = \frac{1.734}{1.416} = 1.225$$

(3) Using the above values in the Maxwell equation leads to the following:

$$k = 0.578 \left[ \frac{1 - \left(1 - 1.225 \dfrac{0.26}{0.578}\right) 0.0528}{1 + (1.225 - 1)\, 0.0528} \right]$$

$$k = 0.558\text{ W/m C}$$

(4) In the Kopelman equation for heat conduction parallel to a fibrous product; $N^2 = 0.0528$

Therefore:

$$k_{\parallel} = 0.578\,[1 - 0.0528\,(1 - 0.26/0.578)]$$

$$k_{\parallel} = 0.561 \text{ W/m C}$$

(5) When heat conduction is perpendicular to the beef fibers; $N =$ 0.22978 and

$$Q'' = \frac{0.22978}{1 - 0.26/0.578} = 0.418$$

then:

$$k_{\perp} = 0.578 \left[ \frac{1 - 0.418}{1 - 0.418\,(1 - 0.22978)} \right]$$

$$k_{\perp} = 0.496 \text{ W/m C}$$

A model for predicting thermal conductivity of gas-filled porous solids was proposed by Harper and El Sahrigi (1964). This model requires an evaluation of several constants but may find application in predicting thermal conductivity of freeze-dried products over various pressure ranges. Chen (1969) proposed a model for predicting the thermal conductivity of a powdered food in a packed bed. The following expression accounts for heat transfer through solid and gaseous phases in the system and predicts the thermal conductivity to within 5%:

$$k = k_g\,\epsilon + k_S\,(1 - \epsilon) \tag{3.43}$$

where $\epsilon$ is the void fraction for the particle bed. This error is created by the fact that heat transfer through contact points of the powdered particles accounts for 2 to 4% of the heat flow; therefore, equation (3.43) would predict thermal conductivities which are lower than actual experimental data.

Harper (1976) has given the following model to estimate thermal conductivities of foods

$$\frac{k_{\text{food}}}{k_{\text{water}}} = 1.0 - 0.5\,X_S \tag{3.44}$$

Similarly Earle (1966) has presented the following expression for foods above freezing point

$$k = \frac{0.55}{100}\,(\%\text{ Water}) + \frac{0.26}{100}\,(100 - \%\text{ Water}) \tag{3.45}$$

A similar expression for foods below freezing was given as

$$k = \frac{2.4}{100}\,(\%\text{ Water}) + \frac{0.26}{100}\,(100 - \%\text{ Water}) \tag{3.46}$$

These simple models should be useful in estimating thermal conductivities of food where high accuracy is not desired. For more rigorous analysis the expressions presented earlier in this chapter are recommended.

## 3.3 STEADY-STATE HEATING AND COOLING

Although the steady-state heat transfer to food products has been discussed and described in the form of several semi-theoretical and empirical expressions, the equations presented up to this point have been developed primarily for Newtonian fluids. As indicated previously, many food products cannot be described by Newtonian flow characteristics, and therefore require an analysis which incorporates non-Newtonian characteristics.

### 3.3.1 Heat Transfer in Laminar Flow

An attempt will be made to present sufficient information on the heat-transfer characteristics of non-Newtonian fluids in tubes along with other flow characteristics which influence heat transfer in specific types of heat exchangers.

A basic differential equation to describe heat transfer during laminar flow in a tube may be developed by conducting an energy balance on an annular ring as shown in Figure 3.6.

Energy in by conduction at $r = (q_r|_r)\ 2\pi r\ \Delta\ z$

Energy out by conduction at $r + \Delta r = (q_r|_{r\ +\Delta r})\ 2\pi(r + \Delta r)\Delta z$

Energy in by conduction at $z = (q_z|_z) 2\pi r \Delta r$

Energy out by conduction at $z + \Delta z = (q_z|_{z\ +\Delta z})\ 2\pi r\ \Delta r$

Energy in with flowing fluid at $z = \rho C_p\ u(T - T_o)|_z 2\pi r \Delta r$

Energy out with flowing fluid at $z + \Delta z = \rho C_p\ u(T - T_o)|_{z\ +\Delta z}\ 2\pi r \Delta r$

Using the first law of thermodynamics

$$\text{energy input} = \text{energy output}$$

and taking limits as $\Delta r$ and $\Delta z$ approach zero.

$$\rho c_p\, u \frac{\partial T}{\partial z} = -\frac{1}{r}\frac{\partial}{\partial r}(r q_r) - \frac{\partial q_z}{\partial z}$$

From Fourier's law $q_r = -k\dfrac{\partial T}{\partial r}\,; q_z = -k\dfrac{\partial T}{\partial z}$

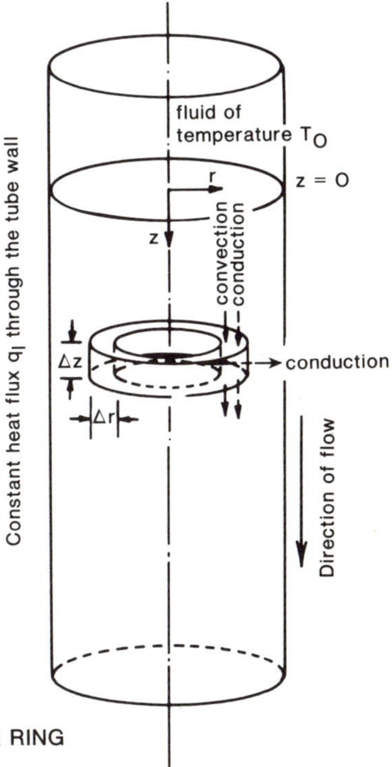

FIG. 3.6. ENERGY BALANCE ON AN ANNULAR RING

$$\rho c_p\, u \frac{\partial T}{\partial z} = k \left( \frac{1}{r} \frac{\partial}{\partial r} \left(r \frac{\partial T}{\partial r}\right) + \frac{\partial^2 T}{\partial z^2} \right) \tag{3.47}$$

Heat conduction in $z$-direction ($\partial^2 T/\partial z^2$) is small compared to heat conductive transfer term ( $\frac{\partial T}{\partial r}$ ) in the radial direction.

$$\therefore \rho c_p\, u \frac{\partial T}{\partial z} = k \left( \frac{1}{r} \frac{\partial}{\partial r} \left(r \frac{\partial T}{\partial r}\right)\right)$$

$$\frac{\partial T}{\partial z} = \frac{\alpha}{u} \left( \frac{\partial^2 T}{\partial r^2} + \frac{1}{r} \frac{\partial T}{\partial r} \right) \tag{3.48}$$

The solution of equation (3.48) depends on the type of velocity profile $u$. In the following, three types of solutions will be discussed.

**3.3.1a Piston Flow.—**Graetz (1883) obtained a solution for a so-called isothermal case in which the fluid properties were assumed to be independent of temperature. This assumption allowed him to utilize the fully-developed velocity profile throughout the entire length of the tube. The following solution was obtained:

$$\frac{T_s - T_\infty}{T_s - T_i} = \sum_{j=1}^{j=\infty} \frac{2}{a_j J_1(a_j)} J_o\left(a_j \frac{r}{R}\right) \exp\left(-a^2 \frac{\pi}{N_{Gz}}\right) \tag{3.49}$$

which is an acceptable solution when the tube wall temperature is uniform. In equation (3.49), $a_j$ is the jth root of $J_o[(a_j)] = 0$, the Graetz number ($N_{Gz}$) is defined in the following way:

$$N_{Gz} = \frac{wc_p}{kL} \tag{3.50}$$

Utilizing equation (3.49), the following expression for the mean Nusselt number can be obtained:

$$N_{\overline{Nu}} = \frac{2}{\pi} \mathrm{Gz} \left[\frac{1 - 4E'(\pi/N_{Gz})}{1 + 4E'(\pi/N_{Gz})}\right] \tag{3.51}$$

where

$$E'\left(\frac{\pi}{N_{Gz}}\right) = \sum_{j=1}^{j=\infty} \frac{1}{a_j^2} \exp\left(-a_j^2 \frac{\pi}{N_{Gz}}\right)$$

Although equation (3.51) cannot be used for $N_{Gz} > 500$, Metzner *et al.* (1959) presented an asymptotic solution which is valid for Graetz number above 100 as follows:

$$\frac{\bar{h}_c D}{k} = \frac{8}{\pi} + \frac{4}{\pi}\left[\frac{wc_p}{kL}\right]^{1/2} \tag{3.52}$$

EXAMPLE 3.4 A tubular heat exchanger is being designed for honey. The equipment will have a $5 \times 10^{-2}$ m diameter and 3-m length. If the heat exchanger is operated at 7 kg/s compute the film heat transfer coefficient.

SOLUTION

(1) The Graetz Number for this problem is larger than 100. By using equation (3.52), the following results: (assume $k$ = 0.42 W/m $C$ and $c_p$ = 2.51 kJ/kg C)

$$\frac{\bar{h}_c(0.05)}{0.42} = \frac{8}{\pi} + \frac{4}{\pi}\left[\frac{7 \times 2.51 \times 1000}{0.42 \times 3}\right]^{1/2}$$

$$\bar{h}_c = 1284.35 \text{ W/m}^2 \text{ C}$$

(2) The value obtained should represent the best estimate of the convective heat transfer coefficient for honey for the given flow conditions.

**3.3.1b Fully developed parabolic velocity profile.**—The second solution to equation (3.48) obtained by Graetz (1883) was for a parabolic velocity distribution for a Newtonian fluid. The expression is:

$$\frac{T_s - T}{T_s - T_i} = \sum_{j=0}^{j=\infty} B_j \phi_j \left(\frac{r}{R}\right) \exp\left[-b_j^2 \frac{\pi}{2N_{Gz}}\right] \tag{3.53}$$

where $B_j$ are coefficients, $\phi_j$ $(r/R)$ are functions of $r/R$, and $b^2_j$ are exponents.

**3.3.1c Fully developed velocity profile for a power-law fluid.**—One of the more commonly used descriptions of non-Newtonian flow characteristics is the power-law expression. By substituting the velocity distribution for the power-law fluid into equation (3.48) the following expression is obtained:

$$u \left[\frac{3n + 1}{n + 1}\right] \left[1 - \left(\frac{r}{R}\right)^{(n+1)/n}\right] \cdot \frac{\partial T}{\partial z} = \alpha \left[\frac{\partial^2 T}{\partial r^2} + \frac{1}{r}\frac{\partial T}{\partial r}\right] \tag{3.54}$$

Solutions to equation (3.54) have been proposed by several investigators, including Lyche and Bird (1956). The latter investigators obtained solutions, but were not able to reduce them to usable forms. Probably the most significant contribution to this particular area of heat transfer has been made by Christiansen and Craig (1962), who solved equation (3.54) for the situations which allowed the variation of physical properties with temperature to be taken into account. These investigators utilized a modified power-law type of equation to describe the velocity distribution, but incorporated a factor which allowed the influence of temperature on the rheological properties to be computed. Christiansen and Craig (1962) solved the resulting expression utilizing a digital computer and standard finite difference techniques along with the following assumptions:

(a) the wall temperature of the tube is constant along the heating section;
(b) the fluid has a uniform temperature at the inlet;
(c) the fluid is fully developed laminar flow at the point it enters the heating section;
(d) the quantities $c_p$, $k$, $E/R_G$, $\rho$, $k_\sigma$, and $n$ are independent of temperature; and
(e) other rheological parameters are temperature-dependent.

Typical results of this solution are presented in Figs. 3.7 and 3.8, where the mean Nusselt number is plotted against the Graetz number. In Fig. 3.7, the influence of the $n$ value is illustrated for isothermal conditions, while in Fig. 3.8, $n$ is constant at a value of 1/3 and the function $\Psi(E)$ is

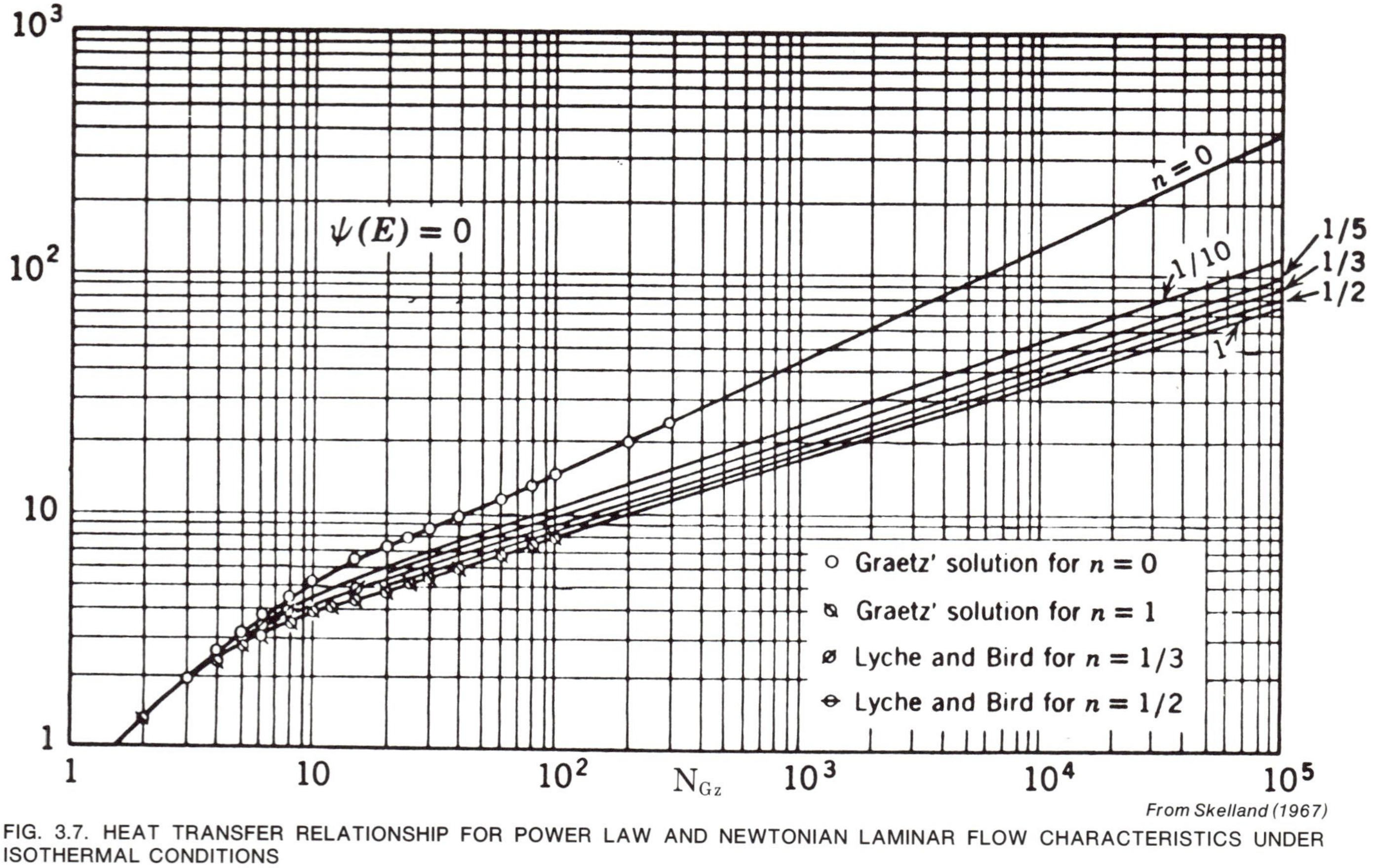


*From Skelland (1967)*

FIG. 3.7. HEAT TRANSFER RELATIONSHIP FOR POWER LAW AND NEWTONIAN LAMINAR FLOW CHARACTERISTICS UNDER ISOTHERMAL CONDITIONS

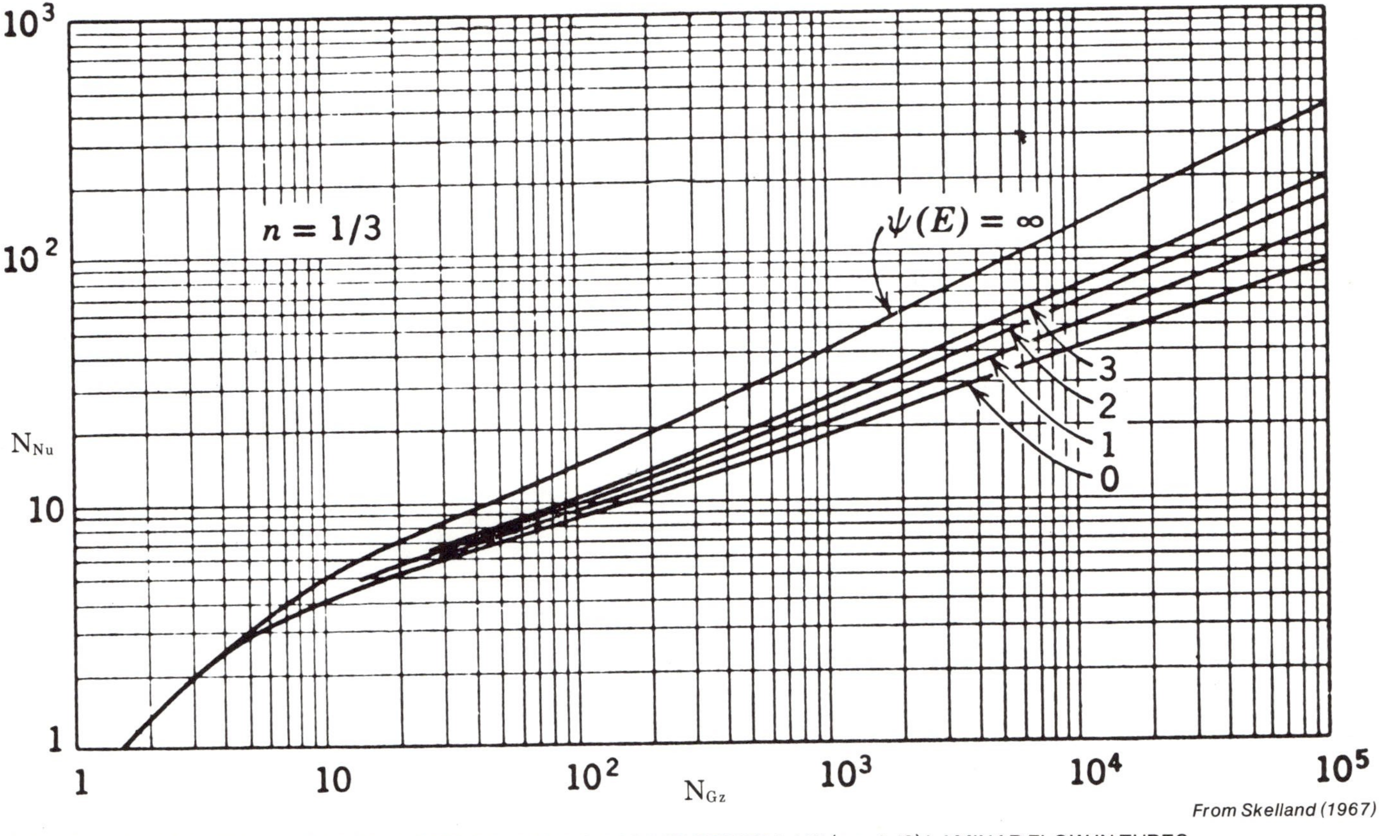


*From Skelland (1967)*

FIG. 3.8. NONISOTHERMAL HEAT TRANSFER RELATIONSHIPS FOR POWER LAW (n = 1/3) LAMINAR FLOW IN TUBES

allowed to vary. The latter parameter is the function which accounts for temperature variations and is defined in the following way:

$$\Psi(E) = \frac{E}{R_G}\left[\frac{1}{T_{Ai}} - \frac{1}{T_{As}}\right] \tag{3.55}$$

Pigford (1955) extended Leveque's approximate solution (Leveque, 1928) for Newtonian flow to non-Newtonian fluids to obtain the following expression:

$$N_{\overline{Nu}} = 1.75\left[\frac{3n+1}{4n}\right]^{1/3}(N_{Gz})^{1/3} \tag{3.56}$$

This particular dimensionless expression is in good agreement with the solution of Christiansen and Craig (1962).

An empirical expression proposed by Charm and Merrill (1959) allows computation of convective heat-transfer coefficients for pseudoplastic fluids. The expression proposed is a modification of the Sieder and Tate (1936) expression for highly viscous Newtonian fluids [equation (3.19)], as follows:

$$\frac{\bar{h}_c D}{k} = 2\left[\frac{wC_p}{kL}\right]^{1/3}\left[\frac{\mu_b}{\mu_s}\right]^{0.14} \tag{3.57}$$

By replacing the viscosity ratio with an apparent viscosity ratio for a power-law fluid, Charm and Merrill (1959) obtained the following expression:

$$\frac{\bar{h}_c D}{k} = 2\left[\frac{wc_p}{kL}\right]^{1/3}\left[\frac{m_b}{m_s}\,\frac{3n+1}{2(3n-1)}\right]^{0.14} \tag{3.58}$$

for which the logarithmic mean temperature difference at the entrance and exit of the tube are suggested for use with the heat-transfer coefficient. Computation of $m_b$ would be at the bulk properties of the fluid and $m_s$ should be evaluated at the mean wall temperature.

EXAMPLE 3.5 Compare the expressions proposed by Pigford [equation (3.56)], Charm and Merrill [equation (3.58)] and the computer solution of Christiansen and Craig (Fig. 3.8). Banana purée is pumped through a $5 \times 10^{-2}$ m diameter uniform wall temperature heat exchanger at a rate of 0.6 kg/s. The heat exchanger length is 3 m and the wall temperature is 49 C. Compute the convective heat-transfer coefficient if the product enters the heat exchanger at 20 C.

SOLUTION

(1) The rheological properties of banana purée (Table A.8) include $n = 0.46$, $m = 6.89$ Pa $s^n$ at 20 C and $m = 4.15$ Pa $s^n$ at 49 C.

(2) A check of the generalized Reynolds number reveals that the value is less than 2100 and flow is laminar.

(3) The thermal conducitivity of banana purée is 0.55 W/m C and specific heat is 3.66 kJ/kg C, (assumed).

(4) Using equation (3.56):

$$h_c = \frac{0.55}{0.05}(1.75)\left[\frac{3(0.46)+1}{4(0.46)}\right]^{1/3}\left[\frac{(0.6)(3.66)(1000)}{(0.55)3}\right]^{1/3}$$

$$h_c = 230.7 \text{ W/m}^2 \text{ C}$$

(5) Using equation (3.58):

$$h_c = \frac{0.55}{0.05}(2)\left[\frac{(0.6)(3.66)(1000)}{(0.55)3}\right]^{1/3}\left[\frac{6.89}{4.15}\;\frac{(3(0.46)+1)}{2(3(0.46)-1)}\right]^{0.14}$$

$$h_c = 304.8 \text{ W/m C}$$

(6) Using equation (3.55); with $E/R_G = 3583/°K$

$$\Psi(E) = \frac{E}{R_G}\left[\frac{1}{T_{Ai}} - \frac{1}{T_{As}}\right] = 3583\left[\frac{1}{293} - \frac{1}{322}\right]$$

$$\Psi(E) = 1.1$$

(7) From Fig. 3.8:
with $N_{Gz} = 1331$; $\Psi(E) = 1.1$ and allowing $n = 1/3$; $N_{Nu} = 23$.

(8) Then:

$$h_c = \frac{0.55}{0.05}(23) = 253 \text{ W/m}^2 \text{ C}$$

(9) The convective heat-transfer coefficient computed using Fig. 3.8 would be expected to be most accurate.

## 3.3.2 Heat Transfer in Turbulent Flow

In the case of heat transfer to non-Newtonian fluids during turbulent flow, published literature are lacking. Utilizing the analogy between heat and momentum transfer processes, Metzner and Friend (1959) developed the following expression:

$$N_{St} = \frac{\bar{h}_c}{c_p \bar{u} \rho} = \frac{f'/2\Theta_m}{\dfrac{1}{\phi_m} + \dfrac{11.8(f'/2)^{1/2}[N'_{Pr} - 1]}{(N'_{Pr})^{1/3}}} \qquad (3.59)$$

where:

$$\Theta_m = \Delta T_m / \Delta T_{max}$$

$$\phi_m = \bar{u}/u_{CL}$$

$$N'_{Pr} = \frac{c_p \mu_{ab}}{k_b}$$

$\mu_{ab}$ = apparent viscosity based on wall shear stress and bulk temperature or $\tau = \mu\,(-du/dr)$.

$$\mu_a = \frac{\tau_w}{(-du/dr)_w},$$

where $\tau_w = \dfrac{\Delta PD}{4L}$ in tube and $\left(-\dfrac{du}{dr}\right)_w = \dfrac{8\bar{u}}{D}\left(\dfrac{3n+1}{4n}\right)$

Since equation (3.59) predicted results to within only 23.6% of experimental data, Christiansen and Peterson (1966) proposed the following improved expression:

$$N_{St} = \frac{(f'/2\Theta_m)(\mu_{as}/\mu_{ab})^{-0.1}}{\dfrac{1}{\phi_m} + \dfrac{11.8(f'/2)^{1/2}[N''_{Pr} - 1]}{(N''_{Pr})^{1/3}}} \tag{3.60}$$

where

$$N''_{Pr} = N'_{Pr}\left(\frac{N'_{Re}}{2100}\right) \tag{3.61}$$

$$N'_{Re} = 1616\,\frac{(n+2)^{(n+2)/(n+1)}}{3n+1} \tag{3.62}$$

Equation (3.60) was considerably better and correlated with experimental data to within 19.2% but is only valid for Reynolds numbers greater than 10,000. Clapp (1961) proposed the following empirical expression:

$$\frac{\bar{h}_c D}{k} = 0.0041\left[\frac{D^n \bar{u}^{2-n}\rho}{m8^{n-1}}\right]^{0.99}\left[\frac{mc_p}{k}\left(\frac{8\bar{u}}{D}\right)^{n-1}\right]^{0.4} \tag{3.63}$$

where the heat-transfer coefficient is based on the logarithmic mean temperature difference and the fluid properties are evaluated at the bulk mean temperature.

### 3.3.3 Heat Exchangers

One of the more important applications of heat transfer in the food industry is in design of heat exchangers for various food products. Numerous types of heat exchangers are used in food-processing plants, including tubular, plate, scraped surface or votator, steam-injection and steam-infusion. The first three of the five listed types may be used for heating or cooling food products, while the last two are used only for heating purposes. In general, the design of any heat exchanger depends on evaluation of the convective heat-transfer coefficient, both for the product and the heating medium. In many situations, the convective heat-transfer coefficient for the product, especially if it tends to have non-Newtonian flow characteristics, may indicate that the product offers considerably more resistance to heat transfer than the heating medium. This situation reduces the influence of the convective heat-transfer coefficient for the heating medium on overall heat transfer coefficient.

**3.3.3a Tubular Heat Exchanger.—**Tubular heat exchangers may have various forms. Probably the most common is the concentric tube heat exchanger illustrated schematically in Fig. 3.9. In this arrangement, the product flows through the inside tube, while the heating medium (usually water) flows through the outer jacket, either in the same direction (called concurrent flow) or in the opposite direction (called countercurrent flow). Probably the most important factor to keep in mind when describing heat transfer in this type of heat exchanger is that the log mean temperature difference defined in equation (3.17) is always utilized as the temperature difference in the heat flux expression. In addition, it seems most likely that the convective heat-transfer coefficient for products flowing in these types of tubular heat exchangers can be described by the most appropriate expression previously presented for heat exchange in tubes.

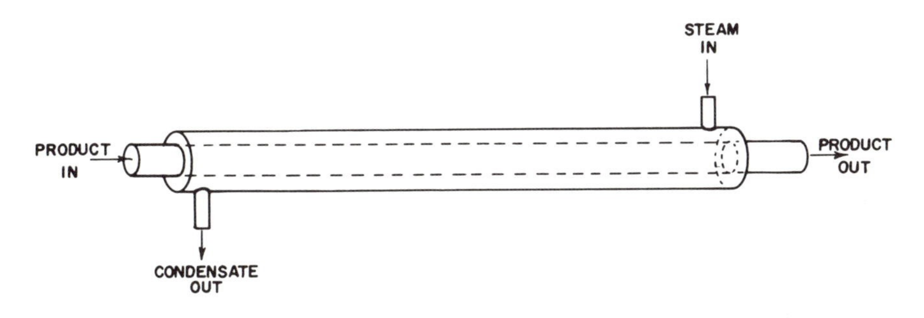


FIG. 3.9. SCHEMATIC ILLUSTRATION OF TUBULAR HEAT EXCHANGER

**3.3.3b Triple Tube Heat Exchanger.—**If, however, the heat exchanger is the triple-tube type, in which the heating medium is flowing through the inner and outer rings of the tubular arrangement and the product flows through the annulus, somewhat different expressions may be required to describe the heat-transfer coefficient. For this situation, Jakob (1949) has suggested the following expression for laminar flow:

$$\frac{\bar{h}_c D}{k} = 1.0(D_2/D_1)^{0.8} \left[\frac{wc_p}{kL}\right]^{0.45} N_{Gr_b}^{0.05} \tag{3.64}$$

where the computed heat-transfer coefficient refers to both the inner and outer heat transfer surfaces. When flow is turbulent in the annulus, Monrad and Pelton (1942) have proposed the following expressions:

$$\frac{h_{co}}{c_p G}\left[\frac{c_p \mu}{k}\right]^{2/3}\left[\frac{\mu_s}{\mu_b}\right]^{0.14} = \frac{0.23}{[(D_2 - D_1)/\mu_b]^{0.2}} \tag{3.65}$$

and

$$\frac{h_{ci}}{c_{pb}G}\left[\frac{c_p \mu}{k}\right]_b^{2/3} = \frac{0.023\,[0.87\,(D_2/D_1)^{0.53}]}{[(D_2 - D_1)G/\mu_b]^{0.2}} \quad (3.66)$$

for the outer and inner heat-transfer surfaces, respectively. Equations (3.64), (3.65) and (3.66) will apply only to fluids which have Newtonian flow characteristics. Tubular heat exchangers may also be of the multiple-tube type, in which several tubes are utilized for product flow and all tubes are surrounded on the exterior surface by the heating medium. In this case, any of the expressions proposed for the single-tube heat transfer should be adequate for computing the convective heat-transfer coefficient.

**3.3.3c Plate Heat Exchanger.**—In a plate heat exchanger, the product flows between two heated surfaces in a very thin film, resulting in a rapid increase in temperature of the product and usually requiring only a single pass over a given heat-transfer surface. Fig. 3.10 schematically illustrates the flow through a plate heat exchanger. In most cases, the developments for predicting convective heat-transfer coefficients in plate heat

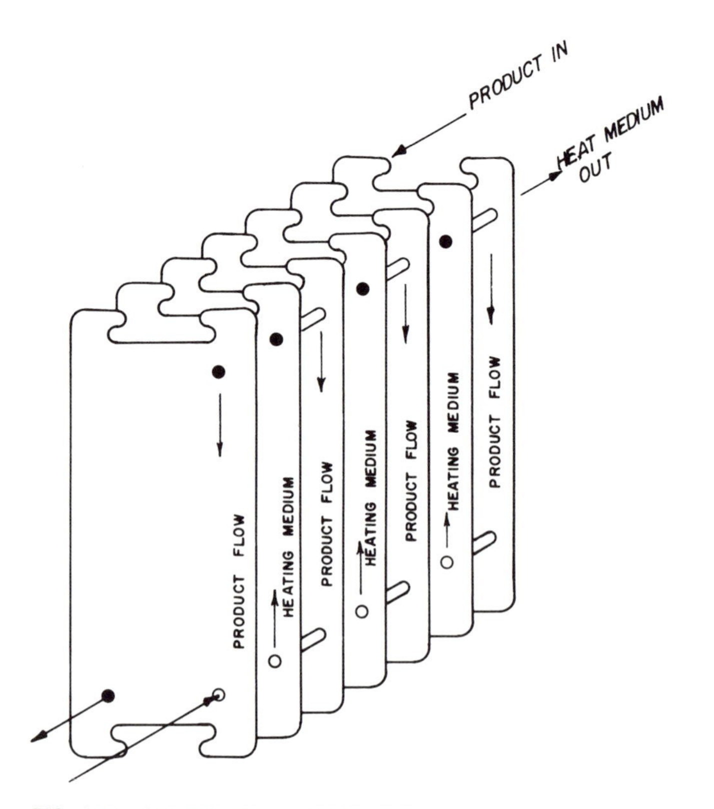


FIG. 3.10. SCHEMATIC ILLUSTRATION OF PLATE HEAT EXCHANGER

exchangers have been based on the heat-transfer characteristics from the heating medium to the product across the plate heating surface. A detailed analysis of heat transfer in both concurrent and counter-current plate heat exchangers is given by Nunge *et al.* (1967). By solving the energy equations for both fluid streams on either side of the heat-transfer surface, the Nusselt numbers (both local and fully developed) have been computed in terms of the Peclet number, which is defined in the following way:

$$N_{Pe} = \frac{2b''\bar{u}c_p\rho}{k} \tag{3.67}$$

for the case of the plate heat exchanger. An example of the results obtained for concurrent flow are presented in Fig. 3.11, where a variety of heat capacity ratios ($H$), thermal conductivity ratios ($K$) and film thickness ratios ($\Delta$) are compared. The results illustrate that when the

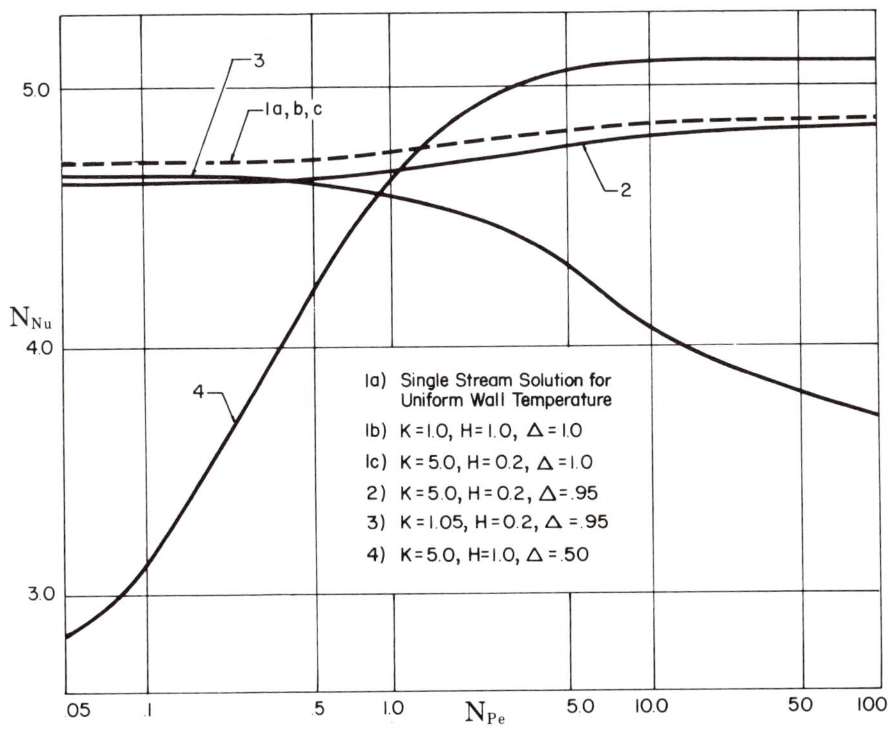


*From Nunge et al. (1967)*

FIG. 3.11. NUSSELT NUMBERS FOR PRODUCT FLOW IN A CONCURRENT PLATE HEAT EXCHANGER WITH ONE ADIABATIC WALL AND ONE UNIFORM WALL TEMPERATURE

fluid streams are equal and the product of the thermal conductivity ratio and heat-capacity ratio are near unity, the results can be predicted with considerable accuracy by single-stream solution with uniform wall temperature. The relationships between the overall Nusselt number and Peclet number for counter-current flow plate heat exchangers are presented in Fig. 3.12. These results illustrate that the overall Nusselt number increases significantly between the Peclet numbers of 10 and 1,000. In general, the results of Nunge *et al.* (1967) indicated that the influence of axial conduction within the fluid streams is significant as long as the Peclet number is below 100. The previous results have been developed for Newtonian fluids only.

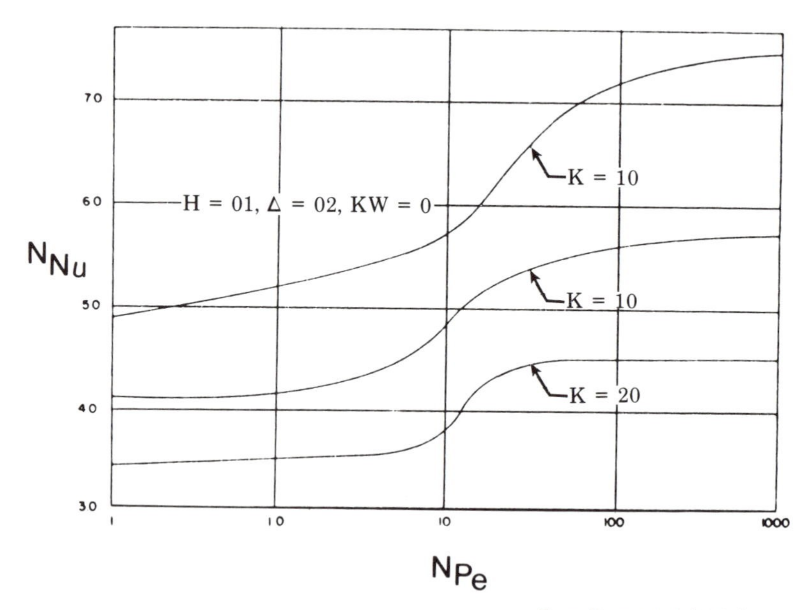


From Nunge et al. (1967)

FIG. 3.12. OVERALL NUSSELT NUMBER RELATIONSHIPS FOR COUNTER-CURRENT FLOW PLATE HEAT EXCHANGER

**3.3.3d Plate Heat Exchangers for Non-Newtonian Fluids.**—For the case of non-Newtonian or power-law fluid characteristics, Skelland (1967) has proposed the following relationship:

$$\frac{\bar{h}_c b''}{k} = N_{Nu} = \frac{6}{\zeta}\left(\frac{n+1}{2n+1}\right) \tag{3.68}$$

where:

$$\zeta = \frac{5}{4} - \frac{2n}{2n+1} + \frac{3n}{4n+1} - \frac{n}{5n+1} \tag{3.69}$$

This relationship has been proposed for the situation in which the cubic polynomial temperature distribution exists in the fluid, which also has constant physical properties. As with most of the relationships presented, Skelland (1967) has assumed thermally and hydrodynamically developed flow in a multistream plate heat exchanger. Blanco and Gill (1967) have used a series expansion method to obtain a solution to the energy equation in a multistream plate heat exchanger under turbulent conditions. This approach led to the prediction of asymptotic Nusselt numbers for the physical situation mentioned, and allowed comparison with previous research, which was done under empirical conditions in most cases. Probably the most significant result was an excellent agreement between the results of Blanco and Gill (1967) and an expression of the type given in equation (3.22). Although the latter expression was confirmed for water in countercurrent flow by Duchatelle and Vautrey (1964), it would appear to be a very acceptable expression to use for predicting a convective heat-transfer coefficient in a plate heat exchanger with turbulent flow of a Newtonian fluid.

EXAMPLE 3.6 Determine the convective heat-transfer coefficient for a plate heat exchanger with a $6.35 \times 10^{-3}$ m distance between plates. The product is non-Newtonian with $n = 0.5$ and thermal conductivity $k = 0.52$ W/m C.

SOLUTION

(1) Using equation (3.69):

$$\zeta = \frac{5}{4} - \frac{2(0.5)}{2(0.5)+1} + \frac{3(0.5)}{4(0.5)+1} - \frac{0.5}{5(0.5)+1} = 1.11$$

(2) Using equation (3.68):

$$N_{Nu} = \frac{6}{1.11}\left[\frac{0.5+1}{2(0.5)+1}\right]$$

$$N_{Nu} = 4.05$$

(3) Then:

$$\bar{h}_c = \frac{0.52}{6.35 \times 10^{-3}}(4.05) = 332 \text{ W/m}^2\text{ C}$$

(4) The heat-transfer coefficient obtained is low in magnitude for a plate heat exchanger, but the non-Newtonian flow characteristics of the fluid reduces the coefficient significantly.

**3.3.3e Scraped Surface Heat Exchanger.**—The scraped-surface heat exchanger or Votator is a widely used type of heat exchanger in the food industry, especially for high-viscosity and non-Newtonian fluids. This heat exchanger appears to be quite similar to the double-tube heat exchanger discussed earlier, but with the addition of a set of scraper blades which continuously move over the heat-transfer surface. The schematic diagram in Fig. 3.13 illustrates this arrangement in more

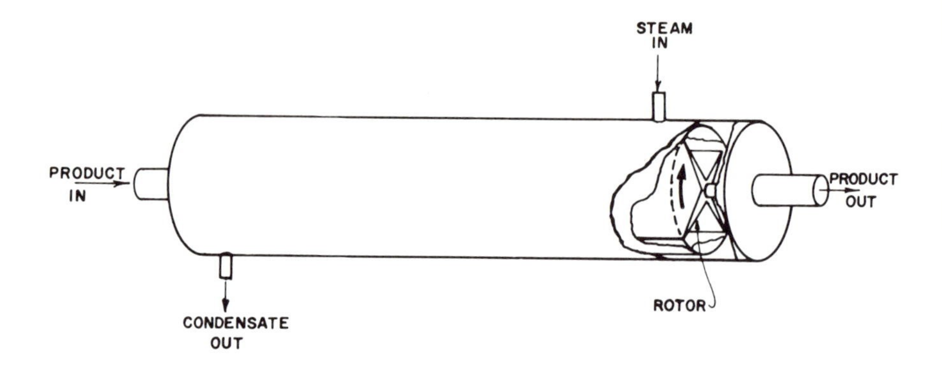


FIG. 3.13. SCHEMATIC ILLUSTRATION OF SCRAPED-SURFACE HEAT EXCHANGER

detail. The only available research conducted to determine the scraped-surface film coefficient for the Votator was conducted by Skelland (1958). After dimensional analysis, he was able to correlate data during cooling of glycerol-water and other glycerol oils to obtain the following:

$$\frac{\bar{h}_c D}{k} = 4.9 \left[\frac{D\bar{u}}{\mu}\right]^{0.57} \left[\frac{\mu c_p}{k}\right]^{0.47} \left[\frac{DN'}{u}\right]^{0.17} \left[\frac{D}{L}\right]^{0.37} \qquad (3.70)$$

where $D$ is internal diameter of the tubular heat exchanger and $N'$ is rotation speed of the Votator shaft. Although no attempt was made to differentiate between flow regimes, Skelland (1958) indicated that there was evidence of laminar, transition and turbulent conditions during flow through the Votator. Under turbulent conditions the exponent of the first and third dimensionless groups on the right side of equation (3.70) would be higher than indicated.

**3.3.3f Direct Contact Heat Exchangers.**—Two heat exchangers which involve direct contact between the product and the heating medium are the steam-injection and steam-infusion systems, illustrated in Figs. 3.14 and 3.15. In the steam-injection system, steam is injected into product flowing through a tubular pipe, resulting in heat exchange from steam bubbles to the product and eventually complete condensation of the steam. In the steam-infusion system, the liquid product is sprayed into a changer containing the steam heating medium. Heat transfer occurs from the steam to the liquid food droplets, resulting in condensation and mixing of condensate with the product. In most cases, both systems utilize vacuum techniques which remove added portions of water. Although there have been no attempts to describe the heat-transfer which occurs from steam to liquid product, both cases must involve heat

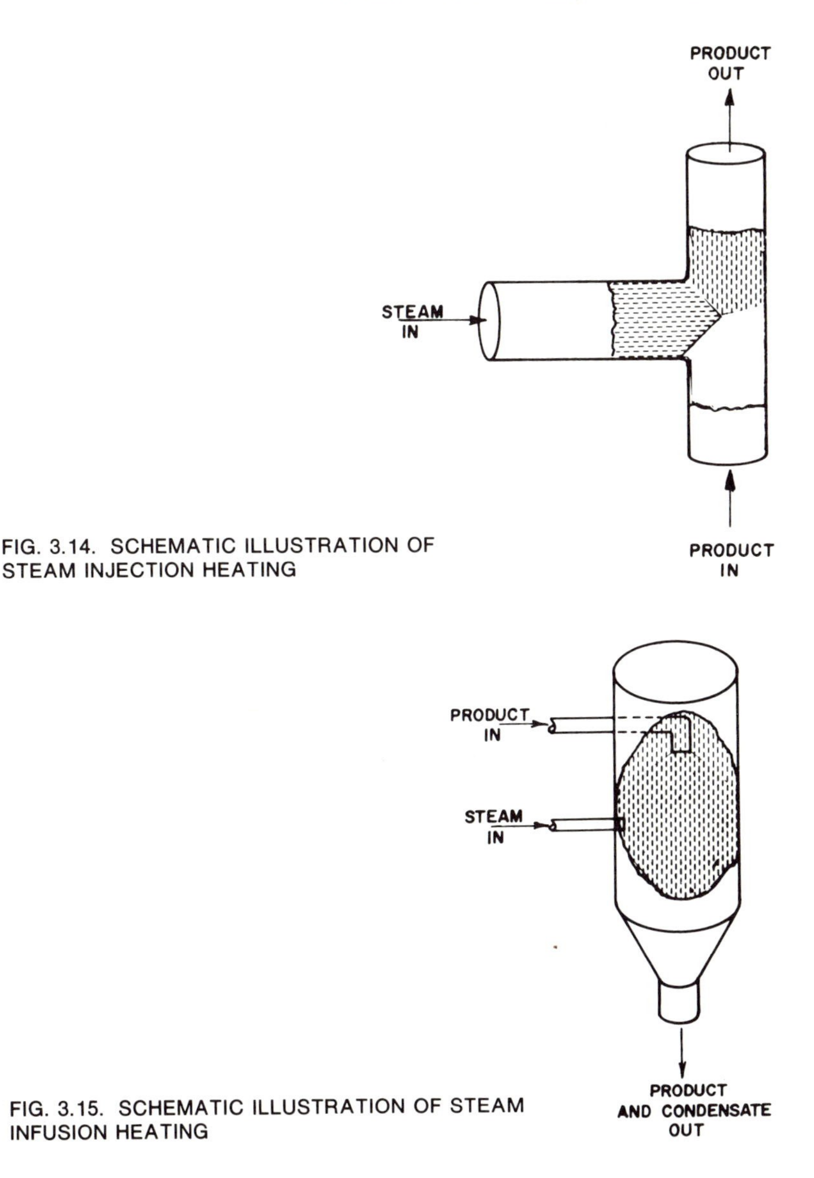


FIG. 3.14. SCHEMATIC ILLUSTRATION OF STEAM INJECTION HEATING

FIG. 3.15. SCHEMATIC ILLUSTRATION OF STEAM INFUSION HEATING

transfer to product through convective heat-transfer films around droplets or bubbles. Ranz and Marshall (1952) have proposed the following expression to describe the heat convective heat-transfer coefficient to or from spherical-shaped object:

$$\frac{\bar{h}_c D}{k_f} = 2 + 0.6 \left[\frac{D\bar{u}\rho_f}{\mu_f}\right]^{0.5} \left[\frac{\mu c_p}{k}\right]_f^{1/3} \tag{3.71}$$

It would seem that this equation could be utilized to estimate convective heat-transfer coefficients from the steam bubbles to the liquid product in the steam-injection system and from the liquid droplets to the surrounding steam in the steam-infusion systems.

EXAMPLE 3.7 Estimate the convective heat-transfer coefficient for liquid droplets of product in a steam-infusion system. Assume the droplets have a $2.54 \times 10^{-3}$ m diameter and an average velocity of 1.52 m/s while moving through the steam. The properties of the steam at 116 C include a thermal conductivity of 0.0275 W/m C, a specific heat of 1674.8 J/kg C, viscosity of $1.488 \times 10^{-5}$ kg/m s and a density of 0.97 kg/m$^3$.

SOLUTION

(1) Using equation (3.71):

$$\frac{\bar{h}_c D}{k_f} = 2 + 0.6\left[\frac{(2.54 \times 10^{-3})(1.52)(0.97)}{1.488 \times 10^{-5}}\right]^{0.5} \left[\frac{(1.488 \times 10^{-5})(1674.8)}{.0275}\right]^{1/3}$$

$$= 2 + 0.6(252)^{0.5}(0.906)^{1/3}$$

$$= 2 + 0.6(15.88)(0.9676) = 2 + 9.21 = 11.21$$

$$\bar{h}_c = \frac{11.21(0.0275)}{2.54 \times 10^{-3}} = 121 \text{ W/m}^2\text{ C}$$

## 3.4 UNSTEADY-STATE HEATING AND COOLING

As indicated earlier in this chapter, a rate of heat transfer which is dependent on time is described as unsteady-state or transient heat transfer. This somewhat more complex type of heat transfer is involved in many of the heating and cooling problems in the food industry. Fortunately many time-saving charts and procedures have been developed to allow some ease in solving these more complex heating and cooling problems.

One of the key factors involved in the evaluation of transient or unsteady-state heat transfer is the relative importance of internal and external resistance to heat transfer. The dimensionless number used in evaluating the importance of these factors is the Biot number, which is defined in the following way:

$$N_{Bi} = \frac{h_c l}{k_s} \tag{3.72}$$

where $l$ is the characteristic dimension of the body involved in the heat-transfer computation. When the Biot number is low ($N_{Bi} < 0.1$), the internal resistance to heat transfer is considerably less than the surface resistance, and the so-called lumped parameter approach can be utilized to describe the heating or cooling characteristics of the body. This approach is normally referred to as negligible internal resistance to heat transfer. When the Biot number is large ($N_{Bi} < 40$), the internal resistance to heat transfer is much greater than the external resistance, and the surface temperature of the object can be assumed to be equal to the heating or cooling medium temperature without significant error. This situation is usually called negligible surface resistance to heat transfer. The heating and cooling problems which are involved between the two extremes are more complex to handle, and are referred to as finite surface and internal resistance to heat transfer. Approaches available to describe these three ranges of transient or unsteady-state heat transfer will be described in the following sections.

### 3.4.1 Negligible Internal Resistance

This situation can be utilized when the surface resistance is considerably greater than the internal resistance. This description implies that the temperature within the object is uniform and that conditions existing at the surface, which determine the magnitude of the convective heat-transfer coefficient, determine the rate at which the object will heat or cool. The expression utilized to describe this type of unsteady-state heat transfer is obtained by conducting an energy balance on the object and setting the change in internal energy equal to the net heat flow to the object in a given period. For cooling of an object this energy balance would then appear as:

$$c_p \rho \, V dT = h_c A_s \, (T - T_M) dt \tag{3.73}$$

where $A_s$ is the surface area of the object which is exposed to the cooling medium. By proper separation of the parameters in equation (3.73) followed by integration, the following expression is obtained:

$$\frac{T - T_M}{T_o - T_M} = \exp(-h_c A_s / c_p \rho V)t \tag{3.74}$$

which describes the temperature history within the object. After some rearrangement, equation (3.74) can be written in completely dimensionless form in the following manner:

$$\frac{T - T_M}{T_o - T_M} = \exp[-(N_{Bi})(N_{Fo})] \tag{3.75}$$

where $N_{Fo}$ is the Fourier Modulus defined in the following way.

$$N_{Fo} = \frac{\alpha t}{l^2} \tag{3.76}$$

This approach represents a relatively simple method of handling heating and cooling problems which meet specifications required, that where $N_{Bi}<0.1$.

EXAMPLE 3.8 A frozen product in a cylindrical can with $5 \times 10^{-2}$ m height and $5 \times 10^{-2}$ m diameter is allowed to thaw in still air at 21 C. The initial product temperature is −18 C. The thermal properties include a thermal conductivity of 2 W/m C and specific heat of 2.51 kJ/kg C, while the density is 961 kg/m³. Estimate the product temperature after 0.5 hr by assuming the product properties do not change significantly.

SOLUTION

(1) Using equation (3.72), the type of unsteady-state heat transfer which is occurring can be characterized. (Still air has $h_c$ = 5.7 W/m² C)

$$N_{Bi} = \frac{(5.7)(2.5 \times 10^{-2})}{2} = 0.07$$

Since the Biot Number is less than 0.1, negligible internal resistance to heat transfer can be assumed.

(2) Using equation (3.74) with:

$$h_c = 5.7 \text{ W/m}^3 \text{ C}$$

$$A_s = \pi(5 \times 10^{-2})(5 \times 10^{-2}) + 2\left[\frac{\pi(5 \times 10^{-2})^2}{4}\right] = 1.18 \times 10^{-2} \text{ m}^2$$

$$c_p = 2.51 \text{ kJ/kg C}$$

$$\rho = 961 \text{ kg/m}^3$$

$$V = \left(\frac{\pi(5 \times 10^{-2})^2}{4}\right)(5 \times 10^{-2}) = 9.82 \times 10^{-5} \text{ m}^3$$

then

$$\frac{T - T_M}{T_o - T_M} = \exp\left[-\frac{(5.7)(1.18 \times 10^{-2})(0.5)(3600)}{2510 \times 961 \times 9.82 \times 10^{-5}}\right]$$

$$= \exp(-0.51) = 0.6$$

$$T = 0.6\,(-18 - 21) + 21$$

$$T = -2.4 \text{ C}$$

(3) The solution indicates that the product temperature increases from −18 C to −2.4 C in 0.5 hr.

### 3.4.2 Negligible Surface Resistance

As indicated previously, when the Biot number is greater than 40, the situation exists when the convective heat-transfer coefficient is sufficiently large to create a situation in which the surface temperature cf the object can be approximately equal to the heating or cooling medium temperature. For these conditions and selected geometrical configurations, series solution to equation (3.1), without the internal energy term, are readily available. The geometric configurations include the infinite slab, infinite cylinder, and sphere. The temperature histories within these geometric configurations can be evaluated directly from the series solutions by assuming the first term approximation. For an infinite slab the series solution is:

$$\frac{T - T_s}{T_0 - T_s} = \frac{4}{\pi} \sum_{n=1}^{\infty} \frac{1}{n} \sin \frac{n\pi x}{L} \exp \left[\left(-\frac{n\pi}{2}\right)^2 \frac{\alpha t}{\delta^2}\right] \tag{3.77}$$

where $T_s$ is the surface temperature which is assumed equal to the cooling or heating medium temperature. For an infinite cylinder the series solution is:

$$\frac{T - T_s}{T_0 - T_s} = \sum_{n=1}^{\infty} \frac{2}{R_n} \frac{J_0(R_n r/R)}{J_1(R_n)} \exp \left(-R_n^2 \frac{\alpha t}{R^2}\right) \tag{3.78}$$

where $J_0$ and $J_1$ are Bessel functions which can be evaluated from tables. The series solution for a sphere is the following:

$$\frac{T - T_s}{T_0 - T_s} = \frac{2}{\pi} \left(\frac{R}{r}\right) \sum_{n=1}^{\infty} \frac{(-1)^{n+1}}{n} \sin n\pi \frac{r}{R} \exp \left(\frac{-n^2 \pi^2 \alpha t}{R^2}\right) \tag{3.79}$$

which will apply at all locations except the geometric center of the sphere.

In addition to direct computation of temperature distribution from any of the equations (3.77), (3.78) and (3.79), the same computations can be accomplished by utilizing charts. These transient heat conduction charts will be described in the following section and will contain a curve corresponding to conditions when the inverse of Biot number is equal to zero.

### 3.4.3 Finite Surface and Internal Resistance

When the Biot number falls between 0.1 and 40, both the internal resistance and convective heat-transfer coefficient at the surface must be accounted for in determining the temperature distribution during heating or cooling of an object. Expressions of the type given by equations (3.77), (3.78) and (3.79) have been obtained for this situation by incorporating a transcendental or root equation which accounts for the influence of the Biot number. The primary assumption made in obtaining these solutions is that the object must have a uniform temperature at $t = 0$, and that the object is exposed instantaneously to the heating or cooling medium temperature at $t = 0$. Utilizing these series solution equations, charts of the type shown in Figures 3.16, 3.17 and 3.18 have been developed. In all three charts, the dimensionless temperature ratio is plotted versus the Fourier Modulus. In the case of the infinite slab,

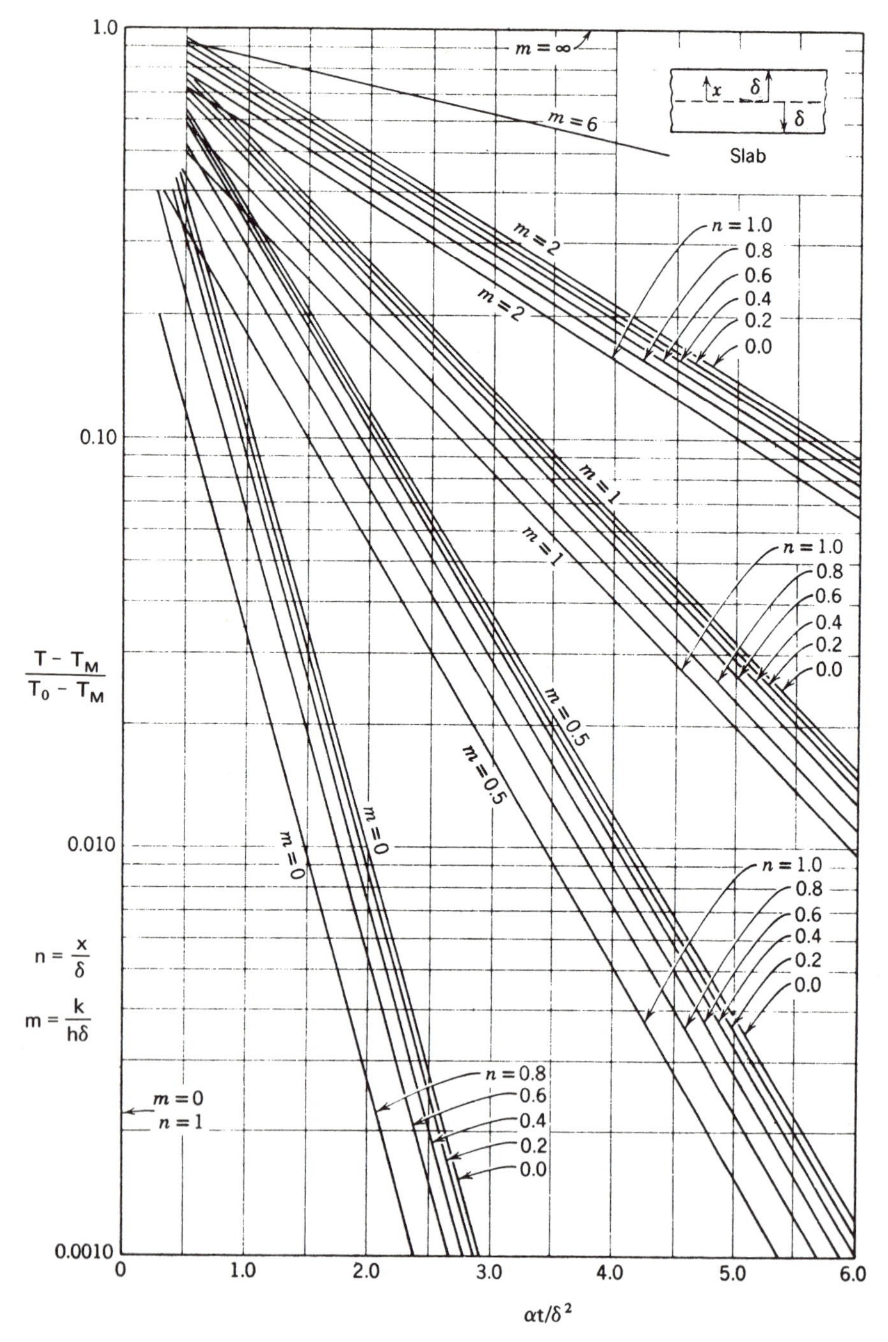


*From Foust et al. (1960)*

FIG. 3.16. UNSTEADY-STATE TEMPERATURE DISTRIBUTIONS IN AN INFINITE SLAB

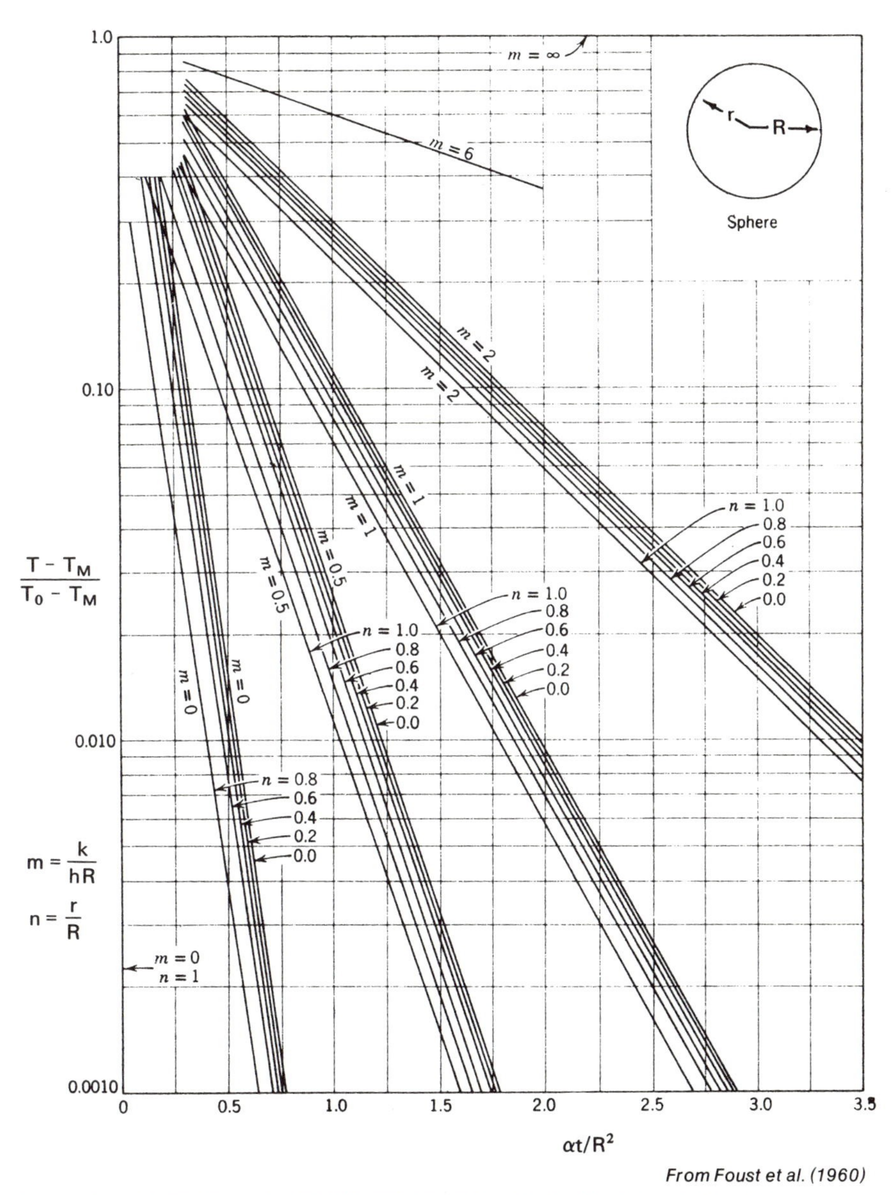


From Foust et al. (1960)

FIG. 3.17. UNSTEADY-STATE TEMPERATURE DISTRIBUTIONS IN A SPHERE

the chart presents six different curves for each value of $m = 1/N_{Bi}$; and each curve represents a different distance from the center of the slab $(x/\delta)$. In each situation, is the half thickness of the slab. In the case of the infinite cylinder, the six curves represent different distances from the

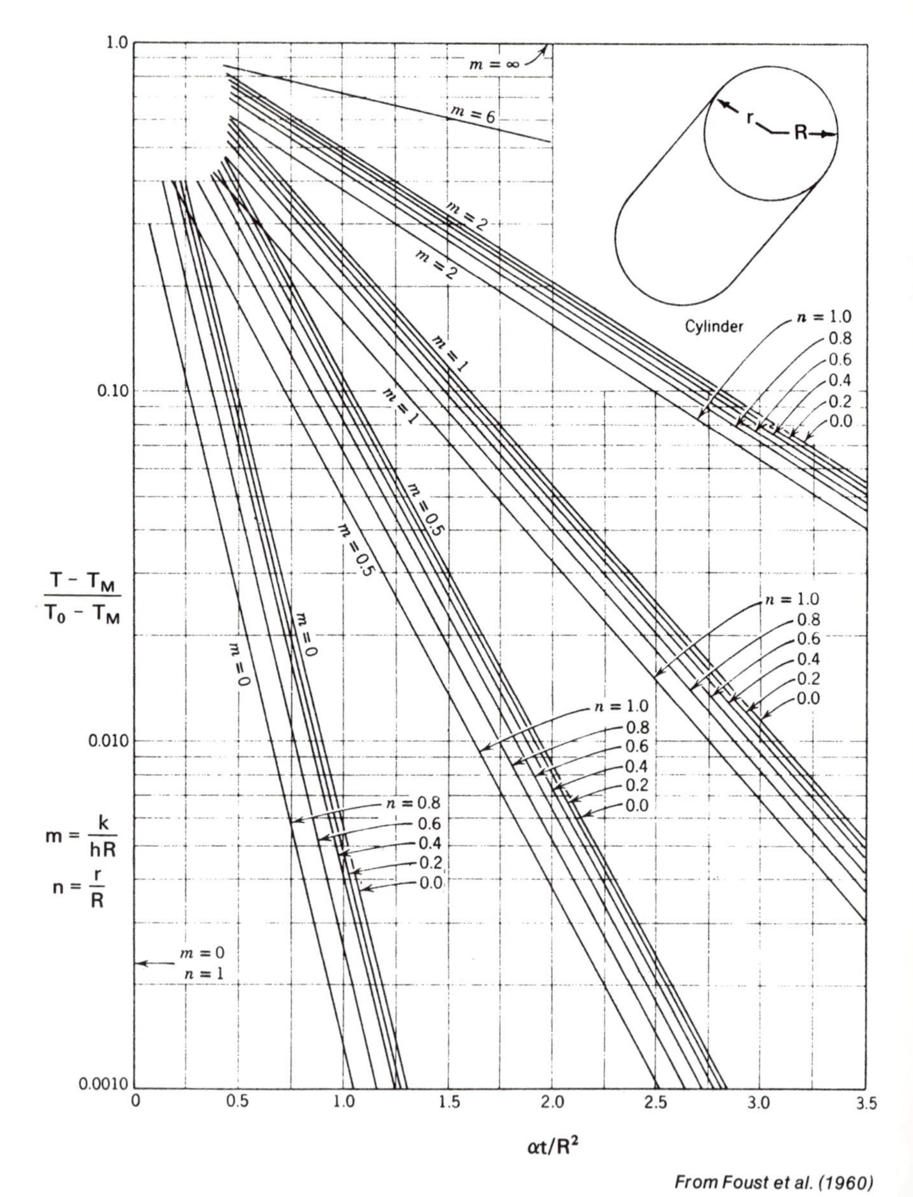


*From Foust et al. (1960)*

FIG. 3.18. UNSTEADY-STATE TEMPERATURE DISTRIBUTIONS IN AN INFINITE CYLINDER

center of the cylinder. The inverse of the Biot number utilizes the radius ($R$) of the cylinder as the characteristic length. The same procedure is utilized for the spherical object.

EXAMPLE 3.9 An apple is being cooled from an initial temperature of 21 C to 4 C in a high velocity water stream at 2 C. Compute the time required for the center of the apple to reach 4 C when the following properties are known; density = 800 kg/m$^3$ specific heat = 3.56 kJ/kg C, thermal conductivity = 0.35 W/m C and radius = 0.03 m. The water stream produced a convective heat transfer coefficient of 3400 W/m$^2$ C, compute time required for location at 0.01 m from surface to reach 4 C.

SOLUTION

(1) Computation of Biot No.:

$$N_{Bi} = \frac{(3400)(.03)}{0.35} = 291$$

Since the Biot number exceeds 40, the curve with $k/hR = 0$ in Fig. 3.17 is available.

(2) In order to use Figure 3.17:

$$\frac{T - T_M}{T_o - T_M} = \frac{4 - 2}{21 - 2} = \frac{2}{19} = 0.11$$

and:

$$\frac{kt}{\rho c_p R^2} = 0.3$$

then tne time required for the center of the apple to reach 4 C becomes:

$$t = \frac{0.3\,(800)\,(3560)\,(.03)^2}{(3600)\,(0.35)} = 0.61\text{ hr} = 36.6\text{ min}$$

(3) At 0.01 m from surface or $r/R = (.02)/(.03) = 0.67$

$$\frac{kt}{\rho c_p R^2} = 0.2; t = \frac{0.2\,(800)\,(3560)\,(.03)^2}{(3600)\,(0.35)}$$

$$t = 0.41\text{ hr} = 24.4\text{ min}$$

## 3.4.4 Use of 'f' and 'j' Parameters

A common procedure in food science literature is to utilize a common equation for heating and cooling of any object with any geometrical configuration. This equation is as follows:

$$\log(T - T_M) = -t/f_h + \log j\,(T_0 - T_M) \tag{3.80}$$

which can be obtained directly from equation (3.74), by letting $f_h$ equal the following equation:

$$f_h = \frac{2.303\, C_p \rho V}{h_c A_s} \tag{3.81}$$

and by letting $j = 1$. The general heating or cooling equation (equation 3.80) can also describe any of the expressions for infinite slab, infinite cylinder, or sphere. In the equations for finite surface resistance or finite surface and internal resistance to heat transfer, equation (3.80) applies when utilizing the first term of the series solution and letting $f_h$ and $j$ equal the appropriate portions of the expressions. Physically, $f_h$ is a time factor representing the rate of heating. On semi-logarithmic coordinates, it becomes equal to the time required for a one log-cycle reduction of the temperature difference between the heating or cooling medium and the initial temperature of the object. The $j$ represents a type of lag factor defined by the following ratio:

$$j = \frac{T_a - T_M}{T_0 - T_M} \tag{3.82}$$

where $T_a$ = the intercept of the heating or cooling curve on semi-log coordinates. Pflug *et al.* (1965) utilized the parameters ($f_h$ and $j$) to develop charts which allowed the determination of these heating and cooling parameters as a function of Biot number. In Fig. 3.19, the relationship of the dimensionless number which includes $f_h$ is presented versus the Biot number. From this figure, it is obvious that the value of $f_h$ would not change at Biot numbers above 40 where negligible surface resistance can be assumed. The influence of Biot number on the lag

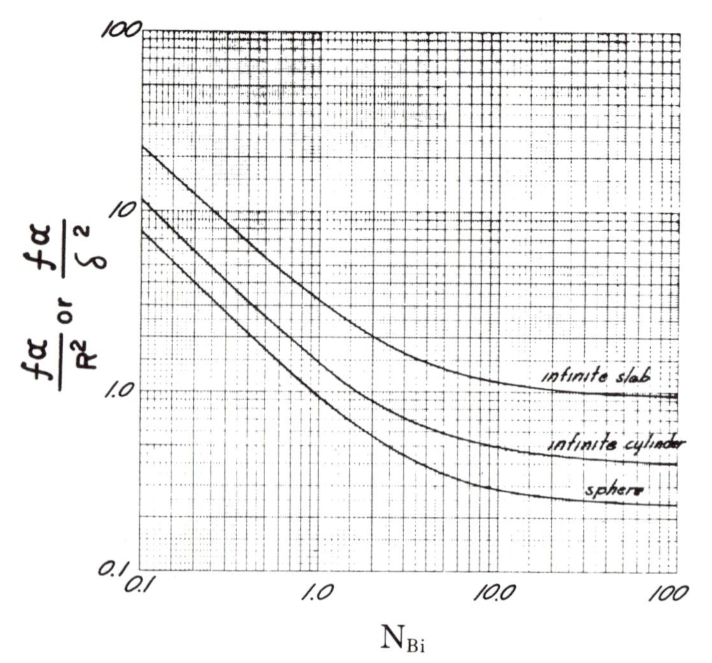


*From Pflug et al. (1965)*

FIG. 3.19. RELATIONSHIP OF THE HEATING OR COOLING MODULUS AND BIOT NUMBER FOR TRANSIENT HEAT TRANSFER

factor ($j$) at the geometric center of the object is presented in Fig. 3.20. Utilizing Figs. 3.19 and 3.20, the heating or cooling characteristics of any of the three basic geometrics can be predicted from knowledge of the Biot number, the thermal properties, and the geometry of the particular object. The only limitation to this approach appears to be the fact that the computations are confined to the geometric center of the object.

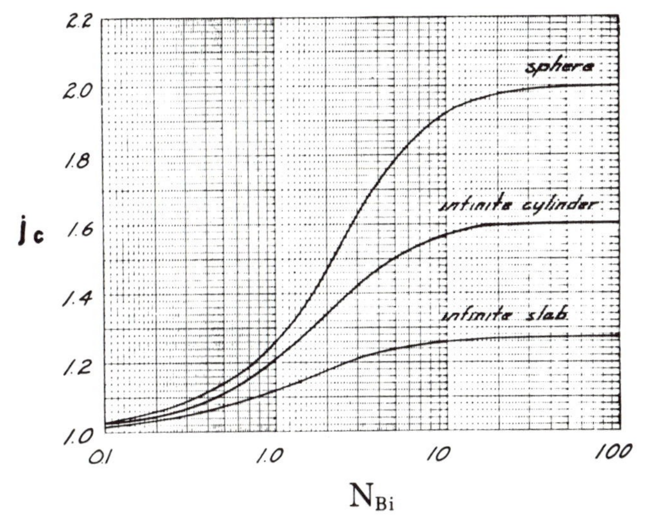


From Pflug et al. (1965)

FIG. 3.20. INFLUENCE OF BIOT NUMBER ON THE LAG FACTOR (j) AT THE GEOMETRIC CENTER OF THE OBJECT

EXAMPLE 3.10 A conduction cooling product in a cylindrical container is being cooled in a 4 C forced-air cooler. A temperature sensor at the geometric center gave the following results:

| Time (min) | Temperature (C) |
|---|---|
| 0 | 38 |
| 5 | 35 |
| 10 | 29 |
| 15 | 24 |
| 20 | 21 |
| 25 | 16 |
| 30 | 13 |
| 35 | 12 |
| 40 | 10 |
| 45 | 9 |
| 50 | 7.2 |
| 55 | 7 |
| 60 | 6.7 |

Determine the $f_c$ and $j$-values for the product in this cooling situation.

SOLUTION

(1) The experimental data are plotted as illustrated in Fig. 3.21.

(2) Since the $f_c$-value is the time required to reduce the product temperature by one log cycle, $f$ = 44 min.

(3) The $j$-value becomes:

$$j = \frac{48 - 4}{38 - 4} = 1.29$$

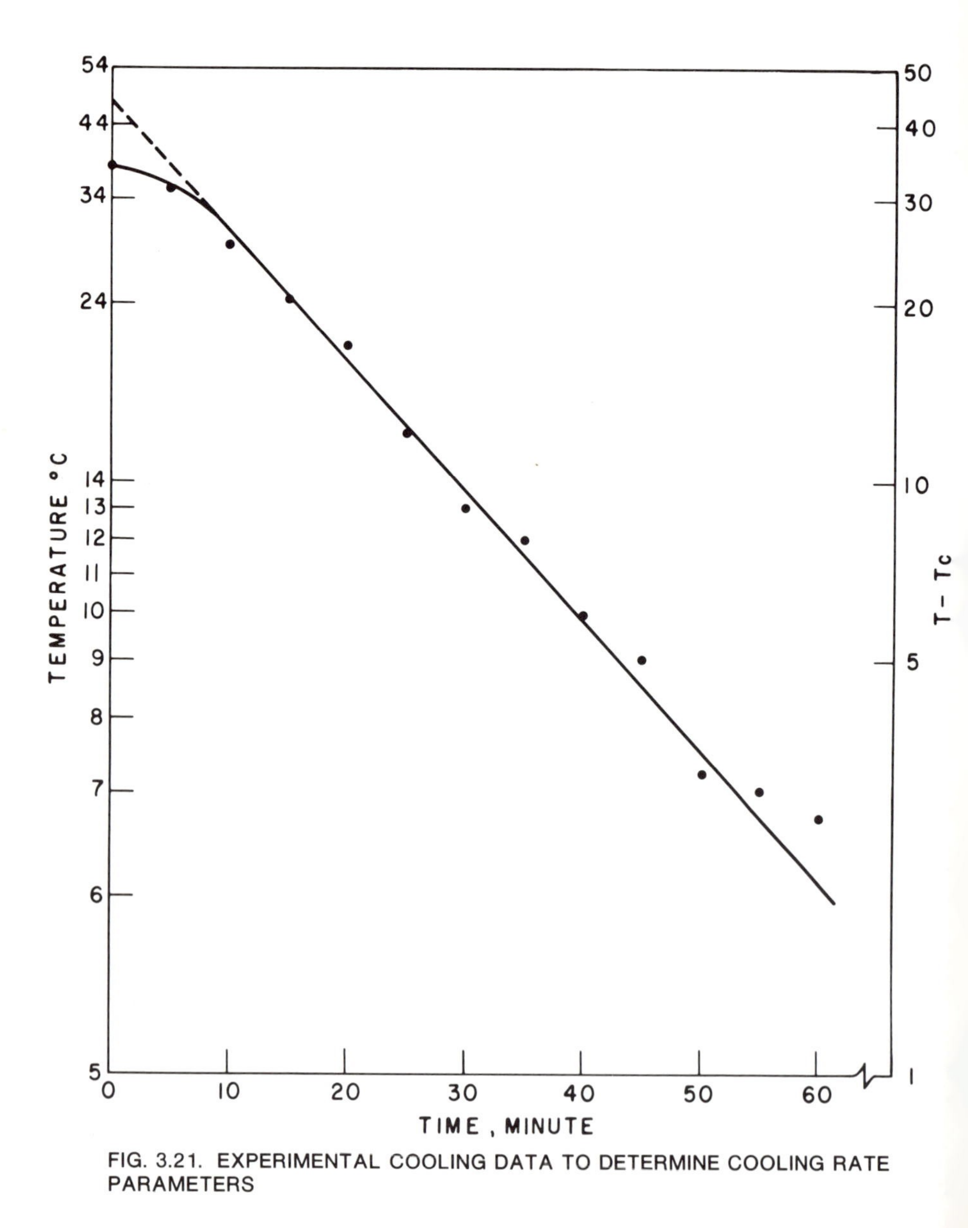


FIG. 3.21. EXPERIMENTAL COOLING DATA TO DETERMINE COOLING RATE PARAMETERS

### 3.4.5 Finite Objects

Although the previously described procedures will allow computation of temperature histories and heating and cooling rates in objects such as infinite slabs, infinite cylinders and spheres, these methods are somewhat limited when trying to describe the conditions in finite slabs and finite cylinders. Since objects such as brick-shapes and can-shapes are so common in food-processing operations, procedures for handling these types of geometric configurations must be investigated. Fortunately, methods have been developed to handle these situations which utilize the charts presented for infinite geometrics. In determining the temperature history within a brick-shaped object the following expression can be utilized:

$$\left[\frac{T - T_M}{T_0 - T_M}\right]_{\text{brick}} = \left[\frac{T - T_M}{T_0 - T_M}\right]_{\text{length}} \times \left[\frac{T - T_M}{T_0 - T_M}\right]_{\text{width}} \times \left[\frac{T - T_M}{T_0 - T_M}\right]_{\text{height}} \quad (3.83)$$

which illustrates that the temperature history within the brick-shaped object is the product of three temperature-history ratios. Each temperature ratio utilized in obtaining the final temperature ratio for the brick-shaped object is obtained utilizing charts available for the infinite slab. The same procedure is applied for each temperature ratio and employs a characteristic dimension, which may be either length, width or height of the brick-shaped object depending on the case being considered. The temperature ratio describing the temperature history within a finite cylinder is obtained in much the same manner. The following expression applies:

$$\left[\frac{T - T_M}{T_0 - T_M}\right]_{\text{finite cylinder}} = \left[\frac{T - T_M}{T_0 - T_M}\right]_{\text{infinite cylinder}} \times \left[\frac{T - T_M}{T_0 - T_M}\right]_{\text{infinite slab}} \quad (3.84)$$

where the temperature ratio for the finite cylinder is the product of two temperature ratios. For this application the temperature ratios are obtained utilizing previously described methods in which the temperature ratio for an infinite cylinder is obtained using the radius of the finite cylinder. The temperature ratio for the length of the finite cylinder is obtained from the chart for an infinite slab in which the half-thickness of the infinite slab is the half-length of the finite cylinder.

Similar procedures can be utilized to obtain heating or cooling rate parameters ($f$) and lag factors ($j$) from the charts proposed by Pflug *et al.* (1965). For these types of applications, it can be shown that the heating or cooling rate constant ($f$) can be obtained for a brick-shaped object from the following expression:

$$\frac{1}{f_{\text{brick}}} = \frac{1}{f_{\text{length}}} + \frac{1}{f_{\text{width}}} + \frac{1}{f_{\text{height}}} \tag{3.85}$$

where each of the $f$ values is evaluated utilizing the chart (Fig. 3.19) in the appropriate manner. In the case of a finite cylinder the following expression would apply:

$$\frac{1}{f_{\text{finite cylinder}}} = \frac{1}{f_{\text{infinite cylinder}}} + \frac{1}{f_{\text{height}}} \tag{3.86}$$

where the $f$ values for infinite cylinder and height of the container or finite cylinder would be obtained from the appropriate curves in Fig. 3.19. The procedure for obtaining the $j$-value in the cases of a brick-shaped object would be described by the following expression:

$$j_{\text{brick}} = j_{\text{length}} \times j_{\text{width}} \times j_{\text{height}} \tag{3.87}$$

where the $j$-values in equation (3.87) would be obtained from Fig. 3.20. An expression for obtaining a $j$-value for a finite cylinder would be written in a similar manner.

EXAMPLE 3.11 A sausage is being heated in a 116 C heating medium with a convective heat-transfer coefficient of 1135 W/m$^2$ C. The sausage has dimensions of 0.1 m diameter and 0.3 m length. Estimate the product temperature at the geometric center after 2 hr when the initial temperature is 21 C and properties include: density = 1041 kg/m$^3$, specific heat = 3.35 kJ/kg C and thermal conductivity 0.48 W/m C.

SOLUTION

(1) Computation of Biot No.

$$N_{Bi} = \frac{1135\,(0.05)}{0.48} = 118 \text{ (based on diameter)}$$

$$N_{Bi} = \frac{1135\,(0.15)}{0.48} = 355 \text{ (based on length)}$$

Since both Biot numbers exceed 40, negligible resistance to heat transfer at the surface can be assumed.

(2) Using Fig. 3.19: with $N_{Bi} = 118$, $f\,\alpha/r^2 = 0.4$
with $N_{Bi} = 355$, $f\,\alpha/\delta^2 = 0.95$

(3) Using Fig. 3.20: for $N_{Bi} = 118$; $j_c = 1.6$
for $N_{Bi} = 355$; $j_c = 1.275$

(4) Since:

$$\alpha = k/\rho c_p\text{, then } f_{\text{infinite cylinder}} = \frac{0.4\,(0.05)^2\,(1041)\,(3350)}{(3600)\,0.48} = 2.02$$

and:

$$f_{\text{infinite slab}} = \frac{0.95\,(0.15)^2\,(1041)\,(3350)}{(3600)\,0.48} = 43\text{ hr}$$

(5) Using equation (3.86):

$$\frac{1}{f_{\text{finite cylinder}}} = \frac{1}{2.02} + \frac{1}{43} = 0.495 + .023 = 0.518$$

$$f_{\text{finite cylinder}} = 1.93\text{ hr}$$

(6) Using an equation of the same form as (3.87):

$$j_{\text{finite cylinder}} = 1.6 \times 1.275 = 2.04$$

(7) Using equation (3.80):

$$\log\,(T - 116) = -2/1.93 + \log\,[2.04\,(21 - 116)]$$

$$\frac{T - 116}{2.04\,(21 - 116)} = \exp[-(2)(2.303)/1.93]$$

$$T = 98\text{ C after 2 hr}$$

### 3.4.6 Anomalous Objects and Ellipsoids

The procedures presented up to this point have described the computation of temperature distributions and heating or cooling rates in many ideal shapes such as a cylinder, a slab, or a sphere. Many of the shapes that occur frequently in the food industry, however, as natural or processed food products, do not conform to the shapes previously discussed. In some cases, computations can be made by approximation of the object by a brick, cylinder or sphere. In other situations, however, the desired accuracy can not be attained without making other assumptions about the geometric configuration. Smith *et al.* (1967) have proposed use of a geometric index ($G'$) to utilize when the object can be approximated more accurately by an ellipsoid. The geometric index can be defined in terms of the following expression:

$$\beta_1^2 = G'\pi^2 \tag{3.88}$$

where $\beta_1$ is the coefficient from the root equation for the first term of the series solution of the heat conduction equation when finite surface and internal resistance to heat transfer is assumed. Equation (3.88) would apply to any geometry between the extremes of $0.25 < G' < 1$ for an infinite slab and for a sphere, respectively. Smith *et al.* (1967) proposed the following expression:

$$G' = \frac{1}{4} + \frac{3}{8A'^2} + \frac{3}{8B'^2} \tag{3.89}$$

where:

$$A' = a'/l \tag{3.90}$$

$$B' = b'/l \tag{3.91}$$

In equation (3.91) $b'$ represents the half-length of the ellipsoidal-shaped object, $a'$ in equation (3.90) represents the half-width of the ellipsoidal-shaped object, and $l$ is described as a characteristics length or minimum half-thickness of the object. These dimensions are described in Fig. 3.22. The coefficients chosen for equation (3.89) by Smith *et al.* (1967), were based on a review of similar research and specifically on approaches proposed by Williamson and Adams (1919).

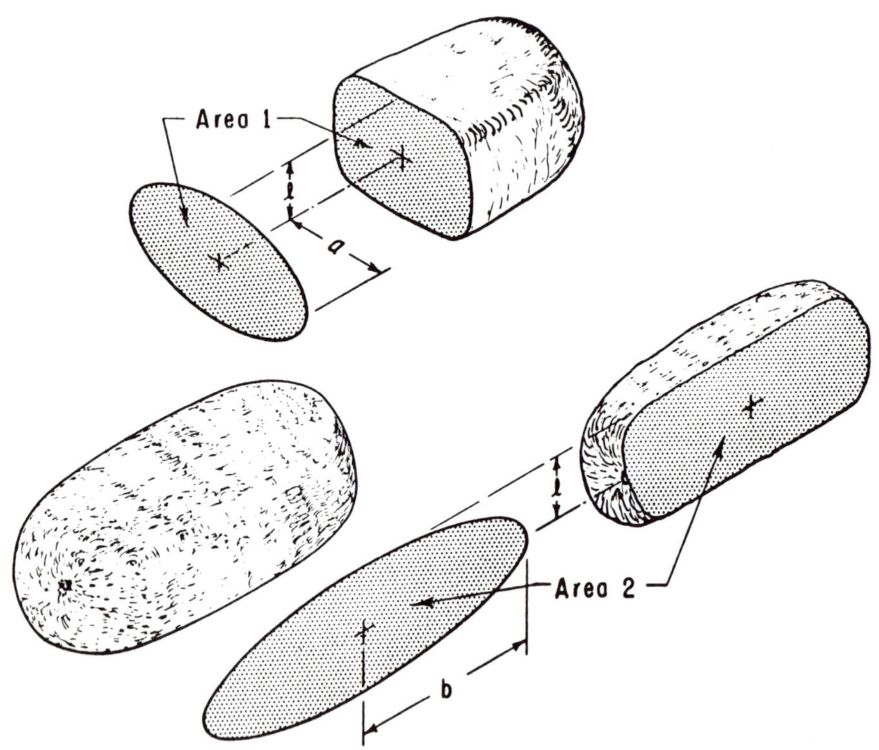


*From Smith et al. (1967)*

FIG. 3.22. DIMENSIONS CONSIDERED IN DEVELOPMENT OF GEOMETRY INDEX FOR ELLIPSOID

EXAMPLE 3.12 The longest overall dimension of an irregular-shaped object is 0.25 m, while the width is 0.13 m and the minimum dimension through the geometric center is .05 m. Compute the geometric index.

SOLUTION

(1) Using equations (3.90) and (3.91):

$$A' = \frac{.0625}{0.025} = 2.5$$

$$B' = \frac{0.125}{0.025} = 5.0$$

(2) Using equation (3.89):

$$G' = \frac{1}{4} + \frac{3}{8(2.5)^2} + \frac{3}{8(5)^2}$$

$$G' = \frac{1}{4} + \frac{3}{50} + \frac{3}{200} = \frac{50 + 12 + 3}{200}$$

$$G' = \frac{65}{200} = 0.3245$$

Smith *et al.* (1967) proposed the nomograph presented in Fig. 3.23, for use in the evaluation of the temperature distribution of an anomalous shape utilizing the geometric index. The nomograph is designed to allow evaluation of the unsteady-state temperature distribution chart for a sphere as given in Fig. 3.17. This chart would be used for all anomalous shapes, although the geometric index has been proved for only ellipsoid shapes. It would appear that this provides a much better approximation of any irregular-shaped object than information previously available.

Smith *et al.* (1967) also determined that the location at which the mass average temperature of the object should be measured would vary with the geometric index ($G'$). Through experimental investigation utilizing ellipsoid shapes, Smith *et al.* (1967) arrived at the following correlation for this location of mass average temperature.

$$L_{ma} = G'^{0.14} - 0.25 \tag{3.92}$$

where $L_{ma}$ is the ratio of any distance involving the characteristic length divided by the characteristic length ($l$). Smith and Nelson (1969) then presented a temperature distribution ratio chart, as illustrated in Fig. 3.24, to illustrate the importance of the geometric index for the condition when negligible resistance to heat transfer exists at the surface. This chart could be utilized directly to determine temperature ratios resulting in computation of the mass average temperature when the specified conditions are met.

EXAMPLE 3.13 Compute the mass average temperature of a product object with a geometric index of 0.45 after 1 hr of exposure to 4 C cooling water with a 5700 W/m$^2$ C convective heat-transfer coefficient. The product has a thermal diffusivity of 11.71 m$^2$/s and the minimum diameter ($l$) through the geometric center is 0.16 m. The initial temperature of the product is 35 C.

SOLUTION

(1) Due to the higher convective heat-transfer coefficient and relatively small characteristic dimension, the condition of negligible surface resistance to heat transfer can be assumed, and Fig. 3.24 will apply.

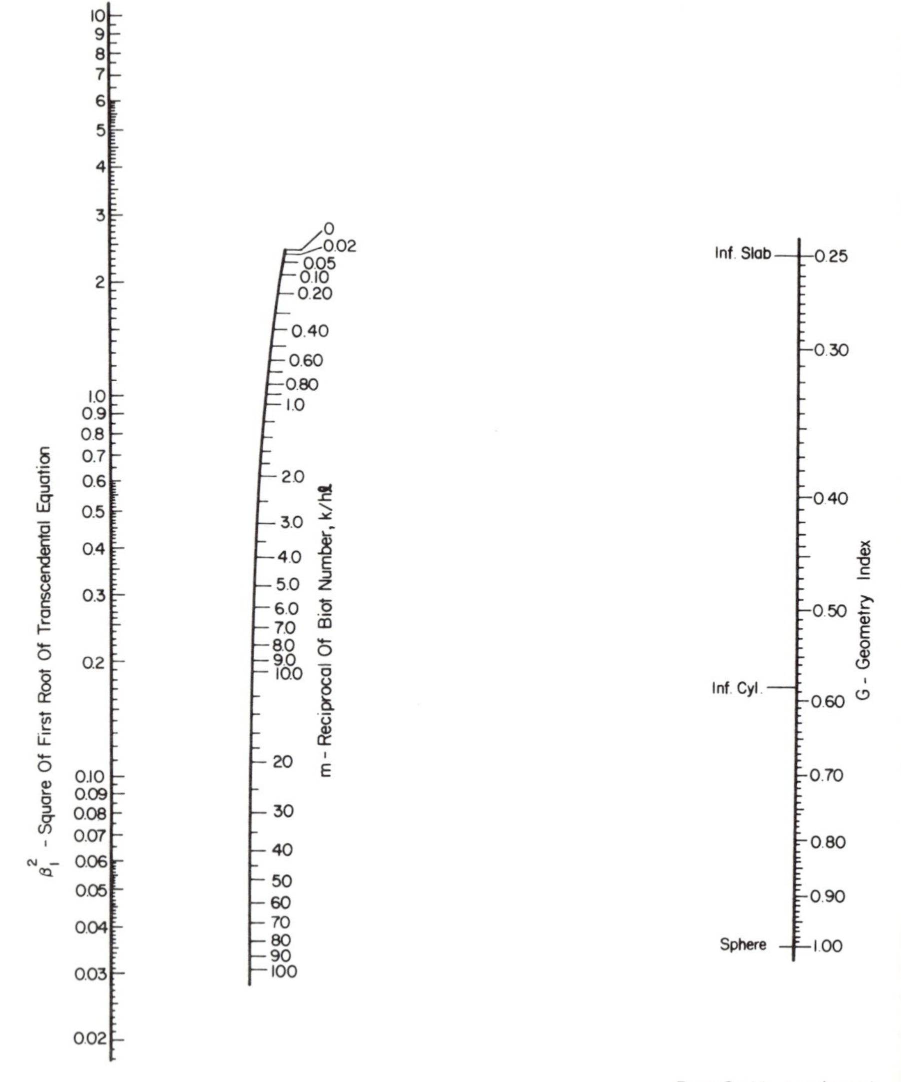


*From Smith et al. (1965)*

FIG. 3.23. NOMOGRAPH FOR EVALUATION OF TEMPERATURE DISTRIBUTIONS IN ANOMALOUS SHAPES

(2) Computation of Fourier number:

$$N_{Fo} = \frac{\alpha t}{l^2} = \frac{(11.71)(1)}{(3600)(.08)^2} = 0.51$$

(3) Use $N_{Fo}$ = 0.51, $G'$ = 0.45 and Fig. 3.24; the mass average temperature ratio becomes:

$$\frac{T_{ma} - T_s}{T_o - T_s} = 0.09$$

(4) The mass average temperature of the product becomes:

$$T_{ma} = 0.09\ (35 - 4) + 4 = 6.8\ \mathrm{C}$$

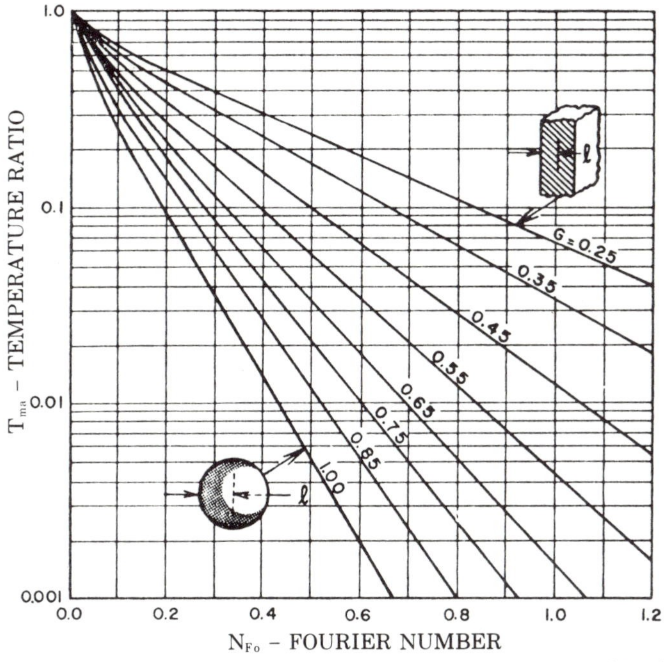


*From Smith and Nelson (1968)*

FIG. 3.24. UNSTEADY-STATE TEMPERATURE DISTRIBUTION FOR MASS AVERAGE TEMPERATURE ILLUSTRATING THE INFLUENCE OF THE GEOMETRIC INDEX WITH NEGLIGIBLE RESISTANCE TO HEAT TRANSFER AT SURFACE

EXAMPLE 3.14 Compare the times required to increase the temperature at the geometric center of two objects with the same geometric index from 21 C to 116 C in a heating medium at 121 C. One object has a minimum diameter of 0.06 m through the geometric center, while the second object has 0.3 m for the same dimension. The geometric index is 0.5, while the properties of the objects include $k$ = 0.48 W/mC, $c_p$ = 3.35 kJ/kg C and $\rho$ = 1121 kg/m³; the convective heat transfer coefficients are 10 W/m² C.

SOLUTION

(1) Computation of Biot numbers:
for 0.03 m object:

$$N_{Bi} = \frac{10\,(.03)}{(0.48)} = 0.63$$

for 0.15 m object:

$$N_{Bi} = \frac{10\,(0.15)}{0.48} = 3.13$$

since both Biot numbers are within the range of 0.1 to 40, internal and surface resistances to heat transfer must be considered.

(2) Computation of temperature ratios:
for 0.03 m and 0.15 m objects:

$$\frac{T - T_\infty}{T_o - T_\infty} = \frac{116 - 121}{21 - 121} = \frac{5}{100} = 0.05$$

(3) Use of Fig. A.1 through A.4:
these figures incorporate influence of Biot number on rate of heating. The value:

$$m = \frac{1}{N_{Bi}} = \frac{1}{0.63} = 1.59 \text{ (for 0.06 m object)}$$

$$m = \frac{1}{N_{Bi}} = \frac{1}{3.13} = 0.32 \text{ (for 0.3 m object)}$$

Since $G' = 0.5$, Fig. A.1 is used to obtain:

$$N_{Fo} = 3.4 \text{ (for 0.06 m object)}$$

$$N_{Fo} = 1.2 \text{ (for 0.3 m object)}$$

(4) Computation of time required:

$$t = \frac{N_{Fo}\, l^2}{\alpha} = \frac{N_{Fo}\, l^2 \rho c_p}{k}$$

for 0.06 m object:

$$t = \frac{3.4\,(0.03)^2\,(1121)\,(3350)}{(3600)\,(0.48)} = 6.7 \text{ hr}$$

for 0.3 m object:

$$t = \frac{1.2\,(0.15)^2\,(1121)\,(3350)}{(3600)\,(0.48)} = 58.7 \text{ hr}$$

### 3.4.7 Numerical Methods

Many unsteady-state heat-transfer problems can be solved with considerable ease using numerical methods. This method offers considerable flexibility in obtaining solutions and does not place restrictions on the use of the solutions, such as the requirement of uniform temperature distribution at the initial time. The numerical or graphical approach was first presented and described by Schmidt (1924). It can best be described by considering the section of an infinitely thick wall, such as presented in Fig. 3.25. The wall section is divided into layers, each with a thickness ($x$). For a *two-dimensional body* the differential equation governing heat transfer is:

$$k\left(\frac{\partial^2 T}{\partial x^2} + \frac{\partial^2 T}{\partial y^2}\right) = \rho c_p \frac{\partial T}{\partial t} \tag{3.93}$$

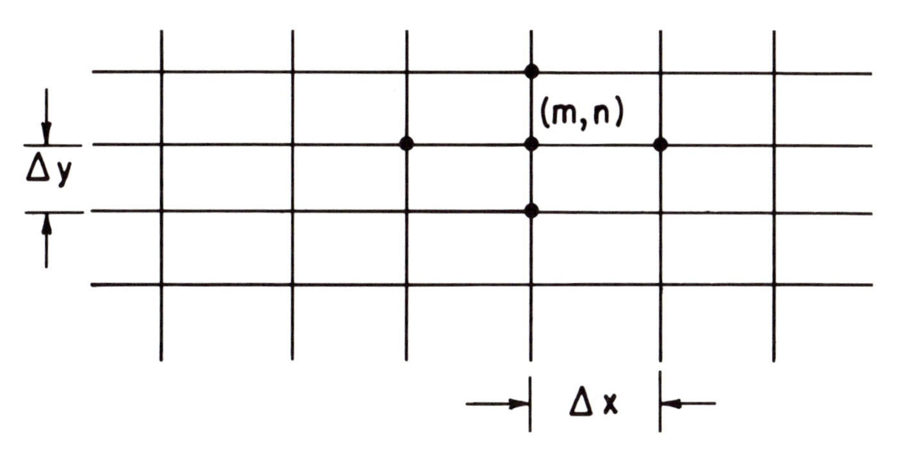


FIG. 3.25. A GRID FOR NUMERICAL APPROXIMATIONS

assuming constant thermal properties. Then second partial derivatives can be approximated by:

$$\frac{\partial^2 T}{\partial x^2} \approx \frac{1}{(\Delta x)^2} \quad (T_{m+1,n} + T_{m-1,n} - 2\,T_{m,n}) \tag{3.94}$$

and:

$$\frac{\partial^2 T}{\partial y^2} \approx \frac{1}{(\Delta y)^2} \quad (T_{m,n+1} + T_{m,n-1} - 2\,T_{m,n}) \tag{3.95}$$

The time derivative is approximated by:

$$\frac{\partial T}{\partial t} \approx \frac{T^{p+1}_{m,n} - T^{p}_{m,n}}{\Delta t} \tag{3.96}$$

Where the superscript, $p$, designate the time increment; then:

$$\frac{T^{p}_{m+1,n} + T^{p}_{m-1,n} - 2\,T^{p}_{m,n}}{(\Delta x)^2} + \frac{T^{p}_{m,n+1} + T^{p}_{m,n-1} - 2\,T^{p}_{m,n}}{(\Delta y)^2} = \frac{1}{\alpha} \; \frac{T^{p+1}_{m,n} - T^{p}_{m,n}}{\Delta t} \tag{3.97}$$

Thus, if the temperatures of the various nodes are known at any time, the temperatures after a time increment $\Delta t$ may be calculated by equation (3.97) for each node and obtaining values of $T^{p+1}_{m,n}$.
If the increments of space coordinates are chosen such that $\Delta x = \Delta y$,

$$T^{p+1}_{m,n} = \frac{\alpha \Delta t}{(\Delta x)^2} \; (T^{p}_{m+1,n} + T^{p}_{m+1,n} + T^{p}_{m,n+1} + T^{p}_{m,n-1}) + \left(1 - \frac{4\,\alpha\,\Delta t}{(\Delta x)^2}\right) T^{p}_{m,n} \tag{3.98}$$

If the time and distance increments are conveniently chosen such that

$$\frac{(\Delta x)^2}{\alpha \Delta t} = 4 \tag{3.99}$$

Temperature of node $(m,n)$ after a time increment is the arithmetic average of the four surrounding nodal temperatures at the beginning of the time increment. When a one-dimensional system is involved

$$T_m^{p+1} = \frac{\alpha \Delta t}{(\Delta x)^2} (T_{m+1}^{\ p} + T_{m-1}^{\ p}) + \left(1 - \frac{2\alpha \Delta t}{(\Delta x)^2}\right) T_m^{\ p} \tag{3.100}$$

If time and distance increments are chosen such that $\frac{(\Delta x)^2}{\alpha \Delta t} = 2$. Temperature at node $m$ after the time increment is arithmetic average of two adjacent nodal temperatures at beginning of time increment.

The selection of the value of parameter $M = \frac{(\Delta x)^2}{\alpha \Delta t}$ governs the case of solution. The larger the values of $\Delta x$ and $\Delta t$ the more rapid the solution proceeds. The smaller the values of $\Delta x$ and $\Delta t$ the more accuracy will be obtained.

If $M < 2$ in equation (3.100), $T_m^p$ becomes negative, so restrict $M \geq 2$ for a one-dimensional case and $M \geq 4$ for a two-dimensional case.

Equations of the forms presented as equation (3.97) can be derived for cylindrical and spherical geometries also. In addition, the influence of varying boundary conditions can be taken into account.

### 3.4.8 Unsteady-state Heat Transfer in Agitated Containers

Many of the heating and cooling processes which occur in the food industry are conducted in containers or vessels that are agitated either by natural convection or by induced mixing. One of the best examples is the agitated, jacketed kettle or tank in which the heating or cooling medium is contained in the jacket.

In order to predict the heating or cooling time for an agitated container or vessel which will occur under unsteady-state conditions, the following expression can be assumed:

$$Wc_p \frac{dT}{dt} = UA\,(T_M - T) \tag{3.101}$$

which assumes that the liquid in the vessel is sufficiently well mixed that the temperature is uniform throughout. In addition, equation (3.101) requires knowledge of the overall heat-transfer coefficient ($U$). This coefficient will depend on film coefficients which exist on the heating medium or cooling medium side of the jacket or tank wall and on the product side of the same wall, in addition to the thermal conductivity of the material used in construction of the tank or vessel. Equation (3.101) can be converted into a more usable form by integration and assuming that the heating medium temperature will be constant at locations inside the tank jacket. This integration results in the following expression:

$$\frac{T - T_M}{T_o - T_M} = \exp\left[-\frac{UAt}{Wc_p}\right] \tag{3.102}$$

where $T_o$ represents the initial uniform temperature distribution of the liquid in the tank.

EXAMPLE 3.15 Cream with 30% butterfat, is being heated in a 0.4 m$^3$ agitated tank. The overall heat transfer coefficient has been computed to be 300 W/m$^2$ C. The tank has 1.4 m$^2$ of heating surface and contains 220 kg product. If the initial product temperature was 21 C and the heating medium was 93 C, compute the product temperature after 30 min. The specific heat of the product is 3.35 kJ/kg C and the density is 996 kg/m$^3$.

SOLUTION

(1) Using equation (3.102):

$$\frac{T - 93}{21 - 93} = \exp\left[-\frac{(300)(1.4)(0.5)(3600)}{(220)(3350)}\right]$$

$$= \exp(-1.03)$$

$$T = (-72)(0.36) + 93$$

$$T = 93 - 26 = 67\ \text{C}$$

In order to evaluate the overall heat-transfer coefficient ($U$), several possibilities exist. It can be assumed that the thermal conductivity of the tank construction material will be known and the film coefficient on the heating-medium side can be predicted in some manner depending on the type of the heating medium. The film coefficient on the product side of the heating surface will probably be of considerable importance. If the tank is agitated, creating induced convection by a horizontal paddle, results obtained by Chilton *et al.* (1944) can be utilized. The expression obtained is as follows:

$$\frac{\bar{h}_c D}{k} = 0.37\left[\frac{D_p^2 N' \rho}{\mu}\right]^{2/3}\left[\frac{c_p \mu}{k}\right]^{1/3}\left[\frac{\mu_w}{\mu}\right]^{0.14} \tag{3.103}$$

where the dimensionless group containing the diameter of the paddle ($D_p$) is a Reynolds number for mixing. Slightly different results related to the influence of the viscosity ratio have been obtained by Uhl and Voznick (1960), who found that the exponent of the ratio should be 0.18. Equation (3.103) is applied to turbulent type mixing in the tank and therefore should best meet the specifications of uniform temperature distribution within the tank during the heating process. Utilizing the convective heat-transfer coefficient determined from equation (3.103), an overall heat-transfer coefficient ($U$) should be computed and utilized in equation (3.102). The temperature increase with time can be computed and an indication of the time required for conducting the heating process is obtained.

Heating and cooling similar to the type described for an agitated tank occurs in a food container during the sterilization process. In this situation, however, the heating or cooling is occurring by natural convection, which is a much more complex type of heating to analyze. To date, no mathematical analysis has been proposed to describe this situation, and therefore the heating and cooling characteristics cannot be predicted. The method normally used involves the experimental determination of the heating or cooling rate constant ($f$) and the lag factor ($j$). These parameters are then utilized in thermal process calculations.

## PROBLEMS

3.1 *Design of insulation:* A cold storage wall (3 m × 6 m) is 0.15 m thick concrete ($k$ = 1.73 W/m C). To maintain a heat transfer rate through the wall at or below 100 W, cork insulation must be provided. If the thermal conductivity of cork is $k$ = 0.04 W/m C, compute the thickness of cork required. The outside temperature is 38 C and inside temperature is 5 C.

3.2 *Heat loss from steampipe:* A .02 m thick steel pipe ($k$ = 43 W/m C) with 0.06 m inside diameter is used to carry 115 C steam from a boiler to a process for a distance of 6 meters. If the pipe is covered with 0.05 m thick insulation ($k$ = 0.07 W/mC), what is the heat loss from steam, if the outside temperature is 24 C?

3.3 *Heat loss from a vessel in ambient environment:* A cylinderical container is being cooled in ambient air at 15 C with no circulation. If the initial temperature of container surface is 95 C, compute the surface heat transfer coefficient due to natural convection during the initial cooling period. The container is 1.2 m diameter and 1.5 m high.

3.4 *Estimate convective heat transfer coefficient:* Steam at 1.22 atm is condensed inside a 9 m length of stainless steel tubing with a 0.025 m outer diameter and 0.019 m inside diameter. The outside of the tube is exposed to 21 C air. If steam is condensing at a rate of

$1.25 \times 10^{-3}$kg/s, estimate the value of the exterior heat transfer coefficient. The surface coefficient for steam on the interior surface is 1136 W/m$^2$ C. The thermal conductivity of stainless steel is 16.3 W/m C.

3.5 *Effectiveness of a heat exchanger:* Effectiveness of a heat exchanger may be defined as a ratio between the temperature difference of inlet and exit of product being heated to the temperature difference of heating medium and the inlet temperature of the product. Using a basic energy balance, derive an expression for the effectiveness of a heat exchanger in which steam is being condensed in the annulus of a concentric heat exchanger to heat a liquid food flowing in the inner pipe. Assuming that the condensing steam remains at a constant temperature throughout the operating time.

3.6 *Selection of double-pipe heat exchangers:* A liquid food ($c_p$ = 3.6 kJ/kg C) is to be heated from 30 to 90 C. The heating medium is oil ($c_p$ = 2.1 kJ/kg C) having an initial temperature of 160 C. The mass flow rate of oil and liquid food is 200 kg/h. Two tubular heat exchangers are available. Exchanger (A) has an overall heat transfer coefficient of 500 W/m$^2$ C with total area of 0.4 m$^2$, whereas, exchanger (B) has an overall heat transfer coefficient of 300 W/m$^2$ C with total area of 1 m$^2$. Which exchanger should be selected for this application?

3.7 *Influence of changing flow rate in a heat exchanger:* Milk is flowing under turbulent conditions in a concentric pipe exchanger. The steam condenses in the outer tube at 110 C. The milk enters at 20 C and exits at 40 C. If the mass flow rate of milk is doubled, estimate the new exit temperature.

3.8 *Comparison of a counter-flow and a concurrent flow heat exchanger:* A tubular heat exchanger is to be used to cool milk ($c_p$ = 4 kJ/kg C) from 65 C to 27 C. Milk flows in the inner tube of internal diameter 0.0229 m and thickness of 0.025 m. Water enters the outer tube at 18 C and exits at 24 C. The flow rate of milk is 3500 kg/h. The overall heat transfer coefficients for the exchanger is 2000 W/m$^2$ C. Should the heat exchanger be operated as counter flow or a concurrent flow system?

3.9 *Heat transfer in a tubular heat exchanger:* A tubular heat exchanger 5 meter long has a 0.05 m inside diameter pipe enclosed in a 0.1m inside diameter outer pipe. A liquid food of specific heat 3.8 kJ/kg C flows at 0.5 kg/s in the inner pipe. In the outer pipe steam condenses at constant temperature of 150 C. The overall heat transfer coefficient based on inner diameter of inside pipe is 500 W/m$^2$ C. What is the product exit temperature if the inlet temperature is 40 C?

3.10 *Transient heat transfer in a sausage:* A sausage 0.4 m long and 0.03 m diameter is being heated in 100 C environment. The initial temperature is 20 C. Estimate the time needed to heat the geometric center of the sausage to 80 C. Assume that convective heat transfer coefficient $h$ = 13 W/m$^2$ C, thermal conductivity $k$ = 0.4 W/m C, specific heat $c_p$ = 3.1 kJ/kg C and density = 950 kg/m$^3$.

3.11 *Cooling of an apple:* An apple is being cooled for an initial temperature of 21 C to 4 C in a high velocity water stream at 1.7 C. Compute the time required for the geometric center of the apple to reach 4 C when the following properties are known, density = 975 kg/m$^3$, specific heat = 3.76 kJ/kg C and thermal conductivity = 0.52 W/m C. The apple is of 0.03 m radius. The water stream produced a convective heat transfer coefficient of 50 W/m$^2$ C.

3.12 *Prediction of temperature at end of heating time in a cylindrical object:* A sausage is being heated in a 115 C heating medium with a convective heat transfer coefficient of 2 W/m$^2$ C. The sausage has following dimension, diameter = 0.1 m, length = 0.3 m. Estimate the temperature at the geometric center after 2 hours of heating if the initial temperature was 21 C. The following properties are assumed, density = 1040 kg/m$^3$, specific heat = 3.43 kJ/kg C and thermal conductivity = 0.4 W/m C.

3.13 *Heating of a liquid food in a steam-jacketed kettle:* A thin consistency liquid food is to be heated with stirring in a steam-jacketed hemispherical kettle. The kettle is 1 meter in diameter. The steam in the jacket is under pressure at 169 kPa. Assume that top of the kettle is well insulated. If the initial temperature of the food is 50 C and overall heat transfer coefficient is 500 W/m$^2$ K; estimate the product temperature after one hour of heating. Assume the following properties for the liquid food $\rho$ = 960 kg/m$^3$, $k$ = 0.6 W/m C and $c_p$ = 4.18 kJ/kg C.

3.14 *Influence of shape and size on mass average product temperature with infinite surface coefficient:* The product being investigated has a thermal conductivity of 0.48 W/m C, specific heat of 3.35 kJ/kg C, and density of 1121 kg/m$^3$. The product shape includes 0.25 m length and 0.12 m width. The third dimension is varied with values of 0.10, 0.15 and 0.20 m. The product is being heated to 115 C in a 121 C heating medium from the initial temperature of 65 C. Develop a plot of product size ($l$) versus time for heating. Discuss the influence of shape and size on the mass average product temperature.

3.15 *Influence of the geometric index and Biot number on temperature history at the geometric center of the product:* Consider a product with same properties as in Problem 3.14. The geometric index is varied to values of 0.3, 0.5, 0.7, and 0.9 corresponding to Figures A.1, A.2, A.3, and A.4. The evaluation is based on the center temperature attained in 5 hr when cooling the product from 71 C in a 15 C cooling medium. Consider surface heat transfer coefficients of 5 and 250 W/m$^2$ C. Evaluate different product sizes expressed by minimum thickness ($l$) of 0.05, 0.10, 0.25, and 0.50 m. Compare the results to values obtained using best traditional geometry for evaluation.

## COMPREHENSIVE PROBLEM III

Numerical Simulation of Heating (or Cooling) a Bed of Potatoes

### Objectives

1) To use numerical solution of a simplified two-equation model to simulate heating (or cooling) of a bed of potatoes with conditioned air.
2) To discuss the influence of heating (or cooling) parameters on the product temperature.

### Procedures

A simplified two-equation model for heating (or cooling) of a stationary bed of biological product will be used in the simulation. Although, due to various assumptions, the results to be obtained from the simulation have poor agreement with the experimental results, the model is a simple illustration of basic concepts used in numerical procedures in heat transfer problems (Bakker-Arkema and Bickert, 1966).

### Assumptions

The assumptions are the following:

(1) no mass transfer
(2) no temperature gradients within the individual particles
(3) no particle-to-particle conduction
(4) plug-type air flow
(5) constant thermal properties of the air and the product in the temperature range considered
(6) adiabatic bed walls of negligible heat capacity

### Energy Balances

The energy balances are written on a differential volume Adx located at an arbitrary location in the stationary bed. The two unknowns are $T_p$

—the product temperature, and $T_g$—the air temperature. Thus, two energy balances result in the following two differential equations:

(1) for the air

energy out = energy in + energy transferred by convection + change in energy in the voids

$$(u\rho c_p)_{air} A T_g dt + (u\rho c_p)_{air} A \frac{\partial T_g}{\partial x} dx dt = (u\rho c_p) A T_g dt +$$

$$haAdx(T_p - T_g)dt + \epsilon Adx(\rho c_p)_{air} \frac{\partial T_g}{\partial t} dt \qquad (1)$$

where 'a' in this problem is product surface area, $m^2/m^3$ and $\epsilon$ is porosity.

Or
$$\frac{\partial T_g}{\partial x} = \frac{ha}{(u\rho c_p)_{air}} (T_p - T_g) + \frac{\epsilon}{u_{air}} \frac{\partial T_g}{\partial t} \qquad (2)$$

(2) for the product

energy transferred by convection = change in internal product energy

$$haAdx(T_p - T_g)dt = (1 - \epsilon)Adx(\rho c_p)_{Product} \frac{\partial T_p}{\partial t} dt \qquad (3)$$

Or
$$\frac{\partial T_p}{\partial t} = \frac{ha}{(1 - \epsilon)(\rho c_p)_{product}} (T_p - T_a) \qquad (4)$$

## Boundary Conditions

Assuming constant, uniform temperature distribution within the bed and constant inlet air conditions, the initial and boundary conditions are:

$$T_p(x,o) = T_g(x,o) = T_p(\text{initial}) \qquad (5)$$

$$T_g(o,t) = T_g \text{ (inlet)} \qquad (6)$$

## Numerical Solution

Before writing the numerical approximations of the above equations, in equation (2), it is observed that the second term on right-hand side containing $(\partial T_g/\partial t)$ is small compared to the term containing $(\partial T_g/\partial x)$. Therefore, equation (2) may be simplified as:

$$\frac{\partial T_g}{\partial x} = \frac{ha}{(u\rho c_p)_{air}} (T_p - T_g) \qquad (7)$$

Equations (4) and (7) can be written in finite difference (forward) formulation as follows:

$$T_p(x,t+\Delta t) = T_p(x,t) - \frac{ha\Delta t}{(1-\epsilon)(\rho c_p)_{\text{product}}}[T_p(x+\Delta x,t) - T_p(x+½\Delta x,t)] \quad (8)$$

and equation (7) becomes:

$$T_g(x+\Delta x,\, t+\Delta t) = T_g(x,t) + \frac{ha\Delta x}{(u\rho c_p)_{\text{air}}}\,[T_p(x+½\Delta x,t) - T_g(x+½\Delta x,t)] \quad (9)$$

Equations (8), (9) and the boundary conditions can be programmed on a digital computer.

For stability reasons, choice of $\Delta t$ and $\Delta x$ should meet the following criteria:

$$\frac{(u\rho c_p)_{\text{air}}}{ha\Delta x} \geq 1/2 \quad (10)$$

and

$$\frac{\Delta x(1-\epsilon)(\rho c_p)_{\text{product}}}{(u\rho c_p)_{\text{product}}\,\Delta t} \geq \frac{2}{1+2(u\rho c_p)_{\text{air}}/ha\Delta x} \quad (11)$$

## Data

A deep bed of potatoes is being *reconditioned* with warm air. It is desired to calculate the temperature of the potatoes within the deep bed as a function of position and time.

The following conditions are given:

1) The initial temperature of potatoes and the air within the bed is 5C
2) The inlet air to the bed is 24C
3) The rate of airflow is $7 \times 10^{-2}$kg/m$^2$/s
4) The heat transfer area per cubic foot of the bed (a) is 130 m$^2$/m$^3$
5) The specific heat of potato flesh ($c_p$) is 3.35 kJ/kgC
6) The density of potato flesh ($\rho$) is 1120 kg/m$^3$
7) The porosity of the bed ($\epsilon$) is 0.40
8) Convective heat transfer coefficient, $h$ = 25 W/m$^2$C
9) The bed depth is 4m
10) The specific heat of air is 1 kJ/kgC

Using numerical simulation, calculate and print in table form the potato tuber and air temperature throughout the bed after 48 hours.

## Programming Notes

a) Set the initial product and air temperature at all positions in the bed equal to 5C or $T_p(x,0) = 5$ and $T_{air}(x,0) = 5$
b) Set the inlet air temperature for all times except time zero equal to 24C or $T\,(0,t) = 24$
c) Set the inlet air temperature at time zero equal to 15 C or $T\,(0,0) = 15$

## NOMENCLATURE

$A$ = area, $m^2$.
$A'$ = constant defined in equation (3.90)
$a$ = constant in equation (3.36).
$a'$ = maximum dimension of geometric shape.
$B'$ = constant defined in equation (3.91).
$b$ = constant in equation (3.36); solids fraction.
$b'$ = width of geometric shape, m.
$b''$ = distance between two parallel heated surfaces, m.
$N_{Bi}$ = Biot number.
$c_p$ = specific heat, kJ/kg C.
$D$ = diameter, m.
$E$ = activation energy, kJ/kg.
$E'$ = function defined in equation (3.51).
$e$ = emissivity.
$f$ = heating or cooling rate constant, hr.
$f'$ = friction factor.
$N_{Fo}$ = Fourier modulus.
$G$ = volumetric flow rate, $m^3/s$.
$G'$ = geometric index.
$g$ = acceleration due to gravity, $m/s^2$.
$N_{Gr}$ = Grashoff number.
$N_{Gz}$ = Graetz number, defined in equation (3.50).
$H$ = heat capacity ratio for heat exchanger.
$h$ = heat transfer coefficient, $W/m^2$ C.
$h_c$ = Convective heat transfer coefficient, $W/m^2$ C.
$J$ = Bessel functions.
$j$ = lag factor defined by equation (3.82).
$K$ = thermal conductivity ratio for heat exchanger.
$k$ = thermal conductivity, W/m C.
$k_\sigma$ = constant used to evaluate influence of temperature on the consistency coefficient (m).
$L$ = length of thickness, m.
$L_{ma}$ = location of mass-average temperature, m.

$l$ = characteristic distance, m.
$M$ = constant in equations (3.37).
$m$ = consistency coefficient, $Pas^n$; $m = 1/N_{Bi}$.
$N'$ = rotation speed, revolutions/min.
$N$ = constant in equation (3.40).
$n$ = flow behavior index.
$N_{Nu}$ = Nusselt number.
$P$ = constant in equation (3.41).
$N_{Pe}$ = Peclet number defined in equation (3.67).
$N_{Pr}$ = Prandtl number.
$Q'$ = constant defined in equation (3.38).
$Q''$ = constant defined in equation (3.40).
$q$ = heat flux, kJ/s.
$q$ = rate of heat generation, $kJ/m^3s$.
$R$ = radius dimension, m.
$R_n$ = variable function in equation (3.78).
$N_{Re}$ = Reynolds number.
$R_G$ = gas constant, kJ/kg K.
$r$ = radial distance, variable.
St = Stanton number.
$T$ = temperature, C.
$T_A$ = absolute temperature, K.
$t$ = time, sec.
$U$ = overall heat transfer coefficient, $W/m^2$ C.
$u$ = fluid velocity, m/s.
$\bar{u}$ = mean fluid velocity, m/s.
$V$ = volume, $m^3$.
$W$ = fluid or solid mass, kg.
$w$ = mass flow rate, kg/s.
$X$ = weight fraction of given component in product.
$x$ = component of cartesian coordinate system or variable distance in $x$-direction.
$y$ = component of cartesian coordinate system or variable distance in $y$-direction.
$Z$ = length.
$z$ = component of cartesian coordinate system or variable distance in $z$-direction.
$\alpha$ = thermal diffusivity, $m^2/s$.
$\beta$ = coefficient of expansion, 1/ C.
$\beta_1$ = function in equation (3.88).
$\delta$ = half-thickness of infinite plate, m.
$\Delta$ = thickness ratio for heat exchanger.
$\epsilon$ = void fraction in particle bed.

$\phi_m$ = constant in equations (3.59) and (3.60); ratio of bulk mean velocity to centerline velocity in tube.
$\mu$ = viscosity, kg/ms.
$\chi$ = characteristic dimension in Grashof number, m.
$\Psi$ = function defined in equation (3.50).
$\rho$ = density, kg/3.
$\sigma$ = Stefan-Boltzmann constant, $W/m^2 K^4$.
$\zeta$ = constant defined by equation (3.69).

## Subscripts

$a$ = ash or mineral component.
$b$ = bulk fluid.
$c$ = carbohydrate component.
$F$ = fat component.
$f$ = fluid property.
$g$ = gas phase.
$i$ = initial condition.
$L$ = liquid component of product.
$L'$ = length.
$m$ = moisture or water component.
$M$ = heating or cooling medium.
$o$ = zero time or location.
$p$ = protein component.
$r$ = radiation heat transfer.
$S$ = component solids.
$s$ = surface location.
$\perp$ = perpendicular analysis.
$\parallel$ = parallel analysis.

## BIBLIOGRAPHY

BAKKER-ARKEMA, F.W. and BICKERT, W.G. 1966. A deep-bed computational cooling procedure for biological products. Trans. ASAE *9*, (6) 834.

BIRD, R.B., STEWART, W.E. and LIGHTFOOT, E.N. 1960. Transport Phenomena. John Wiley & Sons, New York.

BLANCO, J.A. and GILL, W.N. 1967. Analysis of multistream turbulent forced convection systems. Chem. Eng. Prog. Symp. Ser. *77.63,* 67–79.

BLASIUS, M. 1908. Laminar layer in liquids with low friction. Z. Math U. Phys. *56*, No. 1.

CARSLAW, H.W. and JAEGER, J.C. 1959. Conduction of Heat in Solids. Clarenden Press, Oxford, England.

CHARM, S.E. 1978. The Fundamentals of Food Engineering, 3rd Edition. AVI Publishing Co., Westport, Conn.

CHARM, S.E. and MERRILL, E.W. 1959. Heat transfer coefficients in straight tubes for pseudoplastic fluids in streamline flow. Food Res. *24,* 319.

CHEN, A.C. 1969. Mechanisms of heat transfer through organic powder in a packed bed. Ph.D. Thesis. Michigan State University, East Lansing, MI.

CHILTON, T.H., DREW, T.B. and JEBENS, R.H. 1944. Heat transfer coefficients in agitated vessels. Ind. Eng. Chem. *36*, 510.

CHRISTIANSEN, E.B. and CRAIG, S.E. 1962. Heat transfer to pseudoplast fluids in laminar flow. Am. Int. Chem. Engrs. J. *8*, 154.

CHRISTIANSEN, E.B. and PETERSEN, A.W. 1966. Heat transfer to non-Newtonian fluids in transitional and turbulent flow. Am. Inst. Chem. Engrs.

CLAPP, R.M. 1961. International developments in heat transfer. Am. Soc. Mech. Engrs. Publ. Part III, 652–661.

DICKERSON, R.W., JR. 1965. An apparatus for the measurement of thermal diffusivity of food. J. Food Technol. *19*, 880.

DICKERSON, R.W., JR. 1969. Thermal properties of foods. *In*: The Freezing Preservation of Foods, 4th Edition, Vol. 2. D.K. TRESSLER, W.B. VAN ARSDEL and M.J. COPLEY (Editors). AVI Publishing Co., Westport, Conn.

DITTUS, F.W. and BOELTER, L.M.K. 1930. U. California Engr. Publs. *5* No. 2, 23–28. *Cited in:* McAdams, W.H. Heat Transmission. McGraw-Hill Book Co., New York.

DUCHATELLE, L. and VAUTREY, L. 1964. Determination des coefficients de convection d'un alliage Nak en ecoulement turbulent entre plaques planes paralleles. Intern. J. Heat Mass Transfer *7*, 1017.

EARLE, R.L. 1966. Unit Operations in Food Processing. Pergamon Press, New York.

FOUST, A.S., WENZEL, L.A., CLUMP, C.W., MANS, L. and ANDERSON, L.B. 1960. Principles of Unit Operations. John Wiley & Sons, New York.

FURNAS, C.C. 1930. Heat transfer from a gas stream to a bed of broken solids. Ind. Eng. Chem. *22*(8)721.

GRAETZ, L. 1883. On the heat transfer in liquids. Ann. Phys. (Leipzig), *18*, 79–94.

HARPER, J.C. 1976. Elements of Food Engineering. AVI Publishing Company, Inc., Westport, Conn.

HARPER, J.C. and EL SAHRIGI, A.F. 1964. Thermal conductivity of gas-filled porous solids. Ind. Eng. Chem. Fundamentals *3*, No. 4, 318–324.

HOOPER, F.C. and LEPPER, F.R. 1950. Transient heat flow apparatus for determination of thermal conductivities. ASHRAE Trans. *56*, 309.

HURWICZ, H. and TISCHER, R.G. 1952. Heat processing of beef. II. Development of isotherm and isochronal distributions during heat processing of beef. Food Res. *17*, 518.

JAKOB, M. 1949. Heat Transfer, Vol. 1. John Wiley & Sons, New York.

KAYS, W.M. and LONDON, A.L. 1954. Compact heat exchangers—A summary of basic heat transfer and flow friction design data. Tech. Rept. *23*, Stanford University, Stanford, CA.

KOPELMAN, I.J. 1966. Transient heat transfer and thermal properties in food systems. Ph.D. Thesis. Michigan State University, East Lansing, MI.

KREITH, F. 1965. Principles of Heat Transfer, 2nd Edition. International Textbook Co., Scranton, PA.

LENTZ, C.P. 1961. Thermal conductivity of meats, fats, gelatin gels and ice. Food Technol. *15*, 243–247.

LEVEQUE, J. 1928. Ann. Mines Ser. 12, 13, 201, 305, 381. *Cited by* SKELLAND, A.H.P. 1967. Non-Newtonian Flow and Heat Transfer. John Wiley & Sons, New York.

LONG, R.A. 1955. Some thermodynamic properties of fish and their effect on rate of freezing. J. Sci. Food Agr. *6*, 621.

LYCHE, B.C. and BIRD, R.B. 1956. The Graetz-Nusselt problem for a power-law non-Newtonian fluid. Chem. Eng. Sci. *6*, 34.

MAXWELL, J.C. 1904. A Treatise on Electricity and Magnetism, 3rd Edition. Clarendon Press, Oxford, England.

McADAMS, W.H. 1954. Heat Transmission. McGraw-Hill Book Co., New York.

METZNER, A.B. and FRIEND, P.S. 1959. Heat transfer to turbulent non-Newtonian fluids. Ind. Eng. Chem. *51*, 879.

METZNER, A.B., VAUGH, R.D. and HOUGHTON, G.L. 1959. Turbulent flow of non-Newtonian systems. Am. Inst. Chem. Engrs. J. *5*, 189.

MONRAD, C.C. and PELTON, J.F. 1942. *Cited by* W.H. McADAMS. Heat Transmission. 1954. McGraw-Hill Book Co., New York.

NGODDY, P.O., BAKKER-ARKEMA, F.W. and BICKERT, W.G. 1966. Heat transfer in a deep bed of pea beans. Quarterly Bulletin Michigan Agricultural Experiment Station 49(2)132.

NUNGE, R.J., PORTA, E.W. and GILL, W.N. 1967. Axial conduction in the fluid stream of multistream heat exchanger. Chem. Engr. Progr. Symp. Ser. *77.66*, 80–91.

PERRY, R.H. and CHILTON, C.H. 1973. Chemical Engineers Handbook. 5th Edition. McGraw-Hill Book Co., New York.

PFLUG, I.J., BLAISDELL, J.L. and KOPELMAN, I.J. 1965. Developing temperature-time curves for objects that can be approximated by a sphere, infinite plate or infinite cylinder. ASHRAE Trans. *71*, No. 1, 238.

PIGFORD, R.L. 1955. Non-isothermal flow and heat transfer inside vertical tubes. Chem. Eng. Progr. Symp. Ser. *17.51*, 79.

RANZ, W.E. and MARSHALL, W.R., JR. 1952. Evaporation from drops. Chem. Eng. Progr. *48*, 247.

REIDY, G.A. and RIPPEN, A.L. 1971. Methods for determining thermal conductivities of foods. Trans. Am. Soc. Agr. Engrs. *14*, No. 2, 248.

RIEDEL, L. 1949. Measurements of the thermal conductivity of sugar solutions, fruit juices and milk. Chem. Ing-Tech. *21*, 340, 341. (German)

SCHMIDT, E. 1924. Foppls Springer, Berlin. *Cited by* W.H. McADAMS. Heat Transmission. 1954. McGraw-Hill Book Co., New York.

SCHUMAN, T.E.W. 1929. Heat transfer: a liquid flowing through a porous prism. J. Franklin Institute 208(6)408.

SIEBEL, J.E. 1892. Specific heat of various products. Ice Refrig. *2*, 256.

SIEDER, E.N. and TATE, G.E. 1936. Heat transfer and pressure drop in liquids in tubes. Ind. Eng. Chem. *28*, 1429.

SKELLAND, A.H.P. 1958. Correlation of scraped-film heat transfer in the votator. Chem. Eng. Sci. *7*, 166–175.

SKELLAND, A.H.P. 1967. Non-Newtonian Flow and Heat Transfer. John Wiley & Sons, New York.

SMITH, R.E. and NELSON, G.L. 1969. Transient heat transfer in solids: theory vs. experiment. Trans. Am. Soc. Agr. Engrs. *12*, 833.

SMITH, R.E., NELSON, G.L. and HENRICKSON, R.L. 1967. Analysis of transient heat transfer form anomalous shapes. Trans. Am. Soc. Agr. Engrs. *10*, No. 2, 236–245.

SWEAT, V.E. 1974. Experimental values of thermal conductivity of selected fruits and vegetables. J. Food Sci. *39*:1080.

UHL, V.W. and VOZNICK, H.P. 1960. The anchor agitator. Chem. Eng. Prog. *56*, 72.

VOS, B.H. 1955. Measurement of thermal conductivity by a non-steady-state method. Appl. Sci. Res. (Hague) *A5*, 425.

WANG, J.K. and WANG, P.Y. 1968. A computational technique for deep bed forced-air precooling of tomatoes. ASAE Paper No. 68-819.

WILLIAMSON, E.D. and ADAMS, L.H. 1919. Temperature distribution in solids during heating or cooling. Phys. Rev. *14*, 99.

4

# Thermodynamics of Food Freezing

Freezing is one of the more common processes for the preservation of foods. Though it has been recognized as a preservation technique for several hundred years, the major developments in its utilization have occurred only in the last century. It is well known that lowering the temperature reduces the activity of microorganisms and enzyme systems, thus preventing deterioration of the food product. In addition to the influence of temperature reduction on microorganisms and enzymes, crystallization of the water in the product tends to reduce the amount of liquid water in the system and inhibit microbial growth or enzyme activity in the secondary action.

The engineering aspects of food freezing includes several interesting areas. In order to design a refrigeration system that will serve a food-freezing process, some indication of the refrigeration requirements or enthalpy change which occurs during product freezing is required. This aspect is related to the type of product being frozen. The second aspect of food freezing that is closely related to engineering is the rate at which freezing progresses. This area is related to the refrigeration requirement, but the temperature differences existing between the product and freezing medium are also of significance. The rate of freezing is closely related to product properties and quality, also. Product properties resulting from very rapid freezing are significantly different from those obtained by slow freezing. This difference is dependent primarily on the manner in which ice is formed within the product structure. In addition, the rate of freezing will establish the rate of production for a particular food-freezing operation. For this purpose, the most rapid rate of freezing is desirable, provided that product quality is not sacrificed.

# 4.1 PROPERTIES OF FROZEN FOODS

One aspect of food freezing which cannot be over emphasized is the role of the properties of the product. Although the thermal properties of food products, in general, have been discussed in portions of Chapter 3, there are several unique properties of frozen foods which deserve additional consideration. Of particular significance is the role of water and its change of state during the freezing process. Since certain aspects of the thermodynamics, such as phase change by water, were discussed in Chapter 1, some of the more unique features of water as it exists in frozen foods can be presented in more detail.

## 4.1.1 Freezing-point Depression

Probably one of the more revealing properties of water in food is the freezing-point depression. Since all food products contain relatively large amounts of moisture or water in which various solutes are present, the actual or initial freezing point of the water in the product will be depressed to some level below that expected for pure water. The magnitude of this freezing-point depression becomes a direct function of the molecular weight and concentration of the solute in the food product and in solution with the water.

The expression or expressions which predict the extent of freezing-point depression can be derived from thermodynamic relationships based on equilibrium between the states of a system. Based on the situation that the change in free energy for a system in equilibrium must be zero and equation (1.6), which defines chemical potential, it follows that the chemical potential for each phase of a system must be equal, as illustrated by the following equation:

$$\eta^{\alpha} = \eta^{\beta} \tag{4.1}$$

where $\alpha$ and $\beta$ represent two separate phases of the system. Utilizing one of the basic thermodynamic relationships which indicates that the change in chemical potential with pressure at a constant temperature must equal the partial molar volume, it follows that the chemical potential of a solution will be described by the following expression:

$$\eta_A^{\mathrm{soln}} = \eta_A^0 + R_g T_A \ln p_A \tag{4.2}$$

In equation (4.2), the chemical potential at 1 atmosphere pressure has the superscript zero and $T_A$ is absolute temperature. Utilizing Raoult's law, as given by equation (1.10), equation (4.2) becomes:

$$\eta_A = \eta_A^0 + R_g T_A \ln p_A^0 + R_g T_A \ln X_A \tag{4.3}$$

which provides the relationship between mole fraction ($X_A$) and the chemical potential of component $A$ in solution. At a constant temperature and pressure the first two terms of the right-hand side of equation (4.3) are constants, and the expression can be written as follows:

$$\eta_A = \eta^{0\prime} + R_g T_A \ln X_A \tag{4.4}$$

which illustrates the direct relationship between chemical potential of a component ($A$) and the mole fraction of that component in an ideal solution. For the case of solids and liquids at equilibrium, equation (4.4) can be written as follows:

$$\eta_A^s = \eta_A^{0l} + R_g T_A \ln X_A \tag{4.5}$$

providing the relationship between the solid and liquid phases. Utilizing the definition of chemical potential as given by equation (1.6) the following expression is obtained:

$$\frac{\bar{G}_A^{0s} - \bar{G}_A^{0l}}{R_g T_A} = \ln X_A \tag{4.6}$$

which expresses the difference in molar free energies between the liquid and solid phases. Since the free energy can be related to enthalpy by the following expression:

$$\frac{\partial}{\partial T_A}\left[\frac{G}{T_A}\right] = -\frac{H}{T_A^2} \tag{4.7}$$

Equation (4.6) can be written in terms of the differences in enthalpy of the liquid and solid phases or the latent heat of fusion, as illustrated by the following expression:

$$\frac{\bar{H}_A^{0l} - \bar{H}_A^{0s}}{R_g T_A^2} = \frac{\lambda'}{R_g T_A^2} = \frac{d \ln X_A}{d T_A} \tag{4.8}$$

Equation (4.8) relates latent heat of fusion ($\lambda'$) with the mole fraction and temperature. By integration of equation (4.8) the following expression is obtained:

$$\frac{\lambda'}{R_g}\left[\frac{1}{T_{A0}} - \frac{1}{T_A}\right] = \ln X_A \tag{4.9}$$

where $T_{A_0}$ is the freezing point of pure liquid ($A$) and $X_A$ is the mole fraction of water in solution. Although equation (4.9) is in an acceptable form for computation of freezing-point depression, the expression can be simplified for dilute solutions. Equation (4.9) can be modified as:

$$\frac{\lambda'}{R_g}\left[\frac{T_A - T_{Ao}}{T_{Ao}^2}\right] = \ln(1 - X_B) \quad (4.10)$$

where $T_A T_{Ao} \simeq T_{Ao}^2$ for small values of $T_A - T_{Ao}$. By setting $\Delta T_F = T_{Ao} - T_A$ and expanding the log in a power series $[-\ln(1 - X_B) = X_B + 1/2\, X_B^2 + 1/3\, X_B^2 + \ldots]$ results in:

$$\frac{\lambda' \Delta T_F}{R_g\, T_{Ao}^2} = X_B \quad (4.11)$$

when solutions are dilute ($X_B << 1$). By expressing the latent heat of fusion on a per unit mass basis, equation (4.11) becomes:

$$\Delta T_F = \frac{R_g\, T_{Ao}{}^2 W_A\, m}{1000\, L} \quad (4.12)$$

where $m$ is molality in terms of moles of solute per kg solvent. The usual procedure is to incorporate all the parts of the right-hand side of equation (4.12), except molality, into a molal freezing point constant. This constant would have a value of 1.86 for water when the freezing point depression is expressed in C.

EXAMPLE 4.1 Compute the temperature at which ice formation begins in an ice-cream mix with the following composition: 10% butterfat, 12% solids-not-fat, 15% sucrose and 0.22% stabilizer.

SOLUTION

Two assumptions can be made to allow use of existing equations: (a) the sugar in the product is the predominate factor influencing freezing point and (b) the concentration is sufficiently dilute to allow use of equation (4.12).

(1) Equation (4.12) requires knowledge of molality or:

$$m = \frac{M_B \text{ (per 1000 g solvent)}}{W_B}$$

(2) The solute accounted for in the ice-cream mix is sucrose (W = 342) and lactose (W = 342), which represents 54.5% of the solids-not-fat in the mix.

(3) The molality is computed as follows:

Fraction solute = 0.15 + 0.545 (0.12) = 0.2154 g/g product.
When expressed in terms of water fraction (62.78%), then:

$$\frac{0.2154}{0.6278} = 0.3431 \text{ g solute/g solvent}$$

or: 343.1 g solute/1000 g solvent

$$\text{and: } m = \frac{343.1}{342} = 1.003$$

(4) Using equation (4.12):

$$\Delta T_F = \frac{(0.462)(273)^2(18)(1.003)}{1000(333.22)} = 1.86 \text{ K}$$

(5) The computations indicate that the initial ice formation will occur at 271.14 K or −1.86 C.

(6) More accurate computations would account for salts present in the solids-not-fat and would depress the freezing point slightly more than the above value.

The actual freezing process in food products is somewhat more complex than freezing of pure water, as would be expected. This process can be visualized by reference to Fig. 4.1, which compares the freezing curves of water with an aqueous solution containing one solute. In water, the temperature decreases as heat is removed from the system until the freezing point is reached. After the small amount of supercooling, the temperature remains constant as the latent heat is removed from the water system. Following this latent heat removal, the temperature decreases again as energy is removed. In a food product or any solution the removal of heat energy results in a temperature decrease until the initial freezing point is reached, in the same manner as water. The initial freezing point, however, will be depressed to the extent predicted by equation (4.12). The initial freezing results in crystallization of a portion of the water, resulting in a concentration of the remaining solution and further reduction of the freezing point of that unfrozen portion. This results in an additional decrease in temperature before more heat energy is removed, as illustrated in Fig. 4.1. The process continues as a simultaneous crystallization of water, resulting in further depression of

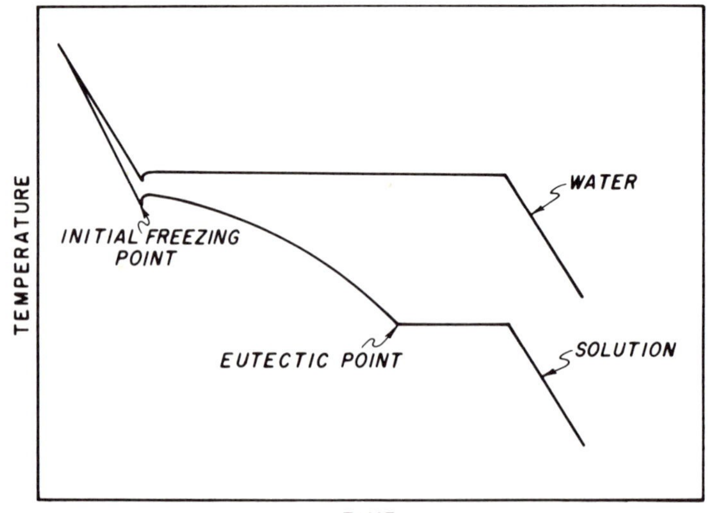


FIG. 4.1. A COMPARISON OF FREEZING CURVES FOR PURE WATER AND AN AQUEOUS SOLUTION CONTAINING ONE SOLUTE

the freezing point of the concentrated solution until the eutectic point of the solute is reached. This point will be unique for each solute present in the system. In a single-solute system, the removal of heat energy beyond the eutectic point does result in temperature decrease, but with a crystallization of the solute as well as ice crystal formation. As would be expected, beyond this point the temperature of the system again decreases. In an actual food product system it is very probable that more than one solute will be present, and therefore several eutectic points may be reached during the freezing process. In fact the temperatures at which eutectic locations are reached during freezing may not be evident because of the presence of so many different solutes in the system.

EXAMPLE 4.2 According to Table A.12, the weight percentage of water in grape juice is 84.7% and the freezing point of grape juice is −1.8 C (271.2 K). Using this experimental value, compute the effective molecular weight ($W_E$) of grape juice to be used in freezing computations.

SOLUTION

(1) Although the solute in grape could be assumed to be sufficiently dilute to use equation (4.12), equation (4.9) can be used as easily and more accurately:

$$\frac{6003}{8.314}\left[\frac{1}{273}-\frac{1}{271.2}\right]=\ln X_A$$

where $\lambda' = 6003$ J/mole. Then:

$\ln X_A = 722.04\,(-2.43 \times 10^{-5}) = -0.01755$

and $X_A = 0.9826$

which represents the effective mole fraction of water in grape juice.

(2) From the definition of mole fraction:

$$0.9826=\frac{84.7/18}{84.7/18+15.3/W_E}$$

$$W_E = 183.61$$

It should be noted that this effective molecular weight value is based only on product components which influence freezing point depression.

## 4.1.2 Ice Crystal Formation

The manner in which ice crystals are formed within a food product during freezing is of considerable interest due to the influence of ice crystal size and configuration on product quality. The physical phenomenon associated with ice crystal formation have been presented by Meryman (1956) and Fennema (1975). The crystallization process occurs in two steps: (a) nucleation or crystal formation and (b) crystal growth. Nucleation is the initiation of freezing and involves the presence or formation of small nuclei which are the centers of the crystals that form. Technically, nucleation may be defined as the generation, within a metastable system or phase, of the smallest particles of a foreign stable phase capable of growing spontaneously. Potentially, two types of nucleation can occur, namely, homogeneous and heterogeneous nucleation. The first case is rather rare and only occurs in systems such as very highly purified water. The nuclei are random accumulations of a sufficient number of water molecules. In heterogeneous nucleation, small particles present in the solution act as nuclei to start the crystal formation. In most cases these particles must have crystal structures similar to that formed by ice. In addition to the presence of these small particles, both mechanical impact and local variations in solute concentration will result in heterogeneous nucleation. Nucleation may be of considerable importance in the overall rate of freezing and the type of crystal structure formed in a food product.

Since nucleation may be of considerable importance in determining the rate of freezing and crystal structure in the food product, it deserves additional study and investigation. Stephenson (1960) proposed an approach based on thermodynamics of phase change, which leads to mathematical expressions describing the rate at which nuclei are formed. Fennema (1975) has summarized the influence of temperature on rate of nucleation as illustrated in Fig. 4.2. This illustration indicates that after some characteristic supercooling, nucleation is initiated and the rate increases rapidly as temperature is decreased.

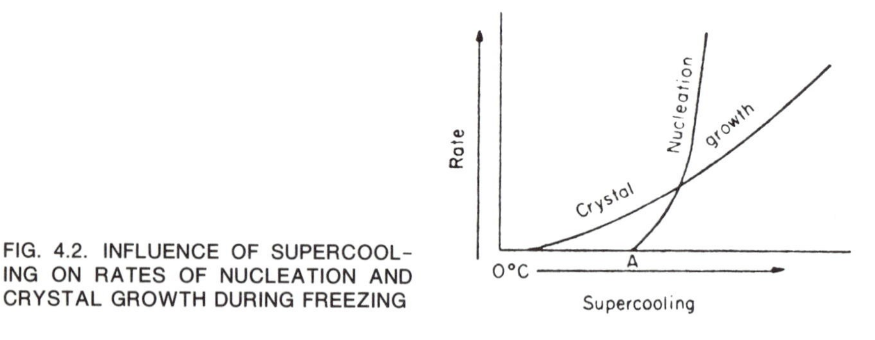


FIG. 4.2. INFLUENCE OF SUPERCOOLING ON RATES OF NUCLEATION AND CRYSTAL GROWTH DURING FREEZING

*From Fennama and Powrie (1964)*

The second step in the formation of ice crystals in a food product is crystal growth. Crystal growth can only occur after the nuclei are formed and exceed a critical size. The rate at which crystal growth may proceed is a function of several factors, including (a) the rate at which the water molecules react at the crystal surface, (b) the diffusion rate of water molecules from the unfrozen solution to the crystal surface, and (c) the rate at which heat is removed. In the latter case, the heat to be removed is the heat of crystallization. An additional factor which influences all the above factors is temperature. As illustrated by Fig. 4.2, crystal growth rate increases moderately as temperature of the product is decreased. Fennema (1975) has emphasized that (a) crystal growth can occur at temperatures very near the melting point and (b) growth rate increases moderately with increasing rates of heat removal until very low temperatures result in high viscosities and crystal growth rates decrease.

The size of the ice crystals which exist at the end of the freezing process or when the product has reached the final temperature of the freezing process is of considerable importance due to the influence on product quality. Fennema (1975) indicated that ice-crystal size is directly related to the number of nuclei which form during freezing. The formation of few nuclei results in small numbers of large crystals, while the development of many nuclei during freezing results in a large number of small crystals. This indicates that the size of the crystals in a product is directly related to the nucleation process. Since nucleation is a function of the extent to which supercooling is attained, the size of the crystals obtained becomes a function of the rate of freezing. As illustrated in Fig. 4.2, nucleation rate increases rapidly after reaching a critical extent of supercooling, while crystal growth rate increases consistently with decreased temperature. If heat removal rate is slow and the product temperature is allowed to reside between 0 C and point A for a significant period, any nuclei that form will be allowed to grow considerably. During rapid heat removal, the product temperature will be decreased to a temperature below point A very quickly and many nuclei will form with limited growth for each crystal. It follows that the mean size of the crystals within the product will vary inversely with the number of nuclei and the number of nuclei can be controlled by the rate of heat removal.

A phenomenon associated with ice crystal formation is recrystallization. This change is of considerable importance to frozen food quality due to the instability of ice crystals formed during freezing and the fluctuations in temperature during product storage. In general, the rate of recrystallization is highly temperature dependent with rates being high at temperatures near the initial freezing point and very slow at very low temperatures. Control of recrystallization can be achieved most effectively by maintaining low, constant storage temperatures for frozen food storage.

## 4.2 ENTHALPY CHANGE DURING FREEZING

One of the basic considerations in the design of a system for the freezing process is the refrigeration requirement for reducing the food-product temperature to the desired level. As indicated earlier in this chapter, the freezing process for food products is relatively complex due to the presence of components which result in depression of the freezing point and a temperature-dependent removal of latent heat. Although the enthalpy change required to reduce the temperature of the food product to some desired level below the freezing point can be measured by calorimetric means, there is considerable advantage in being able to predict these refrigeration requirements. The approach to be presented represents a step-wise description of the freezing process when a solute is present to depress the freezing point and freezing results in concentration of unfrozen portions of the product. The approach is very flexible in that experimental values can be utilized in the prediction equation, or the prediction can be based completely on composition of the product.

The total enthalpy change or change in heat content required to reduce the product temperature from some level above the freezing point to some desired storage temperature can be expressed in the following manner:

$$\Delta H = \Delta H_S + \Delta H_U + H_L + \Delta H_L \tag{4.13}$$

where the terms on the right side of the equation represent the sensible heat removed from product solids ($\Delta H_S$), the sensible heat removed from unfrozen water ($\Delta H_U$), the changes in enthalpy due to latent heat ($\Delta H_L$) and the sensible heat removed from frozen water or ice ($\Delta H_I$). Each part of equation (4.13) should be evaluated separately before obtaining the total enthalpy change.

The sensible heat removed from product solids has two parts as follows:

$$\Delta H_S = M_S \ c_{pS} \ (T_i - T_F) + M_S \ c_{pS} \ (T_F - T) \tag{4.14}$$

where $M_S$ represents the fraction of solids in the product, $c_{pS}$ is the specific heat of the product solids while the temperature difference ($T_i$ – $T_F$) is above the initial freezing point and ($T_F$ – $T$) is the change in temperature below the freezing point. Equation (4.14) can be expressed in difference form as:

$$\int_{H_i}^{H} dH_S = M_S \ c_{pS} \ (T_i - T_F) + \int_{T}^{T_F} M_S \ c_{pS} \ dT \tag{4.15}$$

Evaluation of the other three components in equation (4.13) is somewhat more complex because of the changing state of the product below the initial freezing point. The changes that occur can be visualized by referring to Fig. 4.3, which illustrates schematically the phase changes or the manner in which the phase change in the product occurs during the freezing process. As soon as the product reaches or decreases below the initial freezing point, a portion of the product is

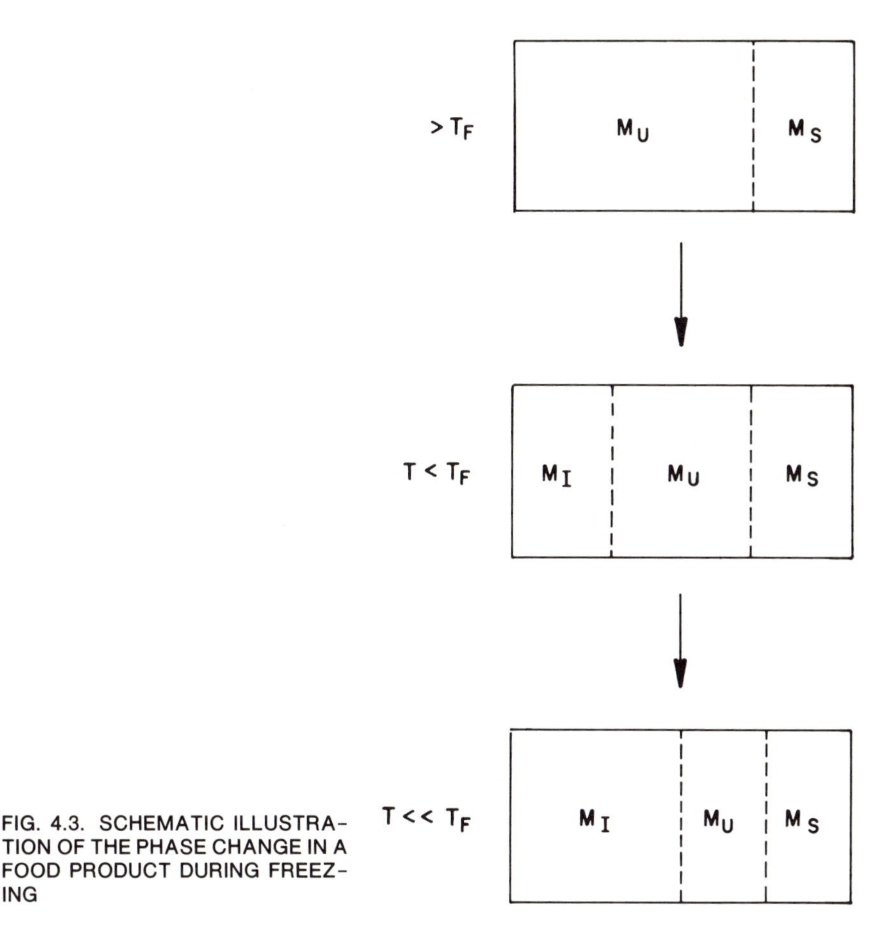


FIG. 4.3. SCHEMATIC ILLUSTRATION OF THE PHASE CHANGE IN A FOOD PRODUCT DURING FREEZING

frozen as ice crystals. This portion of the product, as illustrated in Fig. 4.3, is $M_I$ while the remaining portion of the unfrozen water is $M_U$. Since the product solids and associated solutes are concentrated in the unfrozen water fraction, the freezing point of this fraction is depressed in a step-wise manner as more ice crystals are formed. The process is illustrated in Fig. 4.3 by the increasing size of ice fraction in the product and decreasing fraction of unfrozen water as temperature is decreased.

The changes in frozen and unfrozen water fractions illustrated in Fig. 4.3 and the influence on change in product enthalpy can be described by a series of equations. For the unfrozen water fraction, the following expression will apply:

$$\Delta H_U = M_U\, c_{pU}\, [T_i - T_F] + M_U\, (T)\, c_{pU}\, (T)\, [T_F - T] \qquad (4.16)$$

as long as small temperature increments are considered. In equation (4.16), the mass of unfrozen water ($M_U$) and the specific heat of unfrozen water ($c_{pU}$) are both temperature-dependent at temperatures below $T_F$. Equation (4.16) can be written in differential form as follows:

$$\int_{H_i}^{H} dH_U = M_U \, c_{pU} \, (T_i - T_F) + \int_{T}^{T_F} M_U \, (T) \, c_{pU} \, (T) \, dT \qquad (4.17)$$

Due to the temperature dependent nature of the freezing process, the contribution of latent heat is a function of the magnitude of unfrozen water fraction, as indicated by the following equation:

$$\Delta H_L = M_I \, (T) \, L \qquad (4.18)$$

The expression illustrates that the contribution of latent heat to the total enthalpy change is directly proportional to the mass of water frozen at the appropriate temperature below the freezing point.

The contribution of sensible heat removed from frozen water to total enthalpy change can be expressed as:

$$\Delta H_I = M_I \, (T) \, c_{pI} \, (T) \, (T_F - T) \qquad (4.19)$$

or in differential form as:

$$\int_{0}^{H} dH_I = \int_{T}^{T_F} M_I \, (T) \, c_{pI} \, (T) \, dT \qquad (4.20)$$

where the specific heat of the frozen portion may not be temperature dependent if the temperature range considered is relatively small. Use of equations (4.14), (4.16), (4.18) and (4.19) along with the integrated forms of equations (4.15), (4.17) and (4.20) leads to computation of the total enthalpy change when the results are inserted into equations (4.13). In order to obtain integrated forms of equations (4.15), (4.17) and (4.20), knowledge of the relationship between frozen or unfrozen water fraction and temperature must be established. In addition, specific heats of frozen and unfrozen water as functions of temperature must be obtained for use in equations (4.17) and (4.20).

Information on the portions of frozen and unfrozen water in the food product at any temperature below the initial freezing point can be obtained by the use of equation (4.9). After utilizing this equation to evaluate the initial freezing point, it can be utilized to compute the proportions of frozen and unfrozen water fractions which must exist at various temperatures below the initial freezing point. By referring back to Fig. 4.3, it is obvious that the mole fraction of solvent, which is water in the case of food-product freezing, must decrease as the concentration of solute in the unfrozen fraction increases. The procedure is relatively straightforward in that selecting some temperature below the initial freezing point leads to computation of a new mole fraction of solvent and the amount of frozen water fraction which must exist at that temperature can be computed. The use of this procedure leads to a relationship between unfrozen product fraction and temperature, as illustrated in Fig. 4.4. The relationship between frozen product fraction and temperature would be similar, except that the frozen fraction would increase with decreasing temperature.

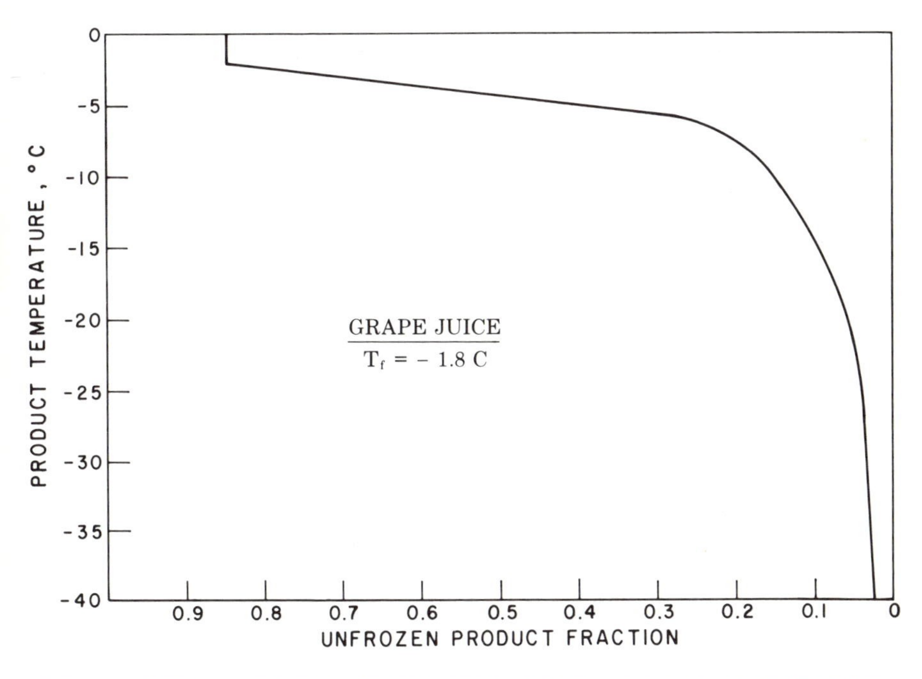


FIG. 4.4. THE CHANGE IN UNFROZEN PRODUCT FRACTION WITH DECREASING TEMPERATURE

After obtaining information on the frozen and unfrozen fractions as a function of temperature and the specific heats of the various components as a function of temperature, equations (4.15), (4.17) and (4.20) can be evaluated by integration. The major limitation to utilizing this procedure for complete prediction of the refrigeration requirements is lack of knowledge concerning the solutes present in various food products, which result in freezing-point depression. Most products contain several components that influence the magnitude of freezing point depression and it is nearly impossible to evaluate which components are contributing most. If experimental data on freezing-point depression are available, equation (4.9) can be utilized to compute an apparent mole fraction of solute and an effective molecular weight which accounts for the freezing-point depression. This effective molecular weight can be utilized to compute the frozen and unfrozen water fractions that exist in the product at various temperatures below the initial freezing point. Lack of this information makes the procedure very inflexible, and unless information on the apparent specific heats of the product during freezing is known, prediction of enthalpy change and refrigeration requirements becomes extremely difficult.

EXAMPLE 4.3 Predict the percent water frozen in grape juice when the temperature has been reduced to −5.5 C

SOLUTION

(1) Using equation (4.9) with $T_A$ = 267.5 K (−5.5 C)

$$\ln X_A = \frac{6003}{8.314}\left[\frac{1}{273} - \frac{1}{267.5}\right]$$

$$\ln X_A = 722.04 \times (-7.53 \times 10^{-5}) = -0.05438$$

$$X_A = 0.947$$

(2) Utilizing the definition of mole fraction and the effective molecular weight of 186.87 from example 4.2:

$$0.947 = \frac{M_U/18}{M_U/18 + 15.3/183.61}$$

$$M_U = 26.8$$

(3) The above value represents the percentage of water which is unfrozen at −5.5 C.

(4) Since the original product contains 84.7% water, the percentage of water frozen will be:

$$\frac{84.7 - 26.8}{84.7} = \frac{57.9}{84.7} = 0.68 \text{ or } 68\%$$

(5) This value indicates that 68.1% of the water is unfrozen, which compares favorably with the value of 65.1% from Fig. 4.8.

This procedure has been used successfully as illustrated by Heldman (1966) to predict the enthalpy change during freezing of ice-cream. The increase in refrigeration requirement for freezing to various levels below the initial freezing point when the initial product temperature was 4.5 C is illustrated in Fig. 4.5. For this particular product it is evident that the refrigeration requirement increases very rapidly as the temperature is decreased to levels between −6 and −10 C. The importance of accounting for all contributions to the total enthalpy change of refrigeration requirement as expressed in equation (4.13) is illustrated in Fig. 4.6. It is evident from this illustration that latent heat is the major contribution to the total enthalpy, but accounts for only 75% of the total, while the sensible heats of the frozen and unfrozen portions increase in relative contribution as the product is decreased to the lower temperatures.

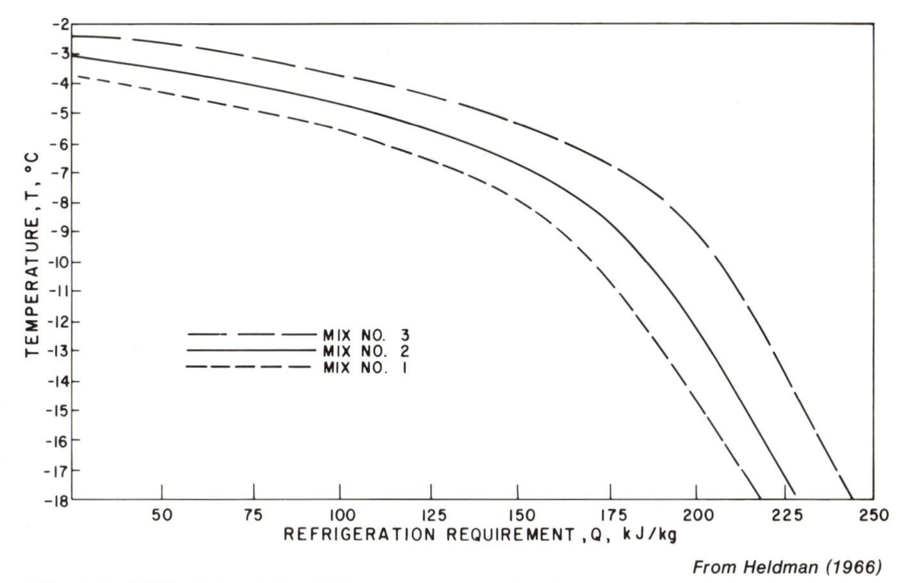


*From Heldman (1966)*

FIG. 4.5. REFRIGERATION REQUIREMENT PREDICTIONS FOR ICE CREAM WITH DIFFERENT COMPOSITIONS

EXAMPLE 4.4 A continuous freezer is being used to freeze a normal composition ice-cream mix to −5 C from an initial temperature of 4.5 C. Determine the refrigeration requirement for a freezing rate of 500 kg mix per hour.

SOLUTION

(1) From Fig. 4.5, the refrigeration requirement for the situation described can be obtained:

108 kJ/kg

(2) Since rate is 500 kg mix per hour, the total refrigeration requirement is 54,000 kJ/hr or 15 W.

Equation (4.13) and the concepts presented for prediction of refrigeration requirements for freezing can be used to predict enthalpy or heat content of a frozen food product above an established reference temperature. Since tables and charts associated with refrigeration use −40 C as a reference temperature, the enthalpy of frozen foods should be established on this reference. Using equations (4.13), (4.14), (4.16), (4.18) and (4.19) and recognition that $H = 0$ at −40 C,

$$\int_{O}^{H} dT = M_S\, c_{pS} \int_{-40}^{T_i} dT + M_U\, c_{pU} \int_{T_F}^{T_i} dT + \int_{-40}^{T_i} M_U\,(T)\, c_{pU}\,(T)\, dT + M_U\,(T)\, L + \int_{-40}^{T_F} M_I\,(T)\, c_{pI}\,(T)\, dT \quad (4.21)$$

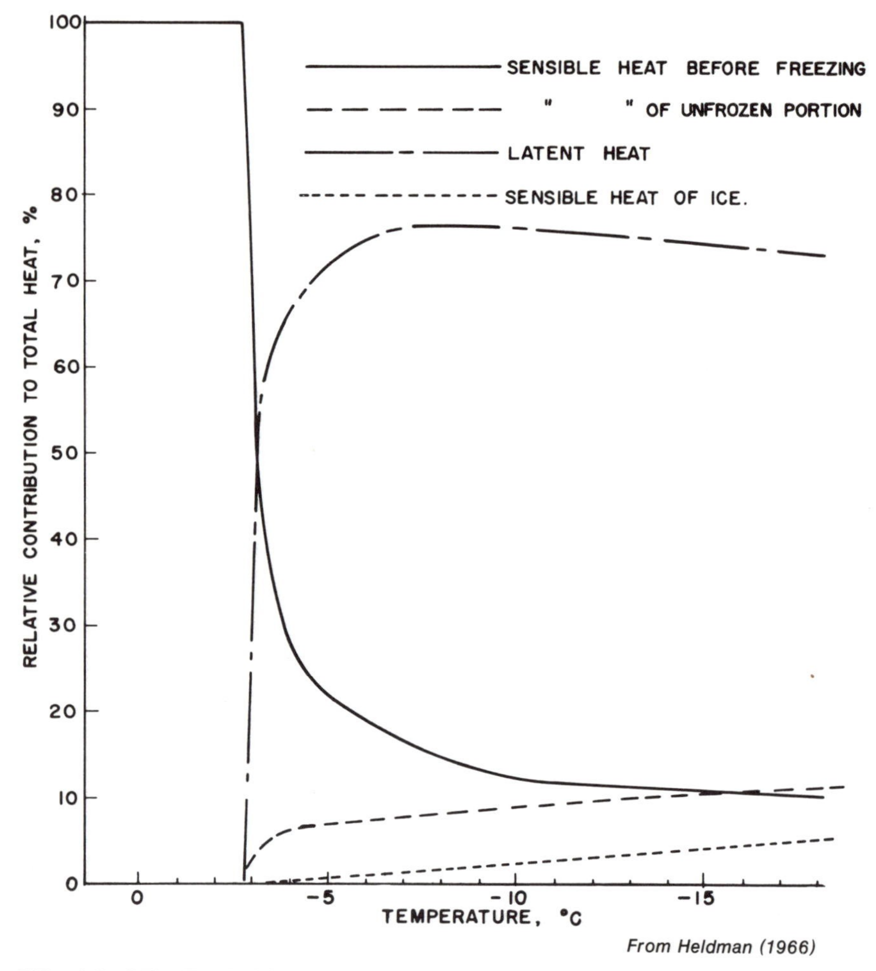


*From Heldman (1966)*

FIG. 4.6. RELATIVE CONTRIBUTIONS TO TOTAL REFRIGERATION REQUIREMENT FOR FREEZING ICE CREAM

Equation (4.21) accounts for the fact that heat added to the product results in increased enthalpy as the temperature is increased above −40 C. In addition, the relationship of unfrozen water fraction specific heat ($c_{pU}$) with temperature must account for the influence of freezing point.

### 4.2.1 Experimental Investigations

Probably the most extensive experimental investigations of enthalpy change during freezing of food products have been conducted and reported by Riedel (1956; 1957A, B). Utilizing an adiabatic calorimeter, the enthalpy changes during freezing of various foods were investigated for a temperature range down to −40 C. The results were presented on the enthalpy versus moisture content charts, as illustrated in Fig. 4.7. These results were obtained by direct measurement of enthalpy above −40 C on lean beef dried to appropriate moisture contents in a stream of

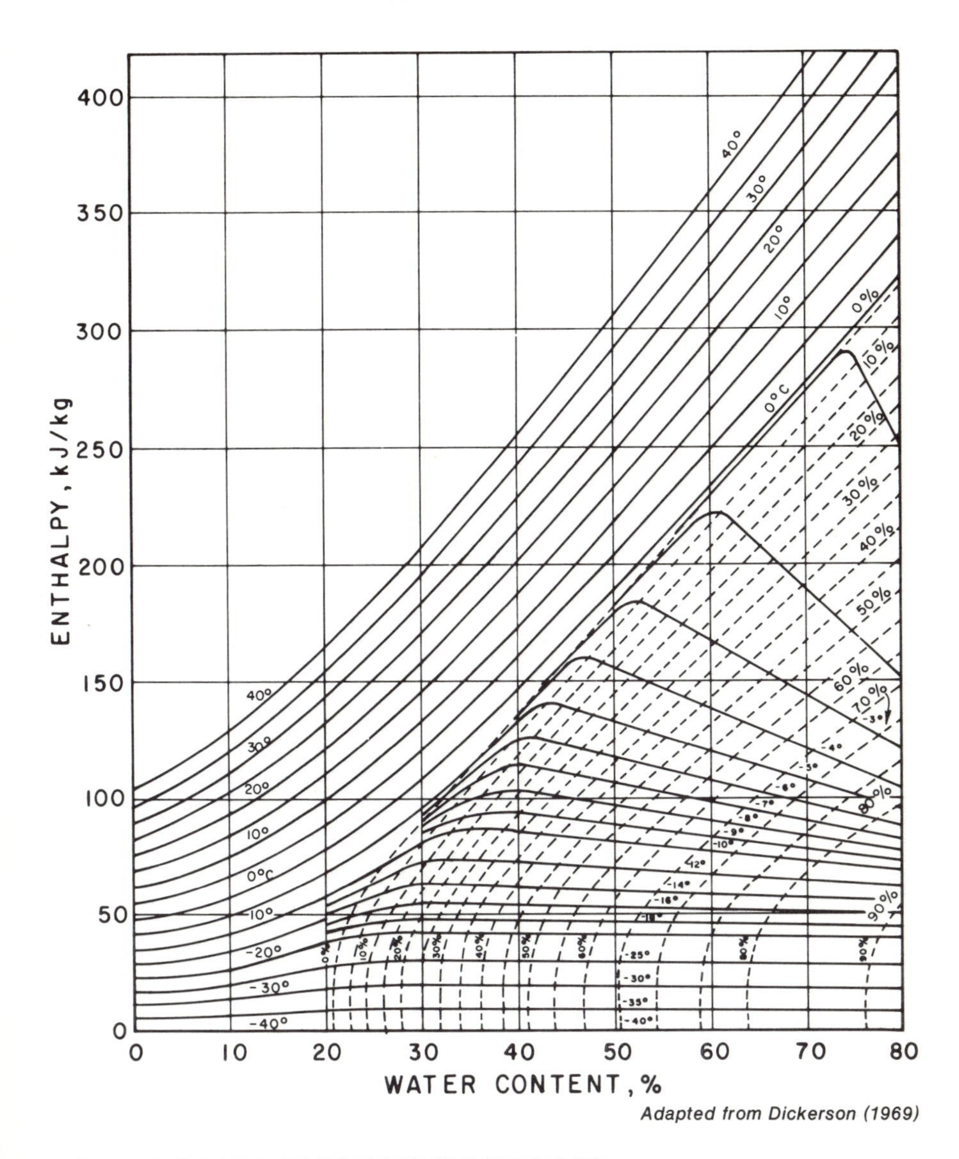

FIG. 4.7. ENTHALPY-COMPOSITION CHART FOR BEEF

air. In addition to the relationship between enthalpy and water content, the influence of percent water unfrozen and temperature is accounted for in the chart presented. The selection of −40 C as the base temperature for zero enthalpy was justified on the basis of findings that negligible amounts of water were frozen below this temperature. There is a certain amount of the water in beef which is unfrozen and apparently remains unfrozen regardless of the temperatures selected below −40 C. This percentage of water (10−12%) is normally referred to as bound water in food products.

A similar enthalpy composition chart was developed for fruit juices and vegetable juices. Although Fig. 4.8 could not be used directly for compu-

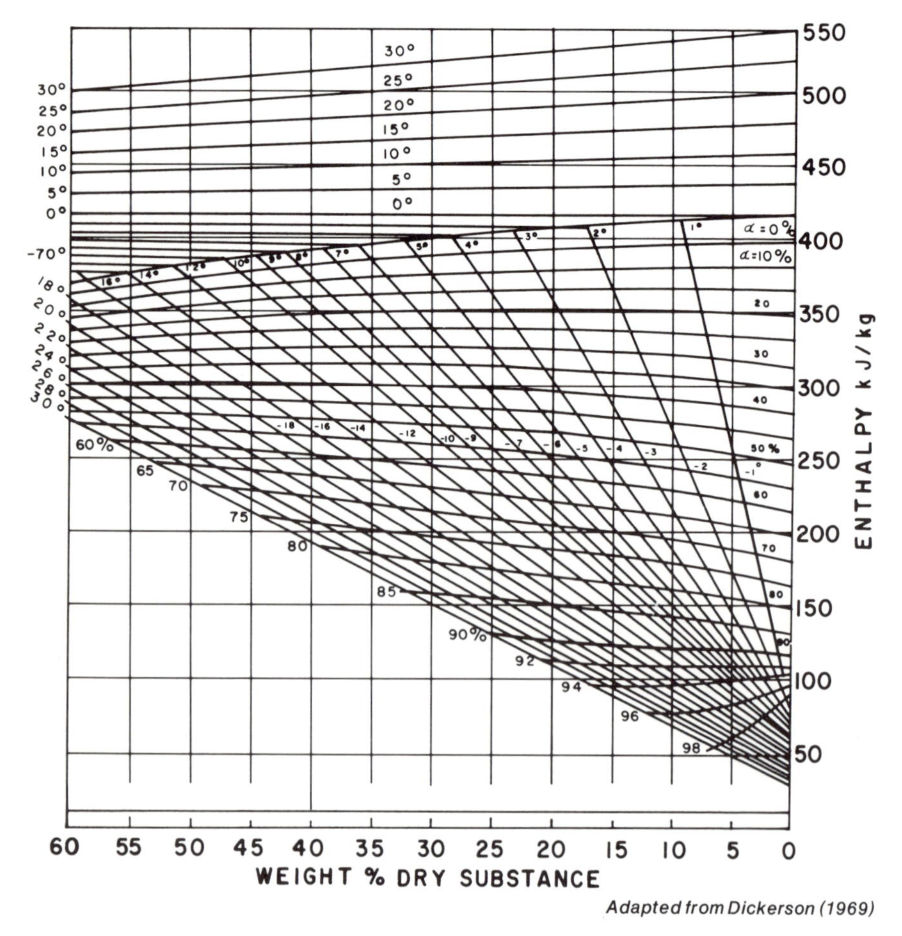


Adapted from Dickerson (1969)

FIG. 4.8. ENTHALPY-COMPOSITION CHART FOR FRUIT AND VEGETABLE JUICES

tation of enthalpy changes for fruits and vegetables, Riedel (1951) reported the following expression:

$$\Delta H = [1 - (\chi_{SNJ}/100)]\ \Delta H_j + 1.21\ (\chi_{SNJ}/100)\ \Delta T \qquad (4.22)$$

which could be used to predict the enthalpy for freezing fruits and vegetables from knowledge of the enthalpy change predicted or obtained from Fig. 4.8 for the vegetable or fruit juice. The term ($\chi_{SNJ}$) refers to the solids content of the product expressed as a percentage of the product solids distinctly different from juice. Dickerson (1969) utilized equation (4.22) to predict the enthalpy changes for several fruits and vegetables and found that, with minor exceptions, the enthalpy changes for these fruits and vegetables could be predicted to within 5% of measured values. In addition, Riedel (1957B) measured and reported the enthalpy change for freezing various components of egg, and predicted refrigeration requirements or enthalpy change for freezing whole egg. These results were presented in chart form similar to Fig. 4.7 and 4.8 and in table form by Dickerson (1969).

The computation of refrigeration requirement or enthalpy change during freezing from charts of the type shown in Fig. 4.7 and 4.8 can be accomplished by a two-step procedure. The first involves determination of the enthalpy content (above −40) of the product in the unfrozen state, and the second step establishes the enthalpy of the frozen product. The information obtained from the chart requires knowledge of the moisture content of the product and the temperature to which the product is to be frozen or the percentage of water unfrozen at the desired temperature. If the water or moisture content is known and the temperature to which the product is to be frozen is selected, this establishes one point on the chart.

From this point the enthalpy content of the product at the selected temperature can be determined, along with the percentage of water unfrozen at that temperature. If the product is to be frozen until a certain portion of the water is frozen in the product, the curve of percentage of water unfrozen along with the total moisture content establishes the point on the chart from which the enthalpy requirement and the temperature of the product under final conditions can be established.

EXAMPLE 4.5 Determine the refrigeration requirement for freezing 50 kg lean beef with 74.5% moisture content. The product is being frozen to −15 C from an initial temperatue of 5 C. What percentage of the product is frozen at −15 C?

SOLUTION

(1) From Fig. 4.7, the enthalpy of lean beef with 74.5% moisture at

−15 C is 58 kJ/kg. Approximately 14% of the water is unfrozen at −15 C.

(2) The enthalpy of lean beef at 5 C is 317 kJ/kg from Fig. 4.7.

(3) The enthalpy change from 5 C to −15 C will be 317 − 58 = 259 kJ/kg.

(4) The refrigeration requirement for freezing 50 kg lean beef will be 259 × 50 = 12,950 kJ.

(5) Since 14% unfrozen water represents 10.43% of the product, the percentage of product unfrozen at −15 C will be the 25.5% solids content in addition to 7.45% of the water. These values indicate that 35.93% of the product is unfrozen or 64.07% is frozen.

EXAMPLE 4.6 Compute the refrigeration requirement for freezing 250 kg strawberries to −10 C when the initial temperature was 15 C. The strawberry fruit solids makes up 24% of the total weight and the strawberry juice has a solids content of 8.3%.

SOLUTION

(1) The enthalpy change for freezing strawberries can be computed from equation (4.22). Since this equation utilizes the enthalpy change for the strawberry juice, Fig. 4.8 must be used initially.

(2) From Fig. 4.8, the enthalpy content of strawberry juice with 91.7% moisture at −10 C is 120 kJ/kg.

(3) The enthalpy content of strawberry juice at 15 C is 476 kJ/kg, so $H_j$ = 476 − 120 = 356 kJ/kg.

(4) Utilizing equation (4.22):

$$\Delta H = \left[1 - \frac{24}{100}\right] 356 + 1.21 \frac{24}{100} 25$$

$$= 0.76\,(356) + 0.2904\,(25) = 270.56 + 7.26$$

$$\Delta H = 277.82 \text{ kJ/kg}$$

(5) The refrigeration requirement becomes:

$$277.82 \times 250 = 69{,}455 \text{ kJ}$$

If, in the case of beef, the product contains a significant amount of fat, a correction presented by Rolfe (1968) can be introduced to account for this factor. The correction involves the use of the following expression:

$$\Delta H = \phi\, \Delta H_f + (1 - \phi)\Delta H_{nf} \tag{4.23}$$

where the fat content of the product ($\phi$) is introduced along with the enthalpy change for the fat which makes up a portion of the product. The enthalpy change for the nonfat portion of the product would be obtained from a chart similar to the Fig. 4.7 for a product with moisture content obtained on a nonfat basis.

## 4.3 PREDICTION OF FOOD PRODUCT FREEZING RATES

The most important consideration associated with food freezing is the rate of the process. This rate not only establishes the structure of the frozen product but the time required for freezing is the basic design consideration for the process. An analysis of current literature indicates significant variations in the definition of freezing rate. Fennema, *et al.* (1973) have identified four methods to describe rate of freezing including: (a) Time-temperature methods, (b) Velocity of ice front, (c) Appearance of specimen and (d) Thermal methods. The most frequently encountered methods are time-temperature including (a) temperature change per unit time or (b) time to transverse a given range of temperatures. The temperature change per unit time is the most appropriate indicator when the primary concern is structure of the frozen product and resulting influence on quality. It must be emphasized that temperature change per unit time will vary significantly during the freezing process and an average value has limited meaning.

The most appropriate indicator of freezing rate for purposes of process design is the time to transverse a given range of temperature. The International Institute of Refrigeration (1971) has proposed the following definition: "The freezing rate of a food mass is the ratio between the minimum distance from the surface to the thermal center and the time elapsed between the surface reaching 0 C and the thermal center reaching 5 C colder than the temperature of initial ice formation at the thermal center. Where depth is measured in cm and time in hour, the freezing rate will be expressed as cm/hr." A variation of the IIR definition is referred to as "thermal arrest time" and represents the time required for the slowest cooling point in the product to decrease from 0 C to 5 C. Long (1955) used thermal arrest time to describe the rate of freezing in fish. The results of this research indicated two significant factors about the use of thermal arrest time. The first factor was the location of the temperature sensor. Small deviations in location of the temperature sensor from the slowest cooling or freezing point in the product resulted in considerable error in determining the thermal arrest time for a given product. The second factor was the influence of initial product temperature. Results reported by Long (1955) indicated that an increase in initial product temperature decreased the thermal arrest time. In other words, the total freezing time was longer when the initial temperature was higher, but the time required to reduce the product temperature from 0° to −5 C was less. As pointed out by Rolfe (1968), this is a significant finding, since literature values of thermal arrest time do not always indicate the initial product temperature. For purposes of the discussion which follows, the time required to reduce the product temperature at the slowest cooling location from the initial freezing point to some desired and specified temperature below the initial freezing point will be utilized as the time to describe freezing rate. Although this definition is not without limitations, it seems to provide the best compromise when considering the advantages and disadvantages of other methods.

Fennema and Powrie (1964) have listed four factors which influence freezing rates: (a) the temperature differential between the product and the cooling medium; (b) the modes of heat transfer to, from and within the product; (c) the size, type and shape of the package containing the product; and (d) the size, shape and thermal properties of the product. Although considerable information is available in heat transfer literature to assist in describing the rates of heat transfer in various shaped packages and products, the major limitation appears to be in the description of transient heat transfer with thermal properties being a function of temperature. The latter must be the case during the freezing of food products, since the apparent specific heat and thermal conductivity are both significant functions of temperature in the freezing zone or below the initial freezing point of the product. Many of the methods utilized to obtain expressions for freezing time have involved simplifying assumptions which do not account for the thermal diffusivity being a function of temperature, in an effort to obtain a solution to a complex heat conduction problem.

### 4.3.1 Plank's Equation

The most straightforward expression available for computing freezing time was derived by Plank (1913). The equation utilized for computation purposes can be derived for various geometries of product. By reference to Fig. 4.9, the case of one-dimensional freezing of a product slab can be illustrated. The three basic equations utilized in the derivation account for heat transfer in various phases of the product during freezing. The first expression is the basic heat-conduction equation for the frozen product region which has a variable thickness of x as follows:

$$q = A(T_s - T_F)k/x \tag{4.24}$$

where $T_F$ is the initial freezing point of the product and represents the temperature which exists in all unfrozen regions of the product, and $k$ is the thermal conductivity of the frozen material. The second expression describes the heat transfer from the product surface to the surrounding medium can be expressed as:

$$q = h_c A(T_\infty - T_s) \tag{4.25}$$

where $h_c$ is a convective heat transfer coefficient at the product surface. Equations (4.24) and (4.25) can be combined into one expression to account for heat transfer in series as follows:

$$q = \frac{A(T_\infty - T_F)}{1/h_c + x/k} \tag{4.26}$$

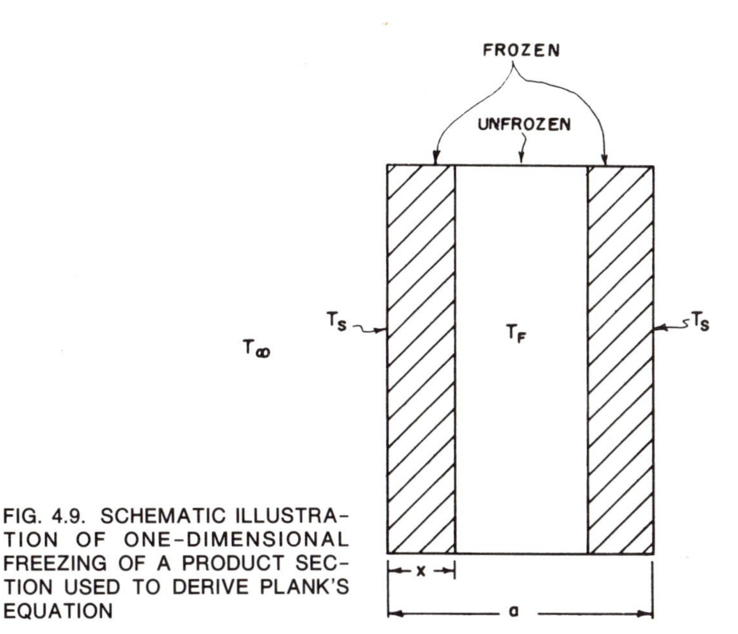


FIG. 4.9. SCHEMATIC ILLUSTRATION OF ONE-DIMENSIONAL FREEZING OF A PRODUCT SECTION USED TO DERIVE PLANK'S EQUATION

and eliminates the need for knowledge of the surface temperature. The third equation, describing the rate at which heat is being generated at the freezing front, is as follows:

$$q = AL\rho \frac{dx}{dt} \tag{4.27}$$

where the differential ($dx/\mathrm{d}t$) represents the velocity of the freezing front. By equating equations (4.26) and (4.27) and by integration between the appropriate limits, the following expression for freezing time is obtained:

$$t_F = \frac{\rho L}{T_F - T_\infty}\left[\frac{a}{2h_c} + \frac{a^2}{8k}\right] \tag{4.28}$$

where $a$ represents the total thickness of the slab being frozen. By introduction of the appropriate constants, the most general form of Plank's equation is obtained as follows:

$$t_F = \frac{\rho L}{T_F - T_\infty}\left[\frac{Pa}{h_c} + \frac{Ra^2}{k}\right] \tag{4.29}$$

where $P$ and $R$ are constants which will vary depending on the geometry of the material being frozen. As is obvious, these constants are ½ and ⅛ for $P$ and $R$, respectively, in the case of an infinite slab. For a sphere, $P = 1/6$ and $R = 1/24$, while for an infinite cylinder, $P = 1/4$ and $R = 1/16$. The dimension, $a$, which is the thickness of the infinite slab, becomes the diameter of a cylinder and a sphere.

When Plank's equation is applied to a brick or block geometry, a chart of the type illustrated in Fig. 4.10 must be utilized. For this situation the dimension ($a$) in equation (4.29) becomes the smallest dimension of the brick or block. The constant ($\beta_1$) becomes the multiplying factor required to make the *product of* $\beta_1$ *and a equal to the second smallest dimension* of the brick or block geometry. The constant ($\beta_2$) is a similar multiplying factor, which makes the *product of* $\beta_2$ *and a equal to the largest dimension* of the geometry considered. Utilizing these constants ($\beta_1$ and $\beta_2$) and the chart given in Fig. 4. 10, the appropriate values of $P$ and $R$ are obtained and applied in the general form of Plank's equation given as equation (4.29). This chart and procedure allows the use Plank's equation for predicting the freezing times of most geometries considered.

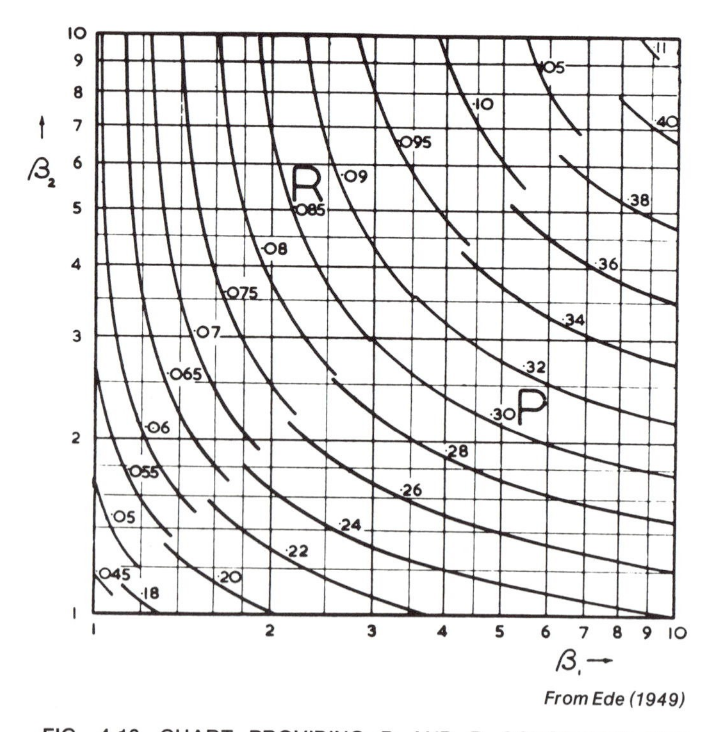


From Ede (1949)

FIG. 4.10. CHART PROVIDING P AND R CONSTANTS FOR PLANK'S EQUATION WHEN APPLIED TO A BRICK OR BLOCK GEOMETRY

EXAMPLE 4.7 A block of lean beef is being frozen in a −30 C convection freezer ($h_c$ = 30 W/m$^2$K). The initial temperature is 5 C and the dimensions of the product are 1 m by 0.25 m by 0.6 m. Compute the time required to freeze the product to −10 C.

SOLUTION

(1) Use of Plank's equation requires knowledge of several product characteristics and related factors:

$\rho$ = 1050 kg/m$^3$ (assumed)

$L$ = 333.22 kJ/kg × 0.745 = 248.25 kJ/kg (based on water content of product—Table A.11)

$k$ = 1.108 W/mK (mean of extreme values for product)
$T_F$ = −1.75 C (assumed)

(2) Determination of $P$ and $R$ values requires use of Fig. 4.10 with:

$$\beta_1 = \frac{0.6}{0.25} = 2.4$$

$$\beta_2 = \frac{1}{0.25} = 4$$

then $P = 0.3$ and $R = 0.085$

(3) Using equation (4.29); recognizing that 1 J/s = 1W:

$$t_F = \frac{(1050\ \text{kg/m}^3)(248.25\ \text{kJ/kg})(1000\text{J/kJ})}{[-1.75\ \text{C} - (-30\ \text{C})]\,(3600\ \text{s/h})}\left[\frac{0.3(0.25\text{m})}{(30\ \text{W/m}^2\text{K})} + \frac{0.085(0.25\text{m})^2}{(1.108\ \text{W/mK})}\right]$$

$$= (2563.05\ \text{Jh/m}^3\text{C})(0.0025\ \text{m}^3\text{K/W} + 0.0048\ \text{m}^3\text{K/W})$$

$$= 18.7\ \text{hr.}$$

The major limitations of Plank's equation for predicting freezing times for food products are obvious. Use of the equation requires the assumption of some latent heat value and does not consider the gradual removal of latent heat over a range of temperatures during the freezing process for food products. In addition, the procedure utilizes only the initial freezing point of the product in the computation equation and neglects the time required to remove sensible heat above the initial freezing point. A third serious assumption is that a constant thermal conductivity must be assumed for the frozen region. As discussed previously in this chapter, the freezing zone has a constantly changing temperature and phase change during the freezing process and therefore has a variable thermal conductivity depending on the temperature. The final limitations is that Plank's equation assumes the product to be completely liquid phase. It would follow therefore that the accuracy of Plank's equation for food products would decrease as the percentage of water in the product decreases. Ede (1949) compared Plank's equation to predictions he obtained using a graphical method for predicting freezing times, and he felt that Plank's equation could be used with reasonable accuracy, even though the assumptions represent rather serious limitations. In addition, Ede (1949) concluded that the benefits of a compact equation of the type shown in equation (4.29) more than compensate for the slight inaccuracies involved in the computations. Modifications of Plank's equation incorporate empirical factors such as Nagaoka *et al.* (1955) based on freezing of fresh fish:

$$t_F = \frac{\Delta H' \rho}{T_F - T_\infty} \left[ \frac{Pa}{h_c} + \frac{Ra^2}{k} \right] \tag{4.30}$$

where

$$\Delta H' = [1 + 0.00445(T_o - T_F)][c_{pU}(T_i - T_F) + L + c_{pI}(T_F - T)]$$

Although the Nagaoka *et al.* (1955) modification does incorporate factors to account for sensible heat above and below the initial freezing point, the equation still assumes that all latent heat is removed at a constant temperature ($T_F$). In addition, applications of the equation establish the desired final temperature of the product ($T$) and adjust the value of the latent heat of fusion ($L$) for the water composition of the product.

EXAMPLE 4.8 Compute the freezing time for the product and conditions in Example 4.7 using the Nagaoka *et al.* equation.

SOLUTION

(1) Using the values presented in the solution of Example 4.7 along with the following information:

$c_{pU}$ = 3.52 kJ/kg K (from Table A.11)
$c_{pI}$ = 2.05 kJ/kg K (for ice; Table A.5)
$k_I$ = 1.108 W/mK

then:

$$H' = [1 + 0.00445\ (5 - (-1.75))][(3.52)(5 - (-1.75)) + 248.25 + (2.05)(-1.75 - (-10)]$$

$$H' = (1.03)(23.76 + 248.25 + 16.91) = 297.59 \text{ kJ/kg}$$

(2) By using equation (4.30):

$$t_F = \frac{(1050)(297.59)(1000)}{[-1.75 - (-30)](3600)} \left[ \frac{0.3(0.25)}{30} + \frac{0.085(0.25)^2}{1.108} \right]$$

$$= (3072.44)(2.5 \times 10^{-3} + 4.795 \times 10^{-3})$$

$$t_F = 22.41 \text{ hr}$$

More significant modifications of Plank's equation have resulted from an in-depth analysis conducted by Cleland and Earle (1976, 1977, 1979a, 1979b). By writing Plank's equation in dimensionless form:

$$N_{Fo} = P \left[ \frac{1}{N_{Bi}\, N_{Ste}} \right] + R \left[ \frac{1}{N_{Ste}} \right] \tag{4.31}$$

where

$N_{Fo}$ = Fourier Number = $\alpha\ t/a^2$

$N_{Bi}$ = Biot Number = $h_c a/k$

$N_{Ste}$ = Stefan Number = $c_{pI}(T_F - T_\infty)/\Delta H$

the relationship among the variables is more evident. By introducing a new dimensionless number ($N_{Pk}$ = Plank's No.), the influence of sensible heat above the initial freezing point can be incorporated:

$$N_{Pk} = \frac{c_{pU}(T_i - T_F)}{\Delta H} \tag{4.32}$$

and

$$N_{Fo} = f(N_{Bi}, N_{Ste}, N_{Pk}) \tag{4.33}$$

Through experimental investigations, Cleland and Earle (1976) were able to establish the following empirical expressions:

$$P = 0.5072 + 0.2018\,N_{Pk} + N_{Ste}\left(0.3224\,N_{Pk} + \frac{0.0105}{N_{Bi}} + 0.0681\right) \tag{4.34}$$

and:

$$R = 0.1684 + N_{Ste}\,(0.274\,N_{Pk} + 0.0135) \tag{4.35}$$

for a slab geometry. These correlations should be accurate to within ±3% for products with moisture contents around 77%. In addition, the correlation should be acceptable for initial temperature up to 40 C, freezing medium temperatures between −15 and −45 C, slab thickness up to 0.12 m and surface heat transfer coefficients between 10 and 500 W/m²K.

Cleland and Earle (1979b) presented correlations for cylindrical geometries:

$$P = 0.3751 + 0.0999\,N_{Pk} + N_{Ste}\left(0.4008\,N_{Pk} + \frac{0.071}{N_{Bi}} - 0.5865\right) \tag{4.36}$$

and

$$R = 0.0133 + N_{Ste}\,(0.0415\,N_{Pk} + 0.3957) \tag{4.37}$$

and for spherical geometries:

$$P = 0.1084 + 0.0924\,N_{Pk} + N_{Ste}\left(0.231\,N_{Pk} - \frac{0.3114}{N_{Bi}} + 0.6739\right) \tag{4.38}$$

and

$$R = 0.0784 + N_{Ste}\,(0.0386\,N_{Pk} - 0.1694) \tag{4.39}$$

The expected accuracy of the freezing time predictions would be ±5.2% when using equations (4.36) and (4.37) and ±3.8% when using equations (4.38) and (4.39); when the following ranges are observed:

$$
\begin{aligned}
0.155 &\le N_{Ste} \le 0.345 \\
0.5 &\le N_{Bi} \le 4.5 \\
0 &\le N_{Pk} \le 0.55
\end{aligned}
\tag{4.40}
$$

These ranges should cover most practical freezing situations, but the moisture content of around 77% for the product should be recognized for cylindrical and spherical geometries as well as slabs.

EXAMPLE 4.9 Compute the freezing time for a 0.025 m thick lamb steak being frozen in an air blast freezer. The initial temperature is 20 C and the freezing medium temperature is −30 C. Use the Cleland and Earle modification of the Plank's equation to compute time required to reduce the product temperature to −10 C.

SOLUTION

(1) In order to use the modified Plank's equation, the following additional information is needed:

$\rho$ = 1050 kg/m$^3$ (assume density of frozen product)

$\Delta H$ = 320 − 80 = 240 kJ/kg (using Fig. 4.7 and assuming that enthalpy of lamb at 20 and −10 C is the same as lean beef with moisture content of 65%)

$c_{pU}$ = 3.0 kJ/kg K (assumed)
$c_{pI}$ = 1.75 kJ/kg K (assumed)
$T_F$ = − 2.75 C (assumed)
$k$ = 1.35 W/mK (assumed)

(2) Using equations (4.34) and (4.35) for slab geometry; the values of the dimensionless numbers must be computed:

$N_{Bi} = h_c a/k = (20)(0.025)/1.35 = 0.37$ (where $h_c$ = 20 W/m$^2$K is assumed for air blast freezer)
$N_{Ste} = c_{pI}(T_F - T_\infty)/\Delta H = [(1.75)(-2.75 + 30)]/240 = 0.199$
$N_{Pk} = c_{pU}(T_i - T_F)/\Delta H = [(3.0)(20 + 2.75)]/240 = 0.284$

then:

$$P = 0.5072 + 0.2018(0.284) + (0.199)[0.3224(0.284) + \frac{0.0105}{0.37} + 0.0681]$$

$$= 0.6019$$

$$R = 0.1684 + (0.199)\,[0.274(0.284) + 0.0135]$$

$$= 0.187$$

(3) By use of equation (4.31):

$$t_F = \frac{(1050)(240)(1000)}{[-2.75-(-30)](3600)}\left[\frac{(0.6019)(0.025)}{20} + \frac{0.187(0.025)^2}{1.35}\right]$$

$$= (2568.81)(7.524 \times 10^{-4} + 8.657 \times 10^{-5})$$

$$t_F = 2.155 \text{ hr}$$

## 4.3.2 Neumann Problem

An alternate approach to solving for freezing times is the solution given by Neumann as published in Carslaw and Jaeger (1959). This approach utilizes one-dimensional heat transfer in a semi-infinite body as illustrated in Fig. 4.11. As in the derivation of Plank's equation, there are

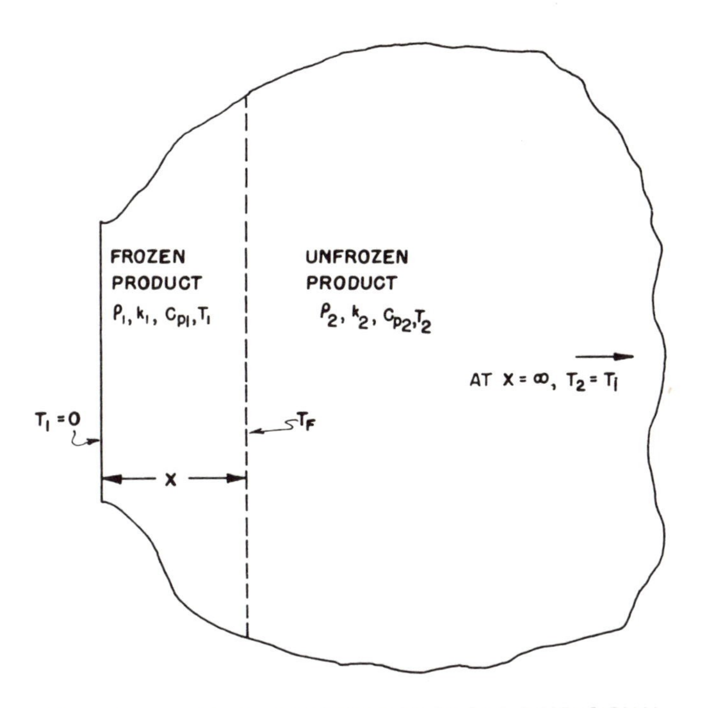


FIG. 4.11. SCHEMATIC ILLUSTRATION OF ONE-DIMENSIONAL FREEZING AS USED IN NEUMANN PROBLEM

three basic equations involved in the development of the equations which describe the Nemann problem. The first two are partial differential equations which describe the temperature distribution in the frozen and unfrozen portions of the semi-infinite body. These can be written as:

$$\frac{\partial^2 T_1}{\partial x^2} = \frac{\rho_1 \; c_{p1}}{k_1} \; \frac{\partial T_1}{\partial t} \tag{4.41}$$

$$\frac{\partial^2 T_2}{\partial x^2} = \frac{\rho_2 \; c_{p2}}{k_2} \; \frac{\partial T_2}{\partial t} \tag{4.42}$$

where subscripts (1 and 2) refer to the frozen and unfrozen portions of the body, respectively. The third equation can be expressed as the difference in the heat flux between the solid and liquid portions of a semi-infinite body which must be equal to the heat generated at the freezing front. This expression can be written as:

$$k_1 \frac{\partial T_1}{\partial x} - k_2 \frac{\partial T_2}{\partial x} = \rho L \frac{dx}{dt} \tag{4.43}$$

where the differential $(dx/dt)$ represents the velocity of the freezing front in the semi-infinite body. The solution of the Neumann problem involves the use of several initial and boundary conditions which may be stated as follows:

$$\text{At } x(t), T_1 = T_2 = T_F \tag{4.44}$$

$$\text{At } x \to \infty, T_2 \to T_i \tag{4.45}$$

$$\text{At } x = 0, T_1 = 0, t > 0 \tag{4.46}$$

$$\text{At } x(t) = 0, T_2 = T_i, t = 0. \tag{4.47}$$

is semi-infinite, solutions to equations (4.41) and (4.42) can be written as follows:

$$T_1 = A' \operatorname{erfc} \frac{x}{2(\alpha_1 t)^{1/2}} \tag{4.48}$$

$$T_2 = T_i - B' \operatorname{erfc} \frac{x}{2(\alpha_2 t)^{1/2}} \tag{4.49}$$

Utilizing the appropriate boundary conditions, the constants $A'$ and $B'$ in equations (4.48) and (4.49) can be evaluated. This procedure leads to the expressions which describe the temperature distribution in the solid and liquid portions of the semi-infinite material. These expressions are as follows:

$$T_1 = \frac{T_F}{\operatorname{erf}\lambda} \operatorname{erf} \frac{x}{2(\alpha_2 t)^{1/2}} \tag{4.50}$$

$$T_2 = T_i - \frac{T_i - T_F}{\operatorname{erfc}[\lambda(\alpha_1/\alpha_2)^{1/4}]} \operatorname{erfc} \frac{x}{2(\alpha_2 t)^{1/2}} \tag{4.51}$$

where $\lambda$ is a numerical constant which must be evaluated by trial and error.

The Neumann problem and the approach used for computation of freezing times may represent slight improvement over Plank's equation since the statement is a more accurate description of the freezing process in foods. The problem statement does allow for different thermal conductivities within the frozen and unfrozen portions of the product but applications are limited by the semi-infinite body geometry. The Neumann problem does assume that latent heat of fusion is removed at a constant temperature ($T_F$) and does not provide for direct incorporation of a surface heat transfer coefficient into the freezing time computation.

The solution to the Neumann problem has been used to compute food freezing times by Charm and Slavin (1962) and Charm (1978) and these computations were discussed by Cowell (1967) and Bakal and Hayakawa (1973). In general, the procedures for calculations are highly complex and involve trial and error evaluation of various constants. It must be concluded that this approach is unlikely to be used when compared to other less complex procedures for calculations of food freezing times.

### 4.3.3 Tao Solutions

Using numerical solutions of the heat conduction equations, Tao (1967) and Joshi and Tao (1974) developed charts to be used for computation of freezing times. These charts for infinite slab, infinite cylinder and sphere are presented as Fig. 4.12, 4.13 and 4.14, respectively. The dimensionless numbers shown on these charts are:

$$\text{dimensionless time} = t_F^* = \frac{t_F\, k_I\, (T_F - T_\infty)}{a^2\, \rho\, L} \tag{4.52}$$

$$\text{inverse Biot No.} = \frac{1}{N_{Bi}} = k_I/h_c a \tag{4.53}$$

$$\text{modified Steffan No.} = N_{Ste}{}^* = \frac{c_{pI}\,(T_F - T_\infty)}{L} \tag{4.54}$$

In addition to the assumption about product shape, use of the Tao charts must account for two additional assumptions: (a) constant ambient temperature at the product surface and (b) initial product temperature is the freezing point with no part of the product frozen.

EXAMPLE 4.10 Compute the time required to freeze a 0.1 m thick slab of lean beef with 73% moisture content. The initial product temperature is 5 C and the product's surfaces are exposed to plates at −42 C that provide a surface heat transfer coefficient of 50 W/m²K.

SOLUTION

(1) The use of the Tao charts requires the following additional information:

$k_I$ = 1.108 W/mK
$\rho_I$ = 970 kg/m³
$c_{pI}$ = 2.05 kJ/kg K (Table A.5)
$L$ = 333.22 × 0.73 = 243.25 kJ/kg
$T_F$ = −2.2 C

(2) By computation of dimensionless numbers:

$$t_F^* = \frac{t_F\,(1.108)(-2.2 + 42)}{(0.1)^2\,(970)\,(243.25)(1000)} = 1.8689 \times 10^{-5}\, t_F$$

$$\frac{1}{N_{Bi}} = \frac{1.108}{(50)(0.1)} = 0.222$$

$$N_{Ste}{}^* = \frac{(2.05)\,(-2.2 + 42)}{243.25} = 0.335$$

(3) Using Fig. 4.12:

$$t_F^* = 0.8 = 1.8689 \times 10^{-5}\, t_F$$

$$t_F = 4.28 \times 10^4 \text{ s}$$

$$t_F = 11.89 \text{ hr}$$

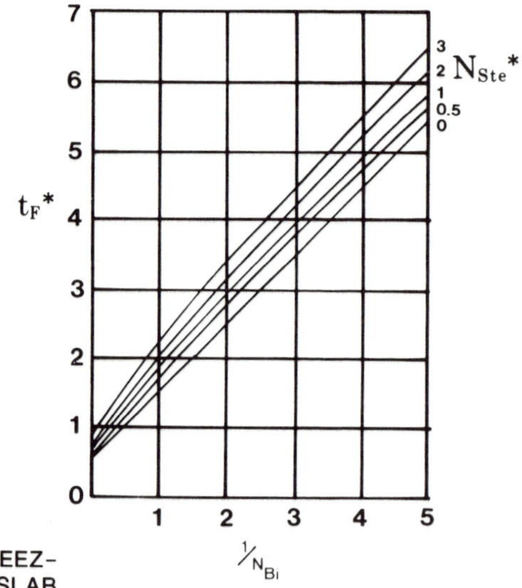


FIG. 4.12. CHART FOR ESTIMATING FREEZING OR THAWING TIME OF AN INFINITE SLAB

*From L.C. Tao (1971, personal communication).*

## 4.3.4 Tien Solutions

Analytical solutions to the heat conduction equation were obtained by Tien *et al.* (1967, 1968a, 1968b, 1969) resulting in expressions similar to Neumann. The assumptions associated with these solutions include: (a) geometric shape is semi-infinite, (b) freezing occurs over a specified range of temperatures below the initial freezing point, (c) the initial temperature is the initial freezing point with the product unfrozen, (d) densities of frozen and unfrozen portions of product are different and (e) the product surface temperature is constant during freezing.

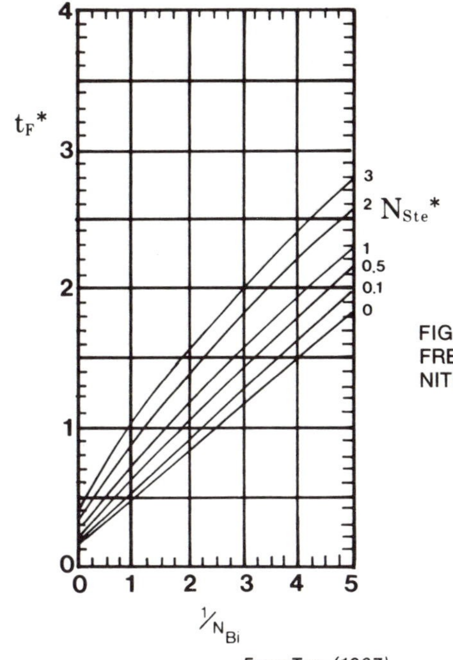


*From Tao, (1967)*

FIG. 4.13. CHART FOR ESTIMATING FREEZING OR THAWING TIME OF AN INFINITE CYLINDER

FIG. 4.14. CHART FOR ESTIMATING FREEZING OR THAWING TIME OF A SPHERE

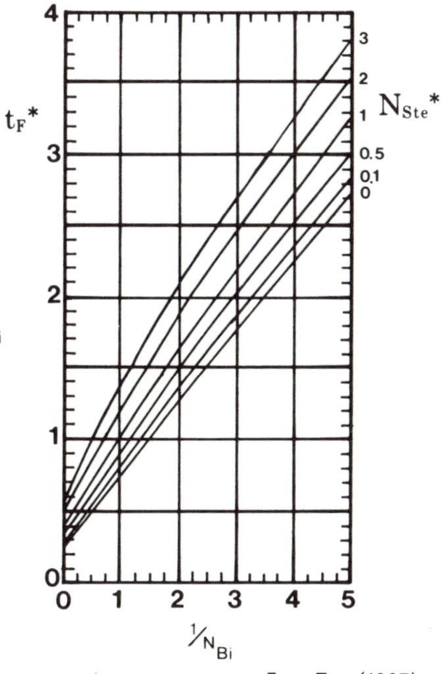


*From Tao, (1967)*

The various equations associated with the Tien solutions have been presented and discussed by Bakal and Hayakawa (1973). In addition, an example illustrating the procedures required in using these expressions are presented by these authors.

### 4.3.5 Mott Procedure

Based on dimensional analysis of experimental data from freezing time experiments, Mott (1964) developed a computation procedure. The procedure is based on the following dimensionless numbers:

$$S = \frac{B+1}{G'} \quad (4.55)$$

$$S = A\ (a/2V) \quad (4.56)$$

$$B = h_c\ a/4\ k_I \quad (4.57)$$

$$G' = \frac{2t_F\ h_c\ (T^* - T)}{\rho_I \Delta H^*\ a} \quad (4.58)$$

The dimensionless numbers are used along with information in Tables 4.1, 4.2 and 4.3. The shape factors in Table 4.1 represent the values of the dimensionless number ($S$) for various frozen food products. The information in Table 4.2 provides property data for various food products with $T^*$ representing the temperature of the product when 60% of the water is frozen. The enthalpy values in Table 4.3 are specific for the products indicated and do vary in magnitude from values presented by Reidel (1956, 1957A, 1957B) for the same products.

TABLE 4.1. SHAPE FACTORS FOR VARIOUS TYPES OF PACKS[a,b]

| Type of pack | Shape factor |
|---|---|
| Flat slab (no transfer at edges) | 1.0 |
| Thin crust freezing | 1.1 |
| Normal crust freezing | 1.2 |
| Small plastic pack, 6.5 × 4 × 1 (in.$^3$) | 1.4 |
| Average pack | 1.5 |
| Average whole fish | 1.5 |
| Square bar or rod (no transfer at ends) | 2.0 |
| Whole chicken | 2.5 |
| Sphere or cube | 3.0 |

[a] From Mott 1964, by permission of Mr. H.G. Goldsten, editor of Aust. Refrig. Air Cond. Heat.
[b] It is assumed that heat transfer is equally effective on all surfaces.

TABLE 4.2. PROPERTIES OF FROZEN FOODS

| Product | Density ($\rho_I$) | Thermal Conductivity ($k_I$) | Temperature ($T^*$) |
|---|---|---|---|
| | (kg/m³) | (W/mK) | (°C) |
| Beef | 1041 | 1.558 | −4.4 |
| Lamb | 1057 | 1.385 | −6.1 |
| Poultry | 1025 | 1.298 | −3.3 |
| Fish | 1009 | 1.125 | −2.8 |
| Beans | 801 | 0.917 | −3.3 |
| Broccoli | 961 | 0.381 | −2.8 |
| Peas | 881 | 0.467 | −3.3 |
| Mashed potatoes | 1089 | 1.091 | −2.8 |
| Cooked rice | 801 | 0.692 | −6.7 |

From Mott (1964) by permission of H.G. Goldsten, editor of Aust. Refrig. Air Cond. Heat.
$T^*$ = temperature of product when 60% of the water is frozen.

EXAMPLE 4.11 Compute the freezing time for the conditions described in Example 4.9 using the Mott Procedures.

SOLUTION

(1) Based on information in Table 4.1, the shape factor(S) would be 1.0 for a flat slab geometry.

(2) In order to determine freezing time at the geometric center; $\frac{a}{2}$ = 0.0125 m.

(3) Assuming an air blast freezer has a surface heat transfer coefficient of 20 W/m²K,

$$B = \frac{(20)(0.025)}{4(1.385)} = 0.09$$

(4) By incorporating the values of $S$ and $B$ into equation (4.55):

$$G' = \frac{0.09 + 1}{1} = 1.09$$

(5) In order to use equation (4.58) a value of $\Delta H^*$ must be obtained from Table 4.3. Using the enthalpy values for lamb product and by linear interpolation:

$$\Delta H^* = 291.94 - 80.56 = 211.38 \text{ kJ/kg}$$

(6) Using equation (4.58) and values from Table 4.2:

$$t_F = \frac{(1050)(211.38)(0.025)(1.09)(1000)}{2(20)(-6.1 + 30)}$$

$$= 6326.5 \text{ s}$$

$$t_F = 1.76 \text{ hr}$$

### 4.3.6 Numerical Solutions

Although the previously described methods for prediction of freezing times can be used for food products with reasonable accuracy under ideal conditions, all of the procedures have limitations. In order to account for all unique features of the food freezing process, the appropriate math-

TABLE 4.3. ENTHALPY OF FROZEN FOODS

| Temperature (°C) | Beef (kJ/kg) | Lamb (kJ/kg) | Poultry (kJ/kg) | Fish (kJ/kg) | Beans (kJ/kg) | Broccoli (kJ/kg) | Peas (kJ/kg) | Mashed Potatoes (kJ/kg) | Cooked Rice (kJ/kg) |
|---|---|---|---|---|---|---|---|---|---|
| −28.9 | 14.7 | 19.3 | 11.2 | 9.1 | 4.4 | 4.2 | 11.2 | 9.1 | 18.1 |
| −23.3 | 27.7 | 31.4 | 23.5 | 21.6 | 16.5 | 16.3 | 23.5 | 21.6 | 31.9 |
| −17.8 | 42.6 | 45.4 | 37.7 | 35.6 | 29.3 | 28.8 | 37.7 | 35.6 | 47.7 |
| −12.2 | 62.8 | 67.2 | 55.6 | 52.1 | 43.7 | 42.8 | 55.6 | 52.1 | 70.0 |
| − 9.4 | 77.7 | 84.2 | 68.1 | 63.9 | 52.1 | 51.2 | 68.1 | 63.9 | 87.5 |
| − 6.7 | 101.2 | 112.6 | 87.5 | 80.7 | 63.3 | 62.1 | 87.5 | 80.7 | 115.1 |
| − 5.6 | 115.8 | 130.9 | 99.1 | 91.2 | 69.8 | 67.9 | 99.1 | 91.2 | 133.0 |
| − 4.4 | 136.9 | 157.7 | 104.4 | 105.1 | 77.9 | 75.6 | 104.4 | 105.1 | 158.9 |
| − 3.9 | 151.6 | 176.8 | 126.8 | 115.1 | 83.0 | 80.7 | 126.8 | 115.1 | 176.9 |
| − 3.3 | 170.9 | 201.6 | 141.6 | 128.2 | 90.2 | 87.2 | 141.6 | 128.2 | 177.9 |
| − 2.8 | 197.2 | 228.2 | 142.3 | 145.1 | 99.1 | 95.6 | 142.3 | 145.1 | 233.5 |
| − 2.2 | 236.5 | 229.8 | 191.7 | 170.7 | 112.1 | 107.7 | 191.7 | 170.7 | 242.3 |
| − 1.7 | 278.2 | 231.2 | 240.9 | 212.1 | 132.8 | 126.9 | 240.9 | 212.1 | 243.9 |
| − 1.1 | 280.0 | 232.8 | 295.4 | 295.1 | 173.7 | 165.1 | 295.4 | 295.1 | 245.6 |
| 1.7 | 288.4 | 240.7 | 304.5 | 317.7 | 361.9 | 366.8 | 304.5 | 317.7 | 254.9 |
| 4.4 | 297.9 | 248.4 | 313.8 | 327.2 | 372.6 | 377.5 | 313.8 | 327.2 | 261.4 |
| 7.2 | 306.8 | 256.3 | 323.1 | 336.5 | 383.3 | 388.2 | 323.1 | 336.5 | 269.3 |
| 10.0 | 315.8 | 263.9 | 332.1 | 346.3 | 393.8 | 398.9 | 332.1 | 346.3 | 277.2 |
| 15.6 | 333.5 | 279.6 | 350.5 | 365.4 | 414.7 | 420.3 | 350.5 | 365.4 | 292.8 |

From Mott (1964) by permission of H.G. Goldstein, editor of Aust. Refrig. Air Cond. Heat.

ematical expressions must be solved numerically using computer simulation. For the case of one-dimensional heat conduction, the following partial differential equation:

$$\frac{\partial T}{\partial t} = \frac{\partial}{\partial x}\left[\alpha \frac{\partial T}{\partial x}\right] \tag{4.59}$$

would be solved numerically using appropriate initial conditions and boundary conditions to account for heat transfer at the product surface.

The numerical solution of equation (4.59) and similar equations can be accomplished as long as a function describing thermal diffusivity ($\alpha$) as a function of temperature is available. This thermal diffusivity can be defined in the following manner.

$$\alpha(T) = \frac{k(T)}{\rho(T)\, c_{pa}(T)} \tag{4.60}$$

where thermal conductivity ($k$), density ($\rho$) and specific heat ($c_p$) are variable with temperature during the freezing process. The specific heat used in equation (4.60) is an apparent property since it incorporates latent heat removed during the freezing process. This approach has lead to the definition of this property as apparent specific heat ($c_{pa}$).

The approach being presented for numerical solution to the heat conduction problem for food product freezing has been explored in-depth by Bonacini and Comini (1973), Bonacini et al. (1973, 1974), Comini and Bonacini (1974), Hohner and Heldman (1970), Heldman (1974a) and Heldman and Gorby (1975a, 1975b). Bonacini *et al.* (1974) have attempted to estimate the function to describe thermal diffusivity during food freezing while Heldman (1974b) has predicted the function from knowledge of the freezing point depression.

The temperature-dependence of thermal conductivity during freezing is a very significant one. Products which have a relatively high moisture content will have thermal conductivities near the thermal conductivity of water as long as they are in the unfrozen state. The change of state of the water in the product results in a significant increase in the thermal conductivity of the product due to the higher thermal conductivity of ice. To describe this relationship, expressions of the type given as equations (3.32) to (3.38) can be utilized, depending on the type of product considered. The influence of freezing of fish muscle on the thermal conductivity of this product was illustrated in Fig. 3.3. In addition, the variations in thermal conductivity as a function of temperature for frozen fruits and vegetables are illustrated in Fig. 4.15 [Hsieh *et al.* (1977)].

Although the frozen food density does not vary significantly with temperature, the difference in density between the unfrozen and frozen states is directly related to fraction of water frozen. The function

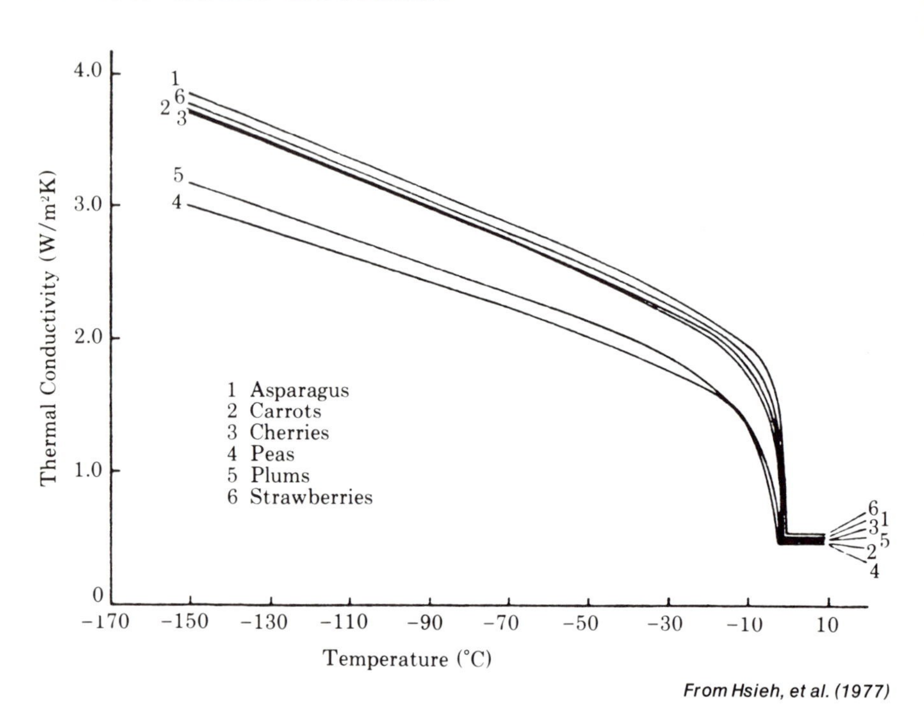


*From Hsieh, et al. (1977)*

FIG. 4.15. PREDICTED THERMAL CONDUCTIVITY FOR DIFFERENT FOOD PRODUCTS

describing product density change during freezing can be predicted from knowledge of fraction water frozen and the following equation:

$$\frac{1}{\rho} = M_U\left(\frac{1}{\rho_U}\right) + M_S\left(\frac{1}{\rho_S}\right) + M_I\left(\frac{1}{\rho_I}\right) \qquad (4.61)$$

where the frozen water fraction ($M_I$) and unfrozen water fraction ($M_U$) can be predicted from equation (4.9). Hsieh *et al.* (1977) used this approach to illustrate the influence of freezing on density of fruits and vegetables in Fig. 4.16.

Evaluation of apparent specific heat as a function of temperature requires use of the expression for determining the refrigeration requirement or enthalpy change of the product at selected temperatures. This leads to the following expression for apparent specific heat of the product in the freezing range of temperatures:

$$c_{pa}(T) = \frac{\Delta H}{\Delta T} = \frac{\Delta H_U + \Delta H_L + \Delta H_I + \Delta H_S}{\Delta T} \qquad (4.62)$$

Equation (4.62) can be evaluated with considerable accuracy by numerical procedures when small increments of temperature change are

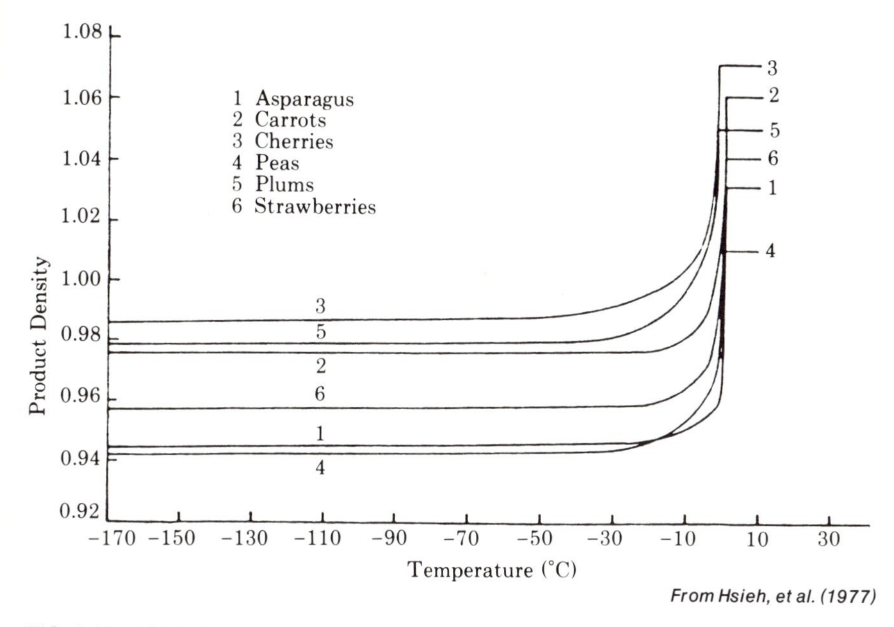


FIG. 4.16. PREDICTED PRODUCT DENSITY FOR DIFFERENT FOOD PRODUCTS

utilized. An example of the variation in apparent specific heat with temperature as evaluated with equation (4.62) is illustrated in Fig. 4.17. Above the initial freezing point, the specific heat is that of the unfrozen product. As the temperature is decreased to the initial freezing point, the apparent specific heat increases considerably to some peak value at the initial freezing point, the apparent specific heat decreases gradually and tends to level off at values just below that of the specific heat for the unfrozen product.

Utilizing the information describing the temperature-dependence of thermal conductivity and density as well as the variation in apparent specific heat with temperature, equation (4.59) can be solved numerically, as illustrated by Hohner and Heldman (1970). The solution obtained leads to information presented in Fig. 4.18, where the time required to freeze the product with various initial temperatures is illustrated. The results indicate that assumptions made in the derivation of Plank's equation can lead to considerable error depending on the product temperature before freezing.

Another factor that may have a dramatic influence on freezing time is shape of the food product. This factor introduces a degree of complexity into all prediction methods including numerical solutions. Comini *et al.* (1978) and Purwadaria (1980) have used finite element analysis to

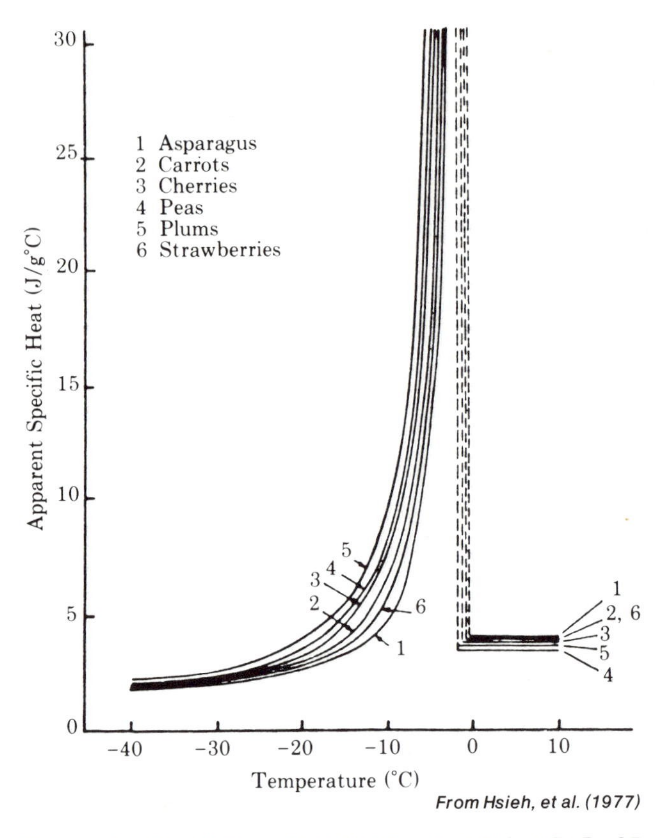


*From Hsieh, et al. (1977)*

FIG. 4.17. PREDICTED APPARENT SPECIFIC HEAT OF DIFFERENT FOOD PRODUCTS

generate solutions for heat conduction during freezing of anomolous shaped foods products. These approaches appear to be quite acceptable for product shapes normally encountered.

Although the influence of the surface heat transfer coefficient on food freezing time is obvious in all procedures, the measurement or estimation of the coefficient magnitude has not been discussed. Typical average surface heat transfer coefficients are presented in Table A.12, but these estimated values do not account for the complexities existing during typical food freezing situations.

Various approaches to measurement of surface heat transfer coefficients during food freezing have been discussed by Cowell and Namor (1974), Earle (1971), Cleland and Earle (1976) and Charm (1978). In these discussions, the attempt is to measure or estimate the average coefficient for the conditions existing in the exact freezing environment

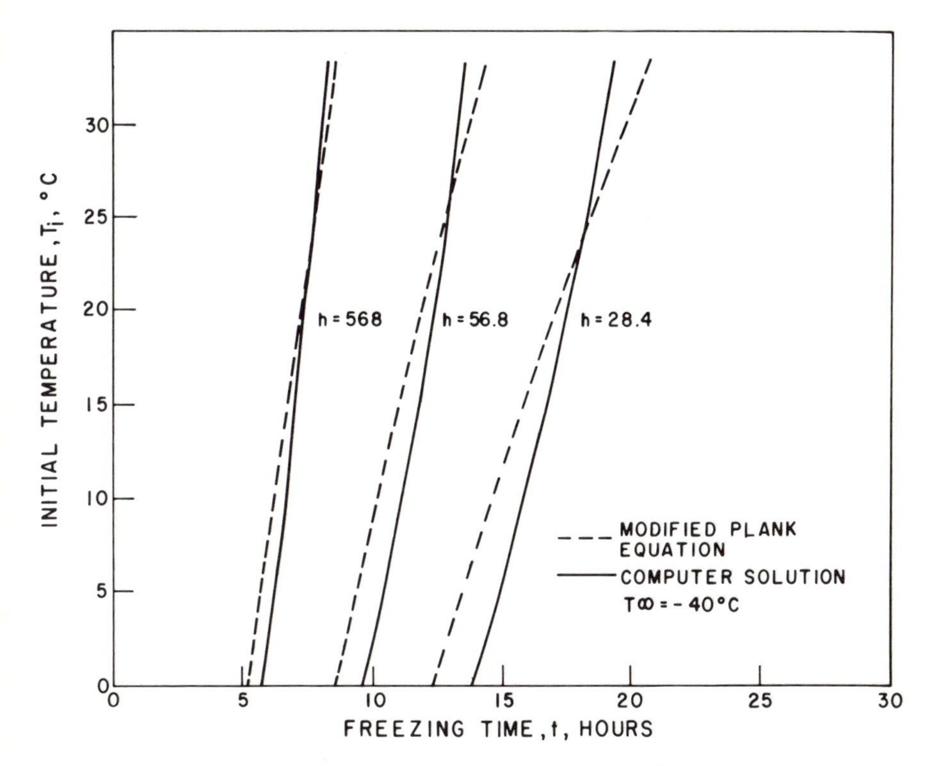


FIG. 4.18. PREDICTED FREEZING TIMES IN 6-INCH PACKS OF CODFISH FOR VARIOUS FREEZING CONDITIONS

of the product. An alternate approach is to establish procedures for measurement or prediction of local surface heat transfer coefficients on the surface of the product during exposure to the freezing environment. As indicated by Chavarria (1978), this approach should lead to improved freezing time predictions since average coefficients may not represent the circumstances existing at the product surface as well as local coefficients. One correlation developed by Chavarria (1978) for a flat slab of ground beef during freezing is:

$$N_{Nu_x} = 0.579\ N_{Re_x}^{0.582} \quad (4.63)$$

Since this expression illustrates that the surface coefficient will vary along the surface in the direction of air flow, the average coefficient must be a function of product length in the direction of air flow. It is difficult therefore, for any single average coefficient to be representative of all products with flat slab geometries.

The prediction and measurement of surface heat transfer coefficients for foods during freezing becomes even more complex for anomolous shapes. It has been concluded by Heldman (1979) that the most accurate approach is the use of a transducer designed to measure local values of the coefficient. In addition, this transducer should have the same geometric shape as the food product for which the freezing time is being predicted.

## 4.4 DESIGN OF FOOD FREEZING EQUIPMENT

In general, freezing equipment for food products can be placed in three categories, as illustrated by Everington (1973): (a) air-blast freezers, (b) plate freezers, and (c) immersion freezers. Only scraped-surface type heat exchangers as used for partial freezing of liquid foods, such as ice-cream, might not be in the indicated categories.

### 4.4.1 Air-blast Freezers

Freezing systems using cold air at high velocities may be designed in any one of several configurations. In most cases, the configurations used will depend on the product being frozen and possibly on the capacity of the system.

Products that are of high density and are frozen in large packages will be placed on trays or conveying systems and exposed to high-velocity cold air. These freezing systems may be batch-type, with the product trays being loaded and removed from a freezing compartment. For such situations, the capacity of the system is established by the size of the freezing compartment and the freezing time computed by methods illustrated in section 4.3.

In freezing systems with the product placed on a conveyor, the product moves through the freezing compartment continuously. In some designs the conveying system may move through the freezing chamber in a spiral manner or in the form of trays which index upward and/or downward from the entrance to exit of the freezing tunnel.

EXAMPLE 4.12 A continuous freezing system for whole chickens in a plastic film is being designed using a spiral conveyor and high-velocity cold air. A schematic cross-section of the system being designed as shown in Fig. 4.19. The initial temperature of the product is 5 C and the freezing point is −2 C. The air used for freezing is at an average of −30 C while in contact with the product, which is being frozen to −18 C. The conveyor that carries the product through the freezing chamber is designed to operate at 3 m/min. Determine the dimensions of the freezing chamber and the capacity of the refrigeration system required.

SOLUTION

(1) Both parts of the solution require the length of conveyor necessary to assure sufficient product exposure to the −30 C air.

(2) The conveyor speed of 3 m/min and freezing time for the product will establish conveyor length.

(3) The freezing time can be computed using a modified form of Plank's equation (4.30).

$$t_F = \frac{\rho \Delta H}{T_i - T_\infty}\left[\frac{Pa}{h_c} + \frac{Ra^2}{k}\right]$$

where $\Delta H$ will be the total enthalpy required to change the product from 5 C to −18 C and was determined from Table 4.3:

$$\Delta H = 315.8 - 37.2 = 278.6 \text{ kJ/kg}$$

The other parameters in the equation will be:

$\rho$ = 1025 kg/m$^3$ (Table 4.2)
$a$ = 0.15 m (approximated diameter for spherical product)
$h_c$ = 22 W/m$^2$K (Appendix Table A.12)
$k$ = 1.298 W/mK (Table 4.2)
$P = 1/6$, $R = 1/24$ } constants for spherical geometry

then:

$$t_F = \frac{(1025)(278.6)(1000)}{[5-(-30)](3600)}\left[\frac{(1/6)(0.15)}{22} + \frac{(1/24)(0.15)^2}{1.298}\right]$$

$$= (2266.4)(1.14 \times 10^{-3} + 7.22 \times 10^{-4})$$

$$t_F = 4.21 \text{ hr} = 252.74 \text{ min}$$

(4) Based on the given conveyor speed and computed freezing time:

Conveyor length = 3 m/min × 252.74 = 758.2m

(5) By establishing the freezing chamber height at 5 m with 0.3 m clearance between sections of the spiral conveyor, approximately 15 complete circular sections of the conveyor could be incorporated.

(6) The diameter of each circular section of the conveyor can be used to assure the desired total length.

(7) Each circular section will have 758.2/15 = 50.55 m of length.

(8) The diameter of each section can be computed:

$$\pi d = 50.55$$

$$d = 16.09 \text{ m}$$

(9) If the width of the conveyor is 0.3 m and the diameter ($d$) represents the distance between conveyor centers, the length and width of the chamber must be 16.09 + 0.3 = 16.39 m.

(10) In order to assure clearance around the conveyor, the chamber dimensions would be 17 × 17 × 5 m with additional space above the chamber for circulation of air from refrigeration coils.

(11) If the whole chickens are spaced at intervals of 0.3 m along the conveyor with 0.1 m between, approximately 2528 chickens will be in the chamber at any time and are moving out of the chamber at a rate of 10 units/min.

(12) Since each chicken will weigh:

$$0.014 \text{ m}^3 \times 1025 \text{ kg/m}^3 = 14.35 \text{ kg}$$

and the enthalpy change is 278.6 kJ/kg, the total refrigeration requirement will be:

$$14.35 \text{ kg/chicken} \times 10 \text{ units/min} \times 278.6 \text{ kJ/kg} = 4 \times 10^4 \text{ kJ/min}$$

Refrigeration requirement = 666 kW

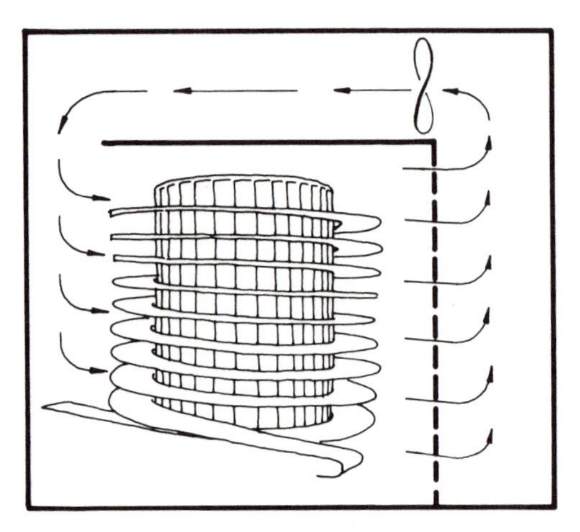

FIG. 4.19. SCHEMATIC OF SPIRAL CONVEYOR AIR-BLAST FREEZING TUNNEL

The modified form of air-blast freezing is the fluidized bed, in which certain products are frozen by being fluidized in the low-temperature air. There is a definite limit to the size (density) of product which can be frozen in a fluidized bed because of the power requirements needed to generate air velocities required for fluidization. The term which has gained commercial acceptance when describing products frozen in a fluidized bed freezer is instant-quick-frozen (IQF). Most fruits and vegetables can be frozen in 3 to 5 min by the fluidized bed or IQF method. Most commercial equipment uses a mesh conveyor to carry the fluidized product through a freezing tunnel.

EXAMPLE 4.13 An IQF tunnel is being used for strawberries. The product conveyor is 1.5 m wide and 6 m long. The air used as a freezing medium is at −34 C and moves through the product bed at a velocity which produces a surface heat transfer coefficient of 85 $W/m^2K$. If the strawberries enter the tunnel at 5 C and are frozen to −20 C, compute the conveyor velocity and estimate the capacity of the freezer.

SOLUTION

(1) In order to compute the conveyor velocity, the product freezing time must be determined.
(2) Using the modified form of Plank's equation (4.30):

$$t_F = \frac{\rho\,\Delta H}{T_i - T_\infty}\left[\frac{Pa}{h_c} + \frac{Ra^2}{k}\right]$$

where:

$\Delta H = 386 - 44 = 342$ kJ/kg (from Table A.11)
$\rho = 960$ kg/m$^3$ (estimated)
$a = 0.013$ m (approximate diameter for sphere)
$k = 2.08$ Wh/mK (estimated)
$P = 1/6$, $R = 1/24$ } constants for spherical geometry

so:

$$t_F = \frac{(960)(342)(1000)}{[5-(-34)](3600)}\left[\frac{(1/6)(0.013)}{85} + \frac{(1/24)(0.013)^2}{2.08}\right]$$

$$t_F = (2338.5)(2.55 \times 10^{-5} + 3.39 \times 10^{-6})$$

$$t_F = 0.0676 \text{ hr} = 4.05 \text{ min}$$

(3) For the strawberries to be exposed to the freezing conditions for 4.05 min, the conveyor must be moving at the following rate:

$$\frac{6 \text{ m}}{4.05 \text{ min}} = 1.48 \approx 1.5 \text{ m/min}$$

(4) The capacity of the system can be estimated by computing the quantity of fruit per meter of conveyor at any time:

(a) Assuming the strawberries are packed tightly on the conveyor:

$$\frac{1.5 \text{ m width}}{0.013 \text{ m/berry}} = 115 \text{ berries across conveyor}$$

(b) The number of strawberries on a 1-meter length of conveyor would be:

$$\frac{1}{0.013 \text{ m/berry}} \times 115 \text{ berries}/1.5 \text{ m width} = 8846 \text{ berries/meter}$$

length.

(c) Using volume and density of strawberries:

$$\frac{4\pi(a/2)\rho}{3} = \frac{4\pi(0.013/2)^3}{3} \times 960 = 1.104 \times 10^{-3} \text{ kg/berry}$$

(d) The quantity of strawberries per foot length:

$$8846 \times 1.104 \times 10^{-3} = 9.77 \text{ kg/m}$$

(5) The capacity of the system will be:

$$9.77 \text{ kg/m} \times 1.5 \text{ m/min} = 14.65 \text{ kg/min}$$
$$= 879 \text{ kg/hr}$$

### 4.4.2 Plate Freezers

By bringing the food product or package into direct contact with plates which are maintained at the desired freezing temperature, rapid freezing of many products can be achieved. Normally, the plate is maintained at the desired temperature by design of the refrigeration system evaporator with the plate in direct contact with the evaporating refrigerant. The systems are designed to bring the product into contact with plates on two sides and with application of pressure to increase the surface heat transfer coefficient as much as possible.

Plate-freezing systems may be designed as batch or continuous units. The batch systems are loaded and unloaded manually; the appropriate freezing time being used to establish frequency of change. Continuous systems are loaded automatically by holding a given freezing station in an open position while product packages are moved into the station from a conveyor (Fig. 4.20). After filling, the station is indexed upward, while a new station is filled. When the product has completed one cycle in the freezing compartment, the frozen product leaves the station as the unfrozen product enters.

Air movement is not required in a plate freezer resulting in less floor area requirement and lower power usage when compared to air-blast freezers. Plate freezers are used widely for freezing of fish and meat products.

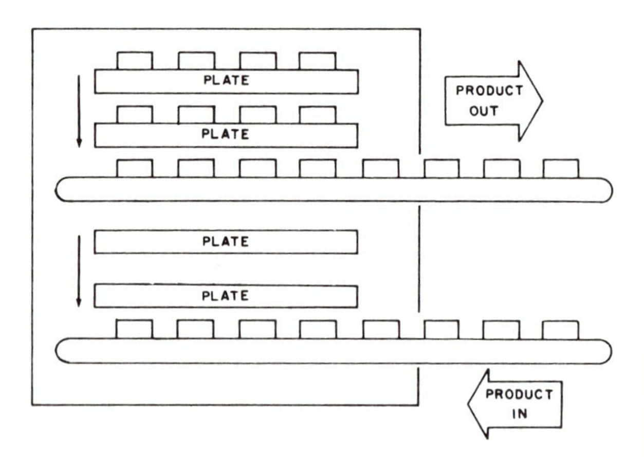


FIG. 4.20. SCHEMATIC OF LOADING AND UNLOADING OF PRODUCT FROM PLATE FREEZER

EXAMPLE 4.14 A continuous plate-freezing system is being designed to freeze 0.5 kg cod fillet packages at a rate of 500 kg/hr. The package dimensions are 0.04 m by 0.1 m by 0.14 m and it enters the freezer at 5 C. The plates for each station are 1 m wide and will accommodate 8 packages of the indicated dimensions. The plates are maintained at a temperature of − 30 C and the surface heat-transfer coefficient is 28 $W/m^2K$. The package material has a thickness of 8

$\times 10^{-4}$ m and a thermal conductivity of 0.05 W/mK. Compute the number of freezing stations required and the approximate size of the freezing compartment needed when freezing the product to −25 C.

SOLUTION

(1) The freezing time for the product can be estimated using the modified form of equation (4.30):

$$t_F = \frac{\rho \Delta H}{T_i - T_\infty}\left[\frac{Pa}{h_c} + \frac{Ra^2}{k}\right]$$

$\Delta H = 341.5 - 30.5 = 311$ kJ/kg (see Table A.7)
$\rho = 880$ kg/m$^3$
$T_i = 5$ C
$T_\infty = -30$ C
$a = 0.04$ m
$k = 2.08$ W/mK
$P = 1/2$, $R = 1/8$ } constants for infinite slab geometry

(a) Before computing $t_F$, the appropriate surface heat-transfer coefficient must be determined. The coefficient should account for contact between plate and package as well as heat transfer through the package material:

$$\frac{1}{h_c} = \frac{1}{28} + \frac{8 \times 10^{-4}}{(0.05)} = 0.0517$$

$$h_c = 19.34 \text{ W/m}^2\text{K}$$

(b) Then:

$$t_F = \frac{(880)(311)(1000)}{[5-(-30)](3600)}\left[\frac{(1/2)(0.04)}{19.34} + \frac{(1/8)(0.04)^2}{2.08}\right]$$

$$t_F = (2172)(1.034 \times 10^{-3} + 9.615 \times 10^{-5})$$

$$t_F = 2.45 \text{ hr}$$

(2) Since the freezing time is 2.45 hr, the freezing system must contain 2.45 × 500 kg/hr = 1225 kg of the product.

(3) Since each freezing station will hold 8 packages of 0.5 kg each, the number of stations required will be:

$$\frac{1225 \text{ kg}}{4 \text{ kg/station}} = 306.25 \text{ or } 307 \text{ stations}$$

(4) The freezing compartment can be arranged in several ways to contain the 307 stations. By assuming that 18 stations can be located in the vertical plane with 0.3 m between the centers of each station, the height of the compartment must be greater than 2.4 m.

(5) Using 8 stations in each vertical plane provides space for 39 vertical planes of freezing stations. Using the width of each station at 0.125 m with 0.025 m between each, the length of the compartment must be at least 5.85 m.

(6) Based on the minimum dimensions from computations, a compartment size of 1.5 m width, 3 m height and 6 m length would be recommended.

### 4.4.3 Immersion Freezers

The concept of immersion freezing involves bringing the product into direct contact with a low-temperature refrigerant. This type of cryogenic freezing has gained popularity due to the improvement in product quality which results as compared to more traditional freezing methods.

Liquid nitrogen is probably the most widely used refrigerant for immersion freezing. Due to an extremely low boiling point (−196 C), freezing rates are very high. In most cases, efficient use of the refrigerant is achieved by moving the product countercurrent to the flow of the nitrogen refrigerant. This brings the product into contact initially with cold gaseous nitrogen and reduces the product temperature considerably before exposure to a spray of liquid nitrogen.

Two other refrigerants which have been used for immersion freezing are liquid carbon dioxide and R-12. Liquid carbon dioxide has a boiling point of −98 C and is brought into contact with the product by spraying in a manner similar to liquid nitrogen systems. R-12 has a much higher boiling point (−30 C) and the product can be introduced directly into a bath of the refrigerant. In general, systems for recovery of the carbon dioxide and refrigerants for reuse have been more successful than those for liquid nitrogen.

EXAMPLE 4.15 Data published by Breyer, Wagner and Ryan (1966a) indicate that a pecan coffee-cake (0.23 m diameter, 0.04 m thick) will freeze in 1.7 min in liquid nitrogen. Additional data indicate that each product unit was 0.372 kg and the liquid nitrogen was used at a rate of 0.665 kg $N_2$/kg product. The coffee-cake was frozen to −18 C from an initial temperature of 22 C. Estimate the surface heat-transfer coefficient for the freezing process.

SOLUTION

(1) In order to use the modified form of Plank's equation (4.43), several parameters must be evaluated.

(a) Product density: based on the known volume of coffee-cake and unit mass, the density will be

$$\rho = \frac{0.372}{\dfrac{\pi (0.23)^2}{4} (0.04)} = 224 \text{ kg/m}^3$$

(b) Enthalpy change for freezing: using the liquid nitrogen consumption rate and the latent heat plus sensible heat change for nitrogen allows computation of product enthalpy change:

$$\begin{aligned} \text{Latent heat} &= 197.98 \text{ kJ/kg } N_2 \\ \text{Sensible heat} &= 1.044 \text{ kJ/kg K } [-18-(-196)] \\ &= 185.83 \text{ kJ/kg } N_2 \end{aligned}$$

where nitrogen is assumed to be heated to lowest temperature of product (−18 C).

$$\text{Enthalpy change} = 0.665 \text{ g } N_2/\text{g product } (197.98 + 185.83)$$
$$= 255.23 \text{ kJ/kg product}$$

(2) Using the modified Plank's equation (4.30):

$$t_F = \frac{\rho \, \Delta H}{T_i - T_\infty} \left[ \frac{Pa}{h_c} + \frac{Ra^2}{k} \right]$$

where:

$t_F = 1.7$ min.
$\rho = 224 \text{ kg/m}^3$
$\Delta H = 255.23$ kJ/kg product
$T_i = 22$ C
$T_\infty = -196$ C
$P = 1/2$, $R = 1/8$ } constants for infinite slab geometry
$a = 0.04$ m
$k = 1.731$ W/m K (estimated)

$$\frac{1.7 \text{ min}}{60 \text{ min/hr}} = \frac{(224)(255.23)(1000)}{22-(-196)\,(3600)} \left[ \frac{1/2(0.04)}{h_c} + \frac{(1/8)(0.04)^2}{1.731} \right]$$

$$0.02833 = (72.85)(0.02/h_c + 1.1554 \times 10^{-4})$$
$$h_c = 73.2 \text{ W/m}^2 \text{ K}$$

# 4.5. STORAGE OF FROZEN FOODS

Although the freezing process has a dramatic influence on the quality of the frozen food, the storage conditions have an equally important influence. In most situations, the inability to maintain adequate storage conditions between the freezing process and the point at which the product is consumed will reduce quality in a significant manner. Singh and Wang (1977) have provided an in-depth review of the storage influences on frozen food quality.

## 4.5.1 Changes During Storage

In general, it must be recognized that the freezing process only reduces the rate at which deterioration reactions occur in the food product. The purpose of frozen storage is to maintain the optimum temperature for minimum rates of product quality reduction. Since several reactions may be occurring within the product, the most critical reactions must be known and controlled during storage.

The deterioration of frozen food quality may be due to physical or chemical changes. The physical changes would include crystallization of ice or desiccation. The crystallization of ice that occurs during fluctuating temperatures results in undesirable texture and appearance. The desiccation of a frozen food occurs due to inadequate packaging and results in reduction of product flavor, texture and color. The most obvious chemical changes during frozen food storage include lipid oxidation, enzymatic browning, flavor deterioration, protein insolubilization and the degradation of chlorophyll and other pigments along with vitamins. In most cases, the rates of deterioration are highly temperature dependent but rates can be controlled through the use of appropriate measures.

The rates of deterioration reactions can be described by equations (1.12) and (1.19). There is an increasing amount of research literature indicating that some rate of quality deterioration may increase at very low storage temperatures. These results would imply that there is some optimum temperature within the frozen food storage temperature region that will maintain maximum quality.

### 4.5.2 Storage Temperature Fluctuations

One of the most obvious influences of storage environment on product quality is the degradation caused by temperature fluctuations. It is generally recognized that small fluctuation of 1−2 C have negligible influence on product quality while fluctuations of 10 C may have definite detrimental effects on quality.

There have been several attempts to develop methods for prediction of quality degradation in frozen foods due to temperature fluctuations. Schwimmer, *et al.* (1955) illustrated that the influence of various shapes and magnitudes of temperature fluctuation could be expressed in terms of "effective" temperature. This effective temperature has been related to rates of reactions causing quality loss. The Time-Temperature Tolerance (T-TT) method was developed by Van Arsdel and Guadagni (1959) and involves the measurement of area under the time-temperature curve when plotted on a transformed coordinate system. The area magnitude is proportional to the total change in frozen product quality.

A computer simulation to predict quality degradation in frozen foods was developed by Singh (1976). The simulation used first-order rate constants as inputs to describe quality degradation. The changes in product quality due to any type of temperature fluctuation could be predicted. The simulation was applied to the oxidation of oxymyoglobin in frozen beef using rate constants from Brown and Dolev (1963) as illustrated in Fig. 4.21. Results of the simulation as presented in Fig. 4.22 illustrate the dramatic influence of the fluctuation period. When the magnitude of fluctuation was −15 ± 10 C and the reaction rate constant was minimum at −5 C, eight-hour cycles caused more rapid quality degradation than 24-hour cycles of temperature fluctuation.

In practice, the stability of frozen foods during storage is expressed in a variety of ways. In most cases, the number of storage days at a given temperature have been measured for the frozen foods having significant market impacts. These types of information are readily available for

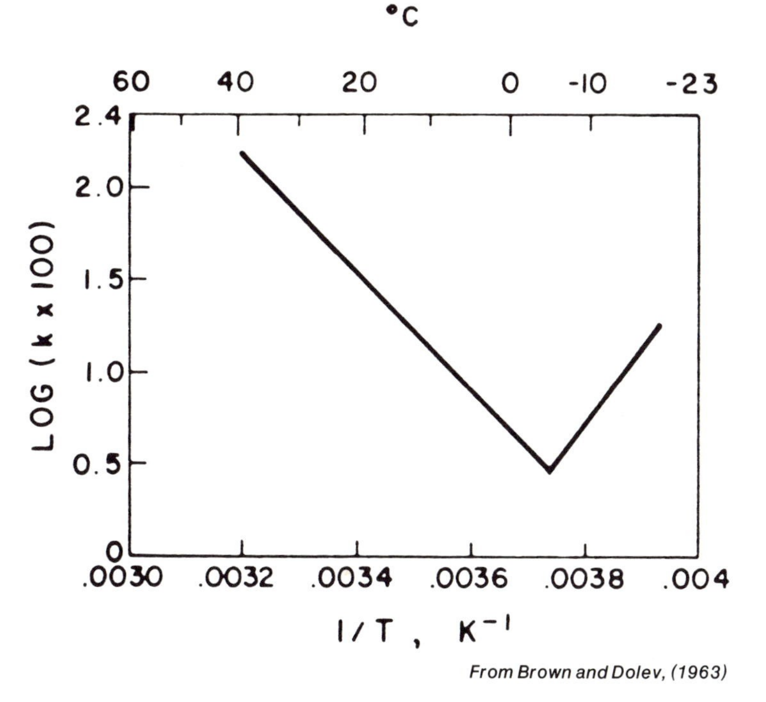


FIG. 4.21. RATES OF OXIDATION OF OXYMYOGLOBIN IN BEEF AT VARIOUS TEMPERATURES

meats, seafoods, vegetables and fruits. In most cases, a uniform storage temperature of −18 C is considered to be desirable.

## PROBLEMS

4.1. *Percent water frozen in a frozen food.* Asparagus has a freezing point of 0.7C and a water content 92.6%. Compute the percent water frozen in the product at −5C.

4.2. *Refrigeration requirements for freezing.* 1000 kg of perch with 79.1% water are being frozen to −10C (approximately 85% of the water frozen). The specific heat of product solids is 1.5 kJ/kgC and 1.9 kJ/kgC for ice and 4.1 kJ/kgC for water. Estimate the refrigeration requirement for freezing the product from an initial temperature of 5C. (Latent heat of fusion for water = 333.22 kJ/kg).

4.3. *Enthalpy change for freezing fruit juice.* Orange juice is being frozen from an initial temperature of 15C to a final temperature of −12C. Compute the enthalpy change required for the process and determine the percent water unfrozen in the final product.

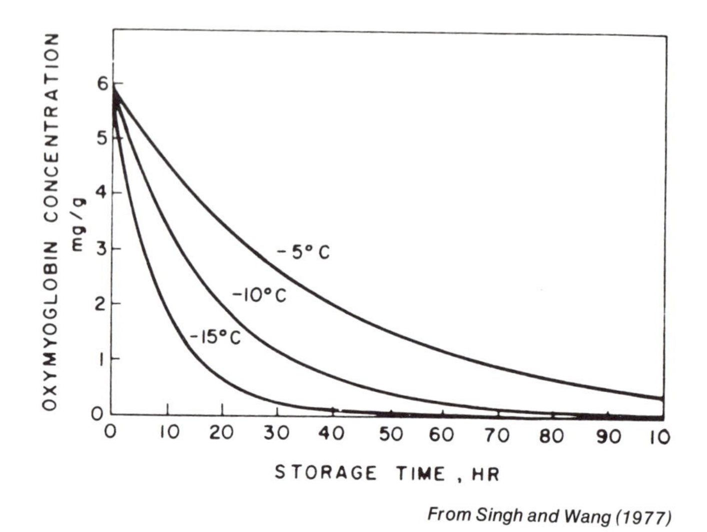


*From Singh and Wang (1977)*

FIG. 4.22. OXYMYOGLOBIN CONCENTRATION IN BEEF STORED AT −15± 10C WITH THREE CYCLIC VARIATIONS IN STORAGE TEMPERATURE

4.4. *Freezing time for fruit juice.* Grape juice is being frozen in a 4 cm diameter by 10 cm tall can in an air blast freezer with 20 $W/m^2K$ as a surface heat transfer coefficient. The initial product temperature is 2C and the air used as a freezing medium is −20C. Estimate the time required to freeze the product to −10C using Plank's equation. (Assume infinite cylinder geometry.)

4.5. *Freezing of sweet cherries using IQF.* Sweet cherries (approximately 1.5 cm diameter) are being frozen in an IQF system with −30C air and a surface heat transfer coefficient of 50 $W/m^2K$. If the initial temperature is 5C, how much time will be required to reduce the center temperature to −15C? Estimate the freezing time using the Cleland-Earle approach.

4.6. *Design of plate freezing system.* A plate freezing system is being designed for freezing of cod packed in packages with 20 cm length, 4 cm width and 2 cm thickness. The system will operate on a continuous basis and will be located in a room with 20 m length, 10 m width and 3 m height. The plate contact will provide a surface heat transfer coefficient of 60 $W/m^2K$. Estimate the capacity of the system that can be conveniently located in the room using the Mott Procedure. The initial product temperature is 3C and the final temperature is −15C.

4.7. *Design of a Spiral Conveyor Freezing System.* Packages of lean beef are to be frozen on a spiral conveyor system. The shape of the packages is nearly spherical with a 15 cm diameter. Freezing will be accomplished with air blast providing a surface heat transfer coefficient of 20 $W/m^2K$. Estimate the size of the freezing tunnel required to freeze 10,000 kg/hr from an inital temperature of 5C to −10C. Use the Tao solutions for the estimation.

## COMPREHENSIVE PROBLEM IV

*Factors Influencing Freezing Rates in Food Products*

### Objectives

1. To become acquainted with the use of a computer program for predicting freezing rates in foods.
2. To evaluate the influence of product thickness, initial product temperature and surface heat transfer coefficient on freezing rate.
3. To use freezing rate information to develop general design specifications for continuous freezing equipment.

### Procedures

A. Computer program
   1. A computer program to predict the temperature history of an infinite slab of food product should be developed.
   2. The input parameters can be selected as appropriate for the product.
   3. The influence of the following factors will be evaluated:
      a. Surface heat transfer coefficients, of 5, 25, and 100 $W/m^2K$
      b. Initial product temperatures of 10, 5, and 2C
      c. Product thickness of 2, 4, and 6 cm.
   4. A consistent criterion for evaluation freezing time should be selected.
   5. The result should be compared to predicted time by Plank's equation.

B. Design specifications
   1. The output from the computer program should be used to develop the physical dimensions of a continuous freezing system.
   2. Factors to be considered:
      a. Capacity of the system.
      b. Dimensions of the freezing system needed to achieve capacity and desired freezing rate.
      c. Magnitude of errors resulting from using Plank's equation for design specifications.

## Discussion

A. Reasons for agreement or lack of agreement between computer prediction and Plank's equation.
B. What is the influence of surface heat transfer coefficient, initial product temperature and product thickness on rates of freezing. Discuss.
C. List and describe factors which influence the design of a continuous belt freezer.

## NOMENCLATURE

$A$ = area, $m^2$.
$A'$ = constant in equation (4.48).
$a$ = thickness of product in equation (4.28), m.
$B$ = dimensionless number in Mott Procedure; equation (4.57).
$B'$ = constant in equation (4.49).
$N_{Bi}$ = Biot Number, dimensionless.
$c_p$ = specific heat, kJ/kgK.
$d$ = diameter, m.
$N_{Fo}$ = Fourier Number, dimensionless.
$G$ = free energy.
$\overline{G}^{ol}$ = free energy of liquid phase at standard state.
$\overline{G}^{os}$ = free energy of solid phase at standard state.
$G'$ = dimensionless number in Mott Procedure; equation (4.58).
$H$ = enthalpy, kJ/kg.
$\overline{H}^{ol}$ = enthalpy of liquid phase at standard state.
$\overline{H}^{os}$ = enthalpy of solid phase at standard state.
$\Delta H$ = enthalpy change during freezing, kJ/kg.
$\Delta H'$ = enthalpy for product freezing, Nagaoka equation, kJ/kg.
$\Delta H^*$ = enthalpy for product freezing, Mott Procedure, kJ/kg.
$h_c$ = convective heat transfer coefficient, $W/m^2K$.
$k$ = thermal conductivity, W/mK.
$L$ = latent heat of fusion, kJ/kg.
$M$ = mass, kg.
$m$ = molality.
$N_{Nu}$ = Nusselt Number, dimensionless.
$P$ = constant for Plank's equation.
$p$ = vapor pressure, Pa.
$p^{\circ}$ = total pressure, Pa.
$p'$ = vapor pressure at standard state, Pa.
$N_{Pk}$ = Plank's Number, dimensionless.
$q$ = heat flux, kJ/hr.
$R$ = constant for Plank's equation.
$R_g$ = gas constant, J/mole K.

$Re$ = Reynold's Number, dimensionless.
$S$ = dimensionless number in Mott Procedure, equation (4.56).
$N_{Ste}$ = Stefan Number, dimensionless.
$T$ = temperature, C.
$T_A$ = absolute temperature, K.
$T^*$ = temperature of product when 60% of the water is frozen, C.
$\Delta T_2$ = freezing point depression, C.
$t$ = time, hr.
$t_F^*$ = dimensionless time; for Tao Solution.
$V$ = product volume for Mott Procedure, $m^3$.
$W$ = molecular weight.
$X$ = mole fraction.
$x$ = variable distance along X-axis, m.
$\alpha$ = thermal diffusivity, $m^2/s$.
$\beta$ = constant used in Fig. 4.10.
$\phi$ = fat content of product expressed as decimal.
$\chi$ = product solids content expressed as percentage in equation (4.22).
$\lambda$ = constant used in Neumann solution; equation (4.50).
$\lambda'$ = latent heat of fusion, kJ/mole.
$\rho$ = density, $kg/m^3$.
$\eta$ = chemical potential.
$\eta^l$ = chemical potential of liquid phase in equilibrium.
$\eta^\beta$ = chemical potential of solid phase in equilibrium.
$\eta^o$ = chemical potential at total pressure.
$\eta^{soln}$ = chemical potential at vapor pressure of solution.
$\eta^{o'}$ = chemical potential at standard state.
$\eta^s$ = chemical potential of solid state.
$\eta^{ol}$ = chemical potential of liquid at standard state.
$\eta^l$ = chemical potential of liquid state.

## Subscripts

$A$ = identification of component in solution.
$a$ = apparent conditions.
$B$ = identification of component in solution.
$E$ = effective.
$F$ = initial freezing point; contribution of frozen product.
$f$ = fat fraction.
$I$ = contribution of frozen water.
$i$ = initial condition.
$j$ = juice fraction.
$L$ = contribution of latent heat.
nf = non-fat fraction.

$o$ = pure substance or liquid.
$S$ = contribution of product solids.
SNJ = solids-not-juice.
$U$ = contribution of unfrozen water.
$x$ = local conditions.
$\infty$ = free stream conditions.

## BIBLIOGRAPHY


BAKAL, A., and HAYAKAWA, K. 1973. Heat transfer during freezing and thawing of foods. Adv. Food Res. *20*, 218.

BONACINA, C., and COMINI, G. 1973a. On a numerical method for the solution of the unsteady state heat conduction equation with temperature dependent parameters. Proceedings of the XIII Int. Congress of Refr. *2*, 329.

BONACINA, C., COMINI, G., FASANO, A. and PRIMICERIO, M. 1973. Numerical solution of phase change problems. Int. J. Heat Mass Transfer *16*, 1825.

BONACINA, C., COMINI, G., FASANO, A., and PRIMICERIO, M. 1974. On the estimation of thermophysical properties in non-linear heat-conduction problems. Int. J. Heat Mass Transfer *17*, 861.

BREYER, F., WAGNER, R.C., and RYAN, J.P. 1966a. Application of liquid nitrogen to the freezing of baked goods. Quick Frozen Foods Intern. *7*, No. 4, 58–62 133–136.

BREYER, F., WAGNER, R.C., and RYAN, J.P. 1966b. Application of cryogenics in the baking industry. Chem. Engr. Prog. Symp. Ser. *62*, No. 69, 93–103.

BROWN, W.D., and DOLEV, A. 1963. Effect of freezing on autoxidation of oxymyoglobin solutions. J. Food Sci. *28*(2), 211–213.

CARSLAW, H.S., and JAEGER, J.C. 1959. Conduction of Heat Solids. Clarendon Press, Oxford, England.

CHARM, S.E. 1978. The Fundamentals of Food Engineering, 3rd Edition. AVI Publishing Co., Westport, Conn.

CHARM, S.E., and SLAVIN, J. 1962. A method for calculating freezing time of rectangular packages of food. Int. Institute of Refr. Meeting Commissions 2, 3, 4, and 6A, Washington, D.C. Bulletin, Annexe: 567.

CHAVARRIA, V.M. 1978. Experimental determination of the surface heat transfer coefficient under food freezing conditions. M.S. Thesis. Department of Agricultural Engineering. Michigan State University, East Lansing, MI.

CLELAND, A.C., and EARLE, R.L. 1976. A new method for prediction of surface heat transfer coefficients in freezing. Bull. I.I.R., Annexe-1, 361.

CLELAND, A.C., and EARLE, R.L. 1977a. A comparison of analytical and numerical methods of predicting the freezing times of foods. J. Food Science, *42*, 1390.

CLELAND, A.C., and EARLE, R.L. 1979a. A comparison of methods for predicting the freezing times of cylindrical and spherical foodstuffs. J. Food Sci. *44*:

CLELAND, A.C., and EARLE, R.L. 1979b. Prediction of freezing times for foods in rectangular packages. J. Food Sci. *44*, 964.

COMINI, G., and BONACINA, C. 1974. Application of computer codes to phase-change problems in food engineering. Int. Institute of Refr. Meeting of Commissions B1, C1, and C2, Bressanone, Italy. Bulletin, Annexe:15.

COMINI, G., DEL GUIDICE, S., STRADA, M., and REBELLATO, L. 1978. The finite element method in refrigeration engineering. Int. J. Refr., *1*, 113.

COWELL, N.D. 1967. The calculation of food freezing times. Proc. 12th Int. Congr. Refrig. pp. 1.

COWELL, N.D., and NAMOR, M.S.S. 1974. Heat transfer coefficients in plate freezing—effect of packaging materials. Proc. of I.I.R. Meeting of Commissions B1, C1 and C2. Bressanone, Italy. Sept. 17–20.

DICKERSON, R.W., JR. 1969. Thermal properties of food. In The Freezing Preservation of Foods, 4th Edition, Vol. 2, D.K. Tressler, W.B. Van Arsdel, and M.J. Copley (Editors). AVI Publishing Co., Westport, Conn.

EARLE, R.L. 1971. Simple probe to determine cooling rates in air. Proc. XIII Int. Congr. Refrig. *2*, 373.

EDE, A.J. 1949. The calculation of the freezing and thawing of foodstuffs. Mod. Refrig. *52*, 52.

EVERINGTON, D.W. 1973. Methods of quick freezing. Food Manufacture *48*, No. 3, 21–25.

FENNEMA, O. 1975. Freezing Preservation. In Physical Principles of Food Engineering. pp. 173. Marcel Dekker, Inc., New York.

FENNAMA, O., and POWRIE, W.D. 1964. Fundamentals of low-temperature food preservation. Advan. Food Res. *13*, 219–347.

FENNEMA, O.R. 1973. Nature of the freezing process. In Low-Temperature Preservation of Foods and Living Matter, Marcel Dekker, New York.

HELDMAN, D.R. 1966. Predicting refrigeration requirements for freezing ice cream. Quart. Bull. Mich. Agr. Expt. Sta. *49*, No. 2, 144–154.

HELDMAN, D.R. 1974a. Predicting the relationship between unfrozen water fraction and temperature during food freezing using freezing point depression. Trans. ASAE *17*, 63.

HELDMAN, D.R. 1974b. Computer simulation of food freezing processes. Proceedings of the VI Int. Congress of Food Science and Technology *IV*, 397.

HELDMAN, D.R., and GORBY, D.P. 1975a. Prediction of thermal conductivity in frozen food. Trans. ASAE 18(1)156–58 & 62.

HELDMAN, D.R., and GORBY, D.P. 1975b. Computer simulation of individual-quick-freezing of foods. ASAE Paper No. 75-6016.

HELDMAN, D.R. 1979. Prediction of food product freezing rates. Presented at 2nd Int. Congr. on Engr. and Food. Helsinki, Finland. Aug. 27–31.

HOHNER, G.A., and HELDMAN, D.R. 1970. Computer simulation of freezing rates in foods. Presented at 30th Ann. Meeting Institute of Food Technologists May 24–27, San Francisco, Calif.

HSIEH, R.C., LEREW, L.E., and HELDMAN, D.R. 1977. Prediction of freezing times for foods as influenced by product properties. J. Food Proc. Engr. *1*, 183.

I.I.R. 1971. International Institute of Refrigeration. Recommendations for the Processing and Handling for Frozen Foods. 2nd Ed. Paris.

JOSHI, C., and TAO, L.C. 1974. A numerical method of simulating the axisymmetrical freezing of food systems. J. Food Science *39*,623.

LENTZ, C.P. 1961. Thermal conductivity of meats, fats, gelatin, gel and ice. Food Technol. *15*, 243–247.

LONG, R.A. 1955. Some thermodynamic properties of fish and their effect on the rate of freezing. J. Sci. Food Agr. *6*, 621.

MERYMAN, H.T. 1956. Mechanics of freezing in living cells and tissues. Science *124*, 515.

MOORE, W.J. 1962. Physical Chemistry, 3rd Edition. Prentice-Hall, Englewood Cliffs, N.J.

MOTT, L.F. 1964. The prediction of product freezing time. Aust. Refrig., Air Cond. Heat *18*, 16.

NAGAOKA, J., TAKAGI, S., and HOTANI, S. 1955. Experiments on the freezing of fish in an air-blast freezer. Proc. 9th Intern. Congr. Refrig., (Paris) *2*, 4.

PLANK, R.Z. 1913. Ges. Kalte-Ind. *20*, 109 (In German). *Cited by* A.J. Ede. 1949. The calculation of the freezing and thawing of foodstuffs. Mod. Refrig. *52*, 52.

PURWADARIA, H.K. 1980. A Numerical Prediction Model for Food Freezing Using Finite Element Methods. Ph.D. Dissertation. Agricultural Engineering Dept. Michigan State University.

RIEDEL, L. 1949. Refractive index and freezing temperatures of fruit juices as a function of concentration. Lebensmittel *89*, 289–299. (German).

RIEDEL, L. 1951. The refrigeration effect required to freeze fruits and vegetables. Refrig. Eng. *59*, 670–673.

RIEDEL, L. 1956. Calorimetric investigations of the freezing of fish meat. Kaltetechnik *8*, No. 12, 374–377. (German).

RIEDEL, L. 1957A. Calorimetric investigations of the meat freezing process. Kaltetechnik *9*, 38–40. (German).

RIEDEL, L. 1957B. Calorimetric investigations of the freezing of egg whites and yolks. Kaltetechnik *9*, No. 11, 342–345. (German).

ROLFE, E.J. 1968. The chilling and freezing of foodstuffs. In Biochemical and Biological Engineering Science, N. Blakebrough (Editor). Academic Press, New York.

SCHWIMMER, S., INGRAHAM, L.L., and HIGHES, H.M. 1955. Temperature tolerance in frozen food processing. Ind. Eng. Chem. *47*(6), 1149–1151.

SINGH, R.P. 1976. Computer simulation of food quality during frozen food storage. Paper presented at the Inter. Inst. of Refrig., Melbourne, Australia.

SINGH, R.P., and WANG, C.Y. 1977. Quality of frozen foods—A Review. J. Food Process Engr. *1*(2), 97.

STEPHENSON, J.L. 1960. Ice crystal formation in biological materials during rapid freezing. Ann. N.Y. Acad. Sci. *85*, 535.

TAO, L.C. 1967. Generalized numerical solutions of freezing a saturated liquid in cylinders and spheres. AICHE J. *13*, 165.

TAO, L.C. 1971. Personal communication, *see* BAKAL, A. and HAYAKAWA, K. 1973. Heat transfer during freezing and thawing of foods. Adv. Food Res. *20*, 218.

TIEN, R.H. and G.E. GEIGER. 1967. A heat transfer analysis of the solidification of a binary eutectic system. J. Heat Transfer *9*:230.

TIEN, R.H. and G.E. GEIGER. 1968a. The unidimensional solidification of a binary eutectic system with a time-dependent surface temperature. J. Heat Transfer. *9C* (1) 27.

TIEN, R.H. and V. KOUMO. 1968b. Unidimensional solidification of a slab-variable surface temperature. Trans. Metallurgical Sec. AIME. *242*:283.

TIEN, R.H. and V. KOUMO. 1969. Effect of density change on the solidification of alloys. ASME Paper No. 69-HT, 45.

VAN ARSDEL, W.B., and GUADAGNI, D.G. 1959. Time-temperature tolerance of frozen foods. XV. Method of using temperature histories to estimate changes in frozen food quality. Food Technol. *13*(1), 14–19.

# 5

# EVAPORATION FOR FLUID FOOD CONCENTRATION

One of the key unit operations for fluid food products is evaporation, as utilized to concentrate or increase the solids concentration of a fluid food. One of the primary objectives of this operation is to reduce the volume of the product by some significant amount without loss of nutrient components. This reduction of volume permits more efficient transportation of the important product components and efficient storage of the solids. An equally important objective of evaporation of moisture from fluid foods is to remove large amounts of moisture effectively and efficiently before the product enters a dehydration process. The evaporation operation may be used for products which vary widely in characteristics, and in many cases these characteristics influence the evaporator design considerably.

The concentration of the solids fraction in a liquid food product is accomplished by evaporation of free moisture present in the product. This is accomplished by raising the product temperature to the boiling point and holding for the time required for the desired concentration. Due to the heat sensitivity of most products, evaporation is usually accomplished under vacuum. By utilizing relatively high vacuum (low pressures), large amounts of moisture can be removed from liquid food products without significant reduction in the quality of heat-sensitive components.

Four basic components are required to accomplish evaporation, as illustrated in Figure 5.1. Any evaporator utilized for fluid food products will contain the following: (a) an evaporation vessel, (b) a heat source, (c) a condenser and (d) a method of maintaining vacuum (removing noncondensable gases). The important design considerations of an evaporator relate very closely to the four components listed. The vacuum system must remove noncondensable gases in order to maintain the desired vacuum in the evaporation vessel. Sufficient heat must be supplied to the product to evaporate the desired amount of moisture, and a condenser is

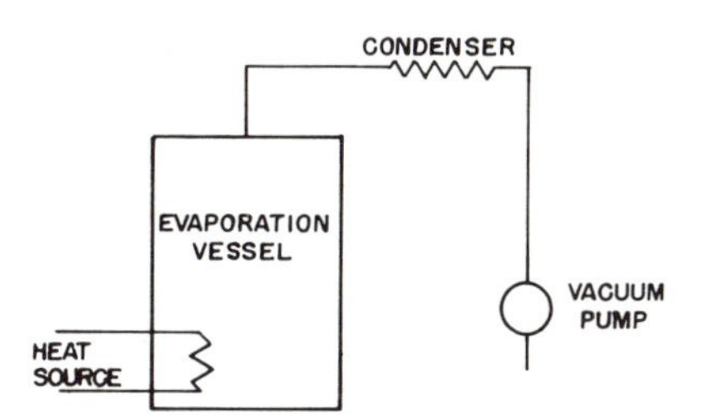


FIG. 5.1. SCHEMATIC DRAWING OF COMPONENTS OF SIMPLE EVAPORATION SYSTEM

provided to condense the vapor produced by evaporation. Basic material balances may be utilized to establish feed rates for given degrees of concentration. These relationships lead to computation of the amounts of heating medium required to provide the desired evaporation of moisture. While the key design consideration is the amount of heat transfer from the heating medium to the product, the required heat-transfer surface area cannot be computed without estimating the overall heat-transfer coefficient for the heating surface.

Although the above design considerations can be predicted with acceptable accuracy, several factors unique to the products result in computations of a more complex nature. Many of the solids that are common to food products result in an elevated boiling point compared to that of water at the same pressure. The boiling point elevation changes as the product becomes more concentrated. The modes of heat transfer from heating medium to product include convection on both the heating-medium side and the product side of the heating surface. The convective heat-transfer coefficient on the product side is a function of viscosity, which changes as the product becomes more concentrated. Finally, the more basic thermal properties of the product change with temperature and moisture content and may have a significant influence on computations relating to evaporator design.

Several factors may influence the design of an evaporation system for liquid foods in an indirect manner. The heat-sensitive characteristics of food components require special attention in order to keep contact time between product and heated surface as short as possible and to prevent product temperature from exceeding predetermined levels. In addition, the characteristics of liquid foods tend to promote scaling or fouling of the heat-transfer surfaces in the evaporator; certain design considerations can minimize the losses in efficiency due to this factor.

## 5.1 THERMODYNAMICS OF EVAPORATION

Thermodynamics plays a very important role in the description of the evaporation process for food products. This importance is particularly evident in the description and evaluation of parameters describing the phase change and in the establishment of boiling-point elevation.

### 5.1.1 Phase Change

During the evaporation process, the phase change is evaporation of water from the liquid state to steam or the vapor state. The latent heat of vaporization is a well-known value for pure water, and can be described as a function of pressure by the Clausius-Clapeyron equation, presented as equation (1.8). Minor variations in equation (1.8) lead to the following expression:

$$\ln p = -\frac{L_v}{R_g T_A} + C \tag{5.1}$$

which clearly expresses the direct relationship between the latent heat of vaporization ($L_v$) and the vapor pressure ($p$). In the case of food products which contain solids and other components that influence the latent heat of vaporization, an expression similar to equation (5.1) can be written as follows:

$$\ln p' = -\frac{L_v'}{R_g T_A} + C \tag{5.2}$$

Equation (5.2) relates the latent heat of vaporization for a fluid food product ($L_v'$) to the vapor pressure ($p'$) at the same temperature ($T_A$). By combining equations (5.1) and (5.2) at equal temperatures, the following expression is obtained:

$$\ln p' = \frac{L_v'}{L_v} \ln p + C' \tag{5.3}$$

to establish the log-log relationship between the pressures of the pure water and the liquid food product. By plotting the logarithm of vapor pressure for the food product versus the logarithm of the pressure for pure water at various temperatures, the relationship between the latent heat of the liquid food product and the latent heat of water is established. Information of this type is necessary to account for the change in latent heat as the concentration of the fluid food occurs during evaporation.

### 5.1.2 Boiling Point Elevation

In addition to the influence of product solids on the latent heat required for evaporation of water from a liquid food product, the influence of solids on the boiling point of the liquid product must be considered. This influence is the result of the product composition containing solutes having molecular weights that cause the boiling point to be elevated above that for pure water. An expression which describes the extent of boiling point elevation can be derived by considering phase equilibrium and the chemical potential of the two phases which exist at equilibrium. This derivation is obtained by utilizing the same procedure as used to derive equation (4.10), which described the freezing point depression as a function of molality. The derivation is almost identical to equation (4.9) and can be restated for the case of boiling-point elevation as follows:

$$\frac{\lambda_v}{R_g}\left(\frac{1}{T_{A0}} - \frac{1}{T_A}\right) = -\ln X_A \tag{5.4}$$

where $\lambda_v$ equals the latent heat of vaporization, $T_{A0}$ represents the boiling point of pure water, and $X_A$ represents mole fraction of water in the solution. By assuming that the boiling point elevation is small and by using only the first term of the logarithmic expansion of equation (5.4), the following expression is obtained:

$$\Delta T_B = \frac{R_g T_{A0}^2}{\lambda_v} X_B \tag{5.5}$$

where $X_B$ is the mole fraction of the solute causing the boiling-point elevation. An additional modification can be accomplished by introducing molality into equation (5.5) to obtain the following expression:

$$\Delta T_B = \frac{R_g T_{A0}^2 W_A}{L_v\, 1000} m \tag{5.6}$$

where $L_v$ is the latent heat of vaporization per unit mass of water. Equation (5.6) can be utilized to compute boiling point elevation as long as the solution is dilute or $X_B$ is small. In situations where the product may become highly concentrated, as will occur in the evaporation process, the assumptions made in obtaining equation (5.6) may create considerable error. Under these conditions, equation (5.4) should be utilized to compute the boiling point of the product.

Use of any of the equations introduced requires knowledge of the specific components of the products that cause changes in the boiling point. Such information may not be readily available for food products due to their complex composition and lack of knowledge about the components that contribute to boiling-point elevation. In most cases, the computations made must be considered estimates based on knowledge of the components which exist in higher concentrations and knowledge about the molecular weight and mole fractions of these components.

EXAMPLE 5.1 Compute the boiling point rise of a 10% NaCl solution at atmospheric pressure (sea-level).

SOLUTION

(1) Since the 10% solution may be assumed to be dilute, equation (5.6) may be used.

(2) Since NaCl has a molecular weight of 59:

$$m = \frac{\text{g NaCl}/1000\,\text{g water}}{W\text{ of NaCl}} = \frac{100}{59} = 1.695$$

(3) Using equation (5.6):

$$\Delta T_B = \frac{(8.314\ \text{J/mole K})(373\ \text{K})^2(18)}{(4.0626 \times 10^4\ \text{J/mole})(1000)}(1.695) = 0.869\ \text{C}$$

(4) The boiling point for the 10% NaCl would be 100.869 C.

EXAMPLE 5.2 The boiling point of a food product would be expected to increase as product solids become more concentrated. Determine the change in boiling point of skimmilk as it is concentrated to 30% T.S.

SOLUTION

(1) The composition of skimmilk is approximately 5.1% lactose, 3.6% protein, 0.8% fat, 0.3% minerals and ash, and 90.2% water. The lactose will be expected to have the predominant influence on boiling point.

(2) The initial boiling point can be computed from equation (5.6) where:

$$5.1\%\ \text{lactose} = 5.1\ \text{g lactose}/100\text{g product} = \frac{5.1}{.902}$$

$$= 5.654\ \text{g lactose}/100\ \text{g water}$$

$$m = \frac{56.54\ \text{g lactose}/1000\ \text{g water}}{342\ (W\text{ of lactose})} = 0.165$$

(3) Using equation (5.6):

$$\Delta T_B = \frac{(8.314)(373)^2(18)(0.165)}{(4.0626 \times 10^4)(1000)} = 0.0846\ \text{C}$$

indicating a boiling point of 100.0846 C.

(4) When the skimmilk is concentrated to 30% T.S., the lactose will represent 15.6% of the product and the molality will be:

$$15.6\% = 15.6\ \text{g lactose}/100\ \text{g product} = \frac{15.6}{0.7}$$

$$= 22.286\ \text{g lactose}/100\ \text{g water}$$

$$m = \frac{222.86}{342} = 0.652$$

(5) If the solute is not considered to be dilute, equation (5.4) should be used with $X_A = 0.9884$

$$\frac{4.0626 \times 10^4}{8.314}\left[\frac{1}{373} - \frac{1}{T_A}\right] = -\ln(0.9884)$$

where

$$X_A = \frac{70/18}{70/18 + 15.6/342} = 0.9884$$

then:

$$13.1 - \frac{4886.4566}{T_A} = 0.01167$$

$$13.1 - 0.01167 = \frac{4886.4566}{T_A}$$

$$13.0887\ T_A = 4886.4566$$

$$T_A = 373.334$$

indicating a boiling point of 100.334 C.

(6) If the solute in the concentrated product is assumed to be dilute, equation (5.6) is used:

$$\Delta T_B = \frac{(8.314)(373)^2(18)(0.652)}{(4.0626 \times 10^4)(1000)} = 0.334C$$

This result can be compared to $\Delta T_B$ = 0.334 C obtained using equation (5.4) and indicates that the equation for dilute solution is acceptable for the situation presented.

The second method that is commonly utilized to estimate boiling point elevation is based on Dühring's rule. The rule states that the temperature at which one liquid exerts a given vapor pressure is a linear function of the temperature at which a reference liquid exerts the identical vapor pressure. The expression utilized to establish the stated relationship is derived from the Clausius-Clapeyron equation given as equation (1.8). Utilizing an integrated form of Clausius-Clapeyron equation, an expression of the type given by equation (5.1) is obtained. By equating expressions of this type for the liquid of interest and a reference liquid at equal vapor pressures, the following expression is obtained:

$$\frac{L_v}{R_g T_A} + C = \frac{L'_v}{R_g T_{A0}} + C_R \tag{5.7}$$

where $T_A$ represents the boiling point of the liquid of interest and $T_{A0}$ represents the boiling point of the reference liquid. A slight rearrangement of equation (5.7) results in the following expression:

$$\frac{L_v}{R_g T_A} = \frac{L'_v}{R_g T_{A0}} + (C_R - C) \tag{5.8}$$

which indicates the linear relationship between the two temperatures being considered. Utilizing an expression of the type given by equation (5.8), a chart presented as Figure 5.2 is obtained. The curves shown on the chart are relationships between the boiling point of a solution and the boiling point of water. The chart illustrates that as the solution becomes more concentrated, the boiling point elevation increases. Although a chart of the type presented in Figure 5.2 may be useful in estimating boiling point elevation for liquid foods, the exact magnitude cannot be computed without more specific information on product composition.

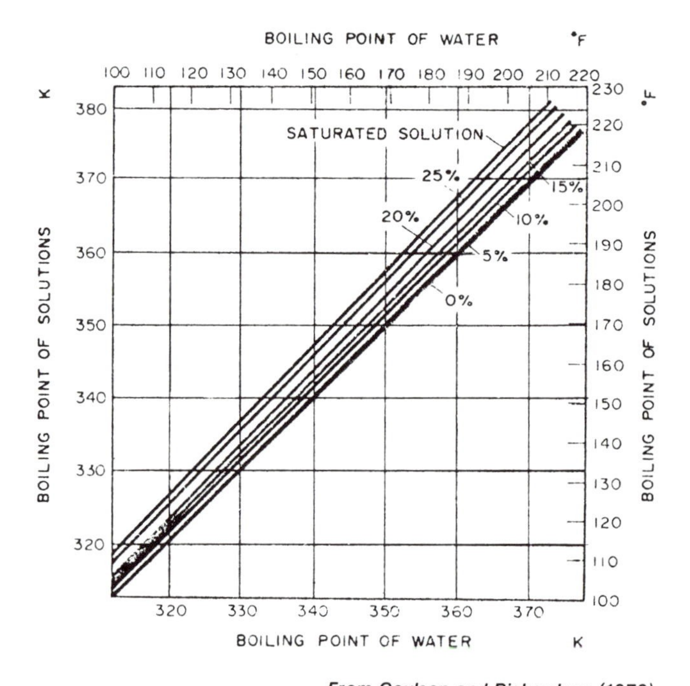


*From Coulson and Richardson (1978)*

**FIG. 5.2. DÜHRING LINES ILLUSTRATING THE INFLUENCE OF SOLUTE CONCENTRATIONS OF BOILING POINT ELEVATION OF NaCl**

EXAMPLE 5.3 A liquid food with composition such that the vapor pressure exerted is similar to sodium chloride is being concentrated in an evaporator at 25 kPa. Determine the initial and final boiling points of the product as concentration is increased from 10% to 25%.

SOLUTION

(1) From Appendix Table A.6, the boiling point of water at 25 kPa is 65 C (338 K).

(2) Using Figure 5.2, the boiling point of the product will be:
339 K (66 C) at 10% total solids concentration.
343 K (70 C) at 25% total solids concentration.

## 5.2 HEAT TRANSFER DURING EVAPORATION

Evaporation of moisture and the resulting concentration of a food product cannot be accomplished without sufficient amounts of heat being transferred to the product. The general equation describing heat transfer from some heating medium to the product is as follows:

$$q = UA(T_s - T_p) \tag{5.9}$$

where heat transfer rate is a function of temperature gradient, heat-transfer surface area and an overall heat transfer coefficient ($U$). The overall heat transfer coefficient is best defined by the following expression:

$$\frac{1}{UA_m} = \frac{1}{h_s A_s} + \frac{x}{kA_m} + \frac{1}{h_p A_p} \tag{5.10}$$

where $x$ is the thickness of material used for the heat-transfer surface, $k$ is the thermal conductivity of the same material and $A_m$ is the average area for the steam and product sides of the heat exchange surface. The most appropriate average would be the logarithmic mean in the case of a tubular heat exchanger. Equation (5.10) describes the three factors involved in determining the resistance to heat transfer during evaporation. Each of the three components of equation (5.10) represents a resistance to heat transfer. These resistances include resistance in the heat-transfer film on the heating-medium side, resistance to heat conduction in the material making up the heat-transfer surface, and resistance in the heat-transfer film on the product side of the heat-transfer surface.

### 5.2.1 Heating Medium

In most evaporators, the heating medium will be steam or some other condensing vapor. Since this type of heating medium is utilized, the resistance to heat transfer is normally created by a condensation film on the heating-medium side of the heat-transfer surface. The expressions available for estimating the heat-transfer coefficients for film condensation are empirical in nature. The basic grouping of parameters can be derived on a theoretical basis, as described by Krieth (1965). Brown *et al.* (1950) presented the following expression:

$$h_s = 1.13 \left[ \frac{k_f^3 \rho_f^2 g L_v}{L \mu_f (T_s - T_w)} \right]^{1/4} \tag{5.11}$$

for prediction of the film coefficient on a vertical surface where $L$ represents the length or height of surface. An expression for identical conditions as presented by Krieth (1965) was:

$$h_s = 0.94 \left[\frac{k_f^3 \rho_f^2 g L_v}{L\mu_f(T_s - T_w)}\right]^{1/4} \tag{5.12}$$

and differs from equation (5.11) only in magnitude of the coefficient. Since the difference between the two coefficients is relatively small, either equation (5.11) or (5.12) should be acceptable for design computations.

The computation of film coefficients for condensate collecting on horizontal tubes can be accomplished using an expression presented by Charm (1978):

$$h_s = 1.18 \left[\frac{k_f^3 \rho_f^2 g \pi D}{\mu_f M}\right]^{1/3} \tag{5.13}$$

where $M$ is the rate at which condensate is collecting. An alternative expression was first presented by Bromley (1950) as:

$$h_s = 0.725 \left[\frac{k_f^3 \rho_f^2 g L_v}{\mu_f D(T_s - T_w)}\right]^{1/4} \tag{5.14}$$

where the amount of heat transfer involved in phase change is accounted for in the same manner as in equations (5.11) and (5.12). The differences in coefficients and exponents in equations (5.13) and (5.14) can be attributed to the difference in the manner of describing phase change.

The property values in equations (5.11) through (5.14) should be evaluated at the mean condensate temperature. The vapor temperature and the mean surface temperature should be utilized in computing the mean condensate temperature.

EXAMPLE 5.4 Compare the heat transfer coefficients for condensation films on vertical and horizontal tubes. Saturated steam at 198.5 kPa is being utilized as a heating medium for evaporation of water. The tube length is 3 m with a 0.05 m outside diameter and an evaporator temperature of 75°C is being used.

SOLUTION

(1) The properties of steam and water are obtained from Tables A.6 and A.4. A mean condensate temperature is computed from 120 C (steam at 198.5 kPa) and 75°C or 97.5 C to be used to select film properties.

$\rho_f = 960.15$ kg/m$^3$ $\quad\quad$ $\mu_f = 0.2849 \times 10^{-3}$ kg/ms
$k_f = 0.68$ W/mK $\quad\quad$ $g = 9.806$ m/s$^2$
$L_v = 2202.59$ kJ/kg (at 198.5 kPa)

(2) For a vertical tube, equation (5.11) will be used:

$$h_s = 1.13\left[\frac{(0.68\ \mathrm{W/mK})^3\,(960.15\ \mathrm{kg/m^3})^2\,(9.806\ \mathrm{m/s^2})\,(2202.59\ \mathrm{kJ/kg})\,(1000\ \mathrm{J/kJ})}{(3\ \mathrm{m})\,(0.2849\times10^{-3}\ \mathrm{kg/ms})\,(393\ \mathrm{K}-348\ \mathrm{K})}\right]^{1/4}$$

$$h_s = 1.13\,(1.628\times10^{14})^{1/4}$$
$$h_s = 1.13\ (3571.9)$$
$$h_s = 4036.2\ \mathrm{W/m^2K}$$

(3) For a horizontal tube, equation (5.14) applies:

$$h_s = 0.725\left[\frac{(0.68)^3\,(960.15)^2\,(9.806)\,(2202.59)\,(1000)}{(0.2849\times10^{-3})\,(0.05)\,(393-348)}\right]^{1/4}$$

$$h_s = 0.725\,(9.767\times10^{15})^{1/4} = 0.725\,(9941.2)$$

$$h_s = 7207.4\ \mathrm{W/m^2K}$$

(4) For the conditions presented, the heat-transfer coefficient through the condensation film on a horizontal tube is nearly double the coefficient for a vertical film.

### 5.2.2 Heating Surface

The second part of the overall resistance to heat transfer described by equation (5.10) is the resistance created by the heating-surface material. Since this surface will normally be constructed of a metal such as stainless steel which has known properties and thickness, this portion of the overall resistance is relatively easy to compute. By knowing the thickness ($x$) and the thermal conductivity ($k$), a direct computation of resistance to heat transfer results.

### 5.2.3 Product

Probably the most complex resistance to heat transfer is created by the resistance film on the product side of the heat-transfer surface. The complexity is due to lack of knowledge about the conditions that exist near the surface during heat transfer to the boiling liquid product. There are some evaporators in which boiling may not occur near the heat-transfer surface and the heat-transfer coefficient can be estimated by empirical equations which describe the fluid flow near the heat-transfer surface. The conditions that exist when heat transfer is occurring to the boiling liquid cannot be overlooked, and these types of evaporators deserve considerable attention.

The heat transfer from a heated surface to a boiling liquid is by convection, but the heat flux varies with the temperature difference between the surface and the liquid, as illustrated for water in Figure 5.3. When the temperature difference is very small (less then 5°C), the heat transfer is by convection in a regime called pool boiling. During this type of heat transfer, the heat is transferred through the liquid by convection to the liquid-vapor interface where evaporation occurs. By increasing the temperature difference between the surface and the liquid, the vaporization begins near the surface, creating bubbles which condense within the liquid before they reach the liquid-vapor interface. These vapor bubbles tend to form at selected locations on the surface, and the regimen is called nucleation boiling. This type of boiling continues until the temperature difference is near 55 C. At the higher temperature differences however, the vapor bubbles that are produced at the surface may be transported to the liquid-vapor interface. As illustrated in Figure 5.3, the heat flux from surface to liquid increases continually up to a point near 55 C temperature difference. This would indicate that the development of vapor bubbles near the surface during nucleation boiling increases the heat transfer coefficient from surface to boiling liquid.

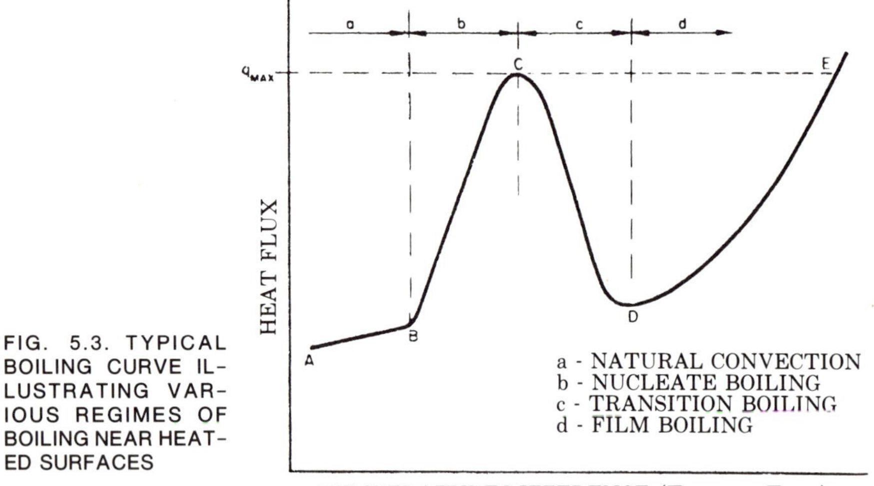


FIG. 5.3. TYPICAL BOILING CURVE ILLUSTRATING VARIOUS REGIMES OF BOILING NEAR HEATED SURFACES

At temperature differences greater than 55 C, a film of vapor bubbles forms and collects near the heat-transfer surface, resulting in a reduced heat flux. Although this film is unstable at temperature differences just slightly greater than 55 C, the film becomes stable at temperature differences of 550 C and results in relatively low heat flux and the corresponding heat-transfer coefficients. At temperature differences greater than 550 C, the film of vapor is very stable, but radiation begins to contribute to the heat flux and results in the increased heat flux and apparent heat-transfer coefficients. There appears to be no evidence to indicate the type of boiling that would exist in an evaporator for liquid food products. It is unlikely, however, that the temperature difference would be greater than 55°C. Either pool or nucleation boiling will probably describe the majority of physical situations that may exist.

Heat transfer to boiling liquids is normally expressed in terms of dimensionless groups in the same manner as other types of heat transfer. The following is the general expression utilized:

$$N_{\overline{Nu}_b} = \frac{h_b D_b}{k_f} = (N_{Re_b})\,\psi(N_{Pr_f}) \tag{5.15}$$

where the characteristic dimension ($D_b$) is the diameter of the bubbles moving through the liquid and the velocity utilized in the Reynolds number is the mass velocity of bubbles leaving the surface. For the case of pool boiling, Rohsenow (1952) derived an expression and correlated experimental data. The derivation accounted for (a) heat transfer to a bubble attached to the heat-transfer surface, (b) the diameter of the bubble that would develop at the surface, (c) the frequency of bubble formation and (d) the fact that the number of bubbles forming at the surface is proportional to the heat-transfer rate. Utilizing this approach, Rohsenow (1952) was able to derive the following expression:

$$N_{\overline{Nu}_b} = \frac{h_b D_b}{k_f} = C_n \beta \frac{L_v}{k_f}\left[\frac{g_c \sigma}{g(\rho_f - \rho_v)}\right]^{1/2} \tag{5.16}$$

where $C_n$ is a constant resulting from the derivation and $\beta$ is the angle between the surface and the bubble forming at the surface. It was found that the best correlation was obtained when the parameters were grouped in the following way:

$$\frac{N_{Re_b}\,N_{Pr_s}}{N_{Nu_b}} = \frac{C_L \Delta T}{L_v} \tag{5.17}$$

which leads to the following empirical expression:

$$\frac{C_L \Delta T}{L\,Pr_f^{\,1.7}} = C_{sf}\left[\frac{q/A}{\mu_f L_v}\;\frac{g_c \sigma}{g(\rho_f - \rho_v)}\right]^{0.33} \tag{5.18}$$

where the constant $C_{sf}$ is a function of the angle $\beta$ and will vary in value from 0.006 for water boiling near heated brass to 0.013 for pool boiling of water on copper and platinum. Equation (5.18) is for clean surfaces, and experimental results indicate that a contaminated surface results in a variation in the exponent of the Reynolds number from 0.8 to 2.0.

For the case of nucleation boiling, Krieth (1965) presented the following empirical expression:

$$\frac{(q/A)_{\max}}{\rho_v L_v} = 143\left[\frac{\rho_f}{\rho_v} - 1\right]^{0.6} \tag{5.19}$$

where $(q/A)_{max}$ is the maximum heat flux occurring during nucleation boiling and is a function of temperature. Equation (5.19) was obtained by correlation of several sets of experimental data collected during nucleation boiling. According to Rohsenow (1952), contamination of the surface results in an increase in heat flux by 15%.

Although film boiling is unlikely to occur in an evaporator, the following expression can be utilized to compute or estimate the heat-transfer coefficient for film boiling on the outside of a horizontal tube:

$$h_b = 0.62 \left[ \frac{k_v^3 \rho_v (\rho_f - \rho_v) g \lambda'}{D \mu_v \Delta T} \right]^{0.25} \tag{5.20}$$

where

$$\lambda' = L_v \left[ 1 + \frac{0.4 \, \Delta T c_{pv}}{L_v} \right] \tag{5.21}$$

Equation (5.21) accounts for the conduction of the heat through the vapor film near the heat-transfer surface and the boiling convection through the surrounding liquid.

Probably the more critical situation in an evaporator is to account for both the heat-transfer coefficient due to boiling and the heat-transfer coefficient due to convection. If the product is flowing through or over the heat-transfer surface, Rohsenow (1952) recommended that the total heat flux can be computed from the following expression:

$$q_{Tot} = q_b + q_c \tag{5.22}$$

which is the sum of the heat flux due to boiling and the heat flux due to pure convection. By utilizing the appropriate heat-transfer coefficients for each situation, the required heat flux components can be computed and summed to obtain the total heat flow.

### 5.2.4 Heat-transfer Coefficients

Although expressions presented up to this point are useful in gaining insight into the heat-transfer mechanisms that may occur during boiling, information may be lacking with regard to making efficient use of these types of expressions. Many investigations have been conducted directly with evaporation systems and have resulted in empirical expressions for convective heat-transfer coefficients to the product. These expressions can be extremely useful as long as the system being considered is similar to the one on which the empirical expression was based.

For the case of natural circulation evaporators, Piret and Isbin (1954) developed and correlated experimental results to obtain the following expression:

$$\frac{h_b D}{k_f} = 0.0086 \left[\frac{u_m D \rho_f}{\mu_f}\right]^{0.8} \left[\frac{c_{pf} \mu_f}{k_f}\right]^{0.6} \left[\frac{\sigma_f}{\sigma}\right]^{0.33} \tag{5.23}$$

where $D$ is the tube diameter, and the influence of a surface-tension ratio has been introduced. The ratio incorporates the surface tension of the product ($\sigma_f$) and the surface tension of water ($\sigma$).

An approach suggested in Rohsenow (1952) might be necessary for natural circulation systems. This approach suggests using a heat-transfer coefficient computed by summation of the coefficient based on natural convection and the appropriate heat-transfer coefficient for boiling.

For the case of forced convection systems, the approach suggested by Rohsenow (1952) has been utilized by Blatt and Adt (1964) and appears to provide very satisfactory results. For the boiling heat transfer occurring in tubes, equation (3.20) can be utilized for the convective portion of the overall heat-transfer coefficient. Coulson and Richardson (1978) suggested the following expression:

$$N_{Nu} = 1.25 \left[0.023\ N_{Re}^{\ 0.8}\ N_{Pr}^{\ 0.4}\right] \tag{5.24}$$

where a factor of 25% has been added to account for the additional contribution of boiling heat transfer during convection in the tube.

Coulson and Richardson (1978) have introduced a new parameter ($X_{tt}$) as developed by Lockhart and Martinelli (1948). The parameter is defined by:

$$\frac{1}{X_{tt}} = \left[\frac{y}{1-y}\right]^{0.9} \left[\frac{\rho_f}{\rho_v}\right]^{0.5} \left[\frac{\mu_v}{\mu_f}\right]^{0.1} \tag{5.25}$$

and provides a mechanism for describing the influence of two-phase turbulent flow on the convective heat transfer coefficient, where $y$ is the mass fraction of vapor. Experimental data have been correlated to within 20% by the expression:

$$\frac{h_b}{h} = 3.5 \left[\frac{1}{X_{tt}}\right]^{0.5} \quad \left(0.25 < \frac{1}{X_{tt}} < 70\right) \tag{5.26}$$

where the convective heat transfer coefficient ($h$) is for convective heat transfer without boiling.

Although other types of heat transfer may occur in various types of evaporators, the description of these types is rather specific and would require discussion of each individual system. Several investigations have been conducted on these individual systems including Coulson and Mehta (1953), Dukler (1960), Sinek and Young (1962), and Kroll and McCutchen (1968). In addition, the ASHRAE Handbook of Fundamentals presents a very complete list of expressions needed in computing overall heat-transfer coefficients for evaporators.

EXAMPLE 5.5 Skim milk is being concentrated in a forced convection evaporator operating at 75 C. Saturated steam at 198.5 kPa is used as the heating medium outside the horizontal tubes with 0.05 m inside diameter. The product is flowing inside the 3 m long stainless steel tubes with 0.08 cm wall thickness at a rate of 385 kg/min. Compute the overall heat transfer coefficient for the heat transfer surface of the evaporator.

SOLUTION

(1) Since the conditions presented are the same as in Example 5.4, the heat transfer coefficient for the heating medium side of the horizontal tube will be $h_s = 7176\ \mathrm{W/m^2K}$.

(2) The convective heat transfer coefficient for the product side of the heat transfer surface can be computed using equation (5.24).

(3) The properties of skim milk can be estimated from properties of water at 75 C (Table A.4):

$$k = 0.671\ \mathrm{W/mK} \qquad \rho = 974.9\ \mathrm{kg/m^3}$$
$$c_p = 4.190\ \mathrm{kJ/kgK} \qquad \mu = 0.3766 \times 10^{-3}\ \mathrm{kg/ms}$$

(4) By using equation (5.24):

$$\mathrm{Nu} = 1.25\left[0.023\left(\frac{974.9 \times 3.35 \times 0.05}{0.3766 \times 10^{-3}}\right)^{0.8}\left(\frac{4.190 \times 0.3766 \times 10^{-3}}{0.671 \times \dfrac{1}{1000}}\right)^{0.4}\right]$$

where:

$$\bar{u} = \frac{\text{mass flow rate}}{\rho A} = \frac{385}{974.9\,\pi(0.025)^2}$$
$$= 201.13\ \mathrm{m/min} = 3.35\ \mathrm{m/s}$$

then:

$$\mathrm{Nu} = 1.25\,[0.023\,(433.605)^{0.8}\,(2.352)^{0.4}]$$
$$= 1.25\,[0.023\,(32335.2)\,(1.408)]$$
$$\mathrm{Nu} = 1308.8$$

(5) The convective heat transfer coefficient on the product side can be computed from:

$$h_p = \frac{(\mathrm{Nu})k}{D} = \frac{(1308.8)(0.671)}{0.05} = 17564\ \mathrm{W/m^2K}$$

(6) The overall heat transfer coefficient can be computed using equation (5.10):

$$\frac{1}{UA_m} = \frac{1}{(7176)\pi(0.0516)(3)} + \frac{0.0008}{(15)A_m} + \frac{1}{(17564)\pi(0.05)(3)}$$

$$= 2.865 \times 10^{-4} + 5.333 \times 10^{-5}/A_m + 1.208 \times 10^{-4}$$

where $k = 15$ W/mK for stainless steel

(7) Since:

$$A_m = 2\pi L \frac{r_2 - r_1}{\ln(r_2/r_1)} = 2\,\pi\,(3)\frac{(0.0258 - 0.025)}{\ln(\frac{0.0258}{0.025})}$$

$$A_m = 0.4787 \text{ m}^2$$

then:

$$\frac{1}{UA_m} = 2.865 \times 10^{-4} + 1.114 \times 10^{-4} + 1.208 \times 10^{-4}$$

$$= 5.187 \times 10^{-4}$$

$$U = 4027 \text{ W/m}^2\text{K}$$

# 5.3 DESIGN OF EVAPORATION SYSTEMS

The design of an evaporation system for liquid food products is a relatively complex procedure in which several factors must be taken into account, including the heat capacity, heat-transfer surface area, heating medium requirements, and several less important factors. The design procedure includes analysis by mass and heat balances on either single or multiple-effect systems.

One indication of the complexity of design procedures is realized when discussing the variety of methods available for evaporation processes. Moore and Hessler (1963) classified evaporators into five different types. The Calandria evaporator includes short vertical tubes through which the product moves by natural circulation. The product passes vertically through the tubes (Fig. 5.4) and moves downward through the central

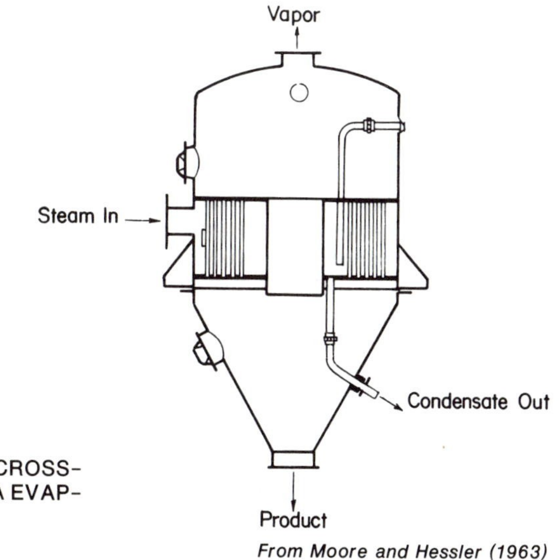


FIG. 5.4. SCHEMATIC CROSS-SECTION OF CALANDRIA EVAPORATOR

portion of the evaporator body. A slightly modified version of this type uses horizontal tubes as heat-transfer surfaces. Both of these evaporator types are batch units.

The second type in Moore and Hessler's classification is the forced-circulation evaporator shown schematically in Fig. 5.5. This evaporator has a bank of horizontal tubes through which the product is pumped for heating purposes. Only limited evaporation occurs in the heating tubes but the vertical chamber provides an opportunity for considerable moisture removal. This evaporator is well suited to products with suspended solids and with high viscosity, and has been used considerably in the evaporation of water from tomato products.

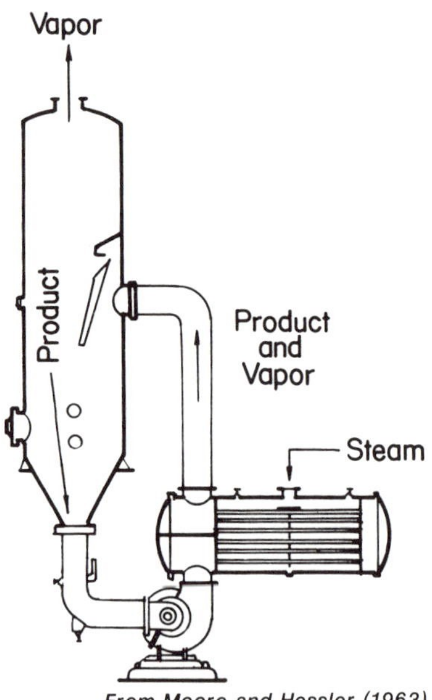


FIG. 5.5. SCHEMATIC CROSS-SECTION OF FORCED CIRCULATION EVAPORATOR

*From Moore and Hessler (1963)*

The third classification includes the long-tube vertical evaporators which operate by natural circulation and a rising or climbing film through the vertical tubes (Fig. 5.6). In this type, the product enters at the bottom of the evaporation chamber and is carried vertically through the tubes by vapor bubbles formed near the bottom of the heating tubes. The vapors and product are separated at the top of the evaporator body. This type of evaporator is limited to products with low to moderate viscosities.

Another approach to handling products with high viscosities is the fourth classification of Moore and Hessler (1963), which includes the use of agitated-film heat exchangers, as illustrated in Fig. 5.7. In these types, the product film near the heat-transfer surface is continuously agitated by a rotor to induce heat transfer and increase the convective heat-

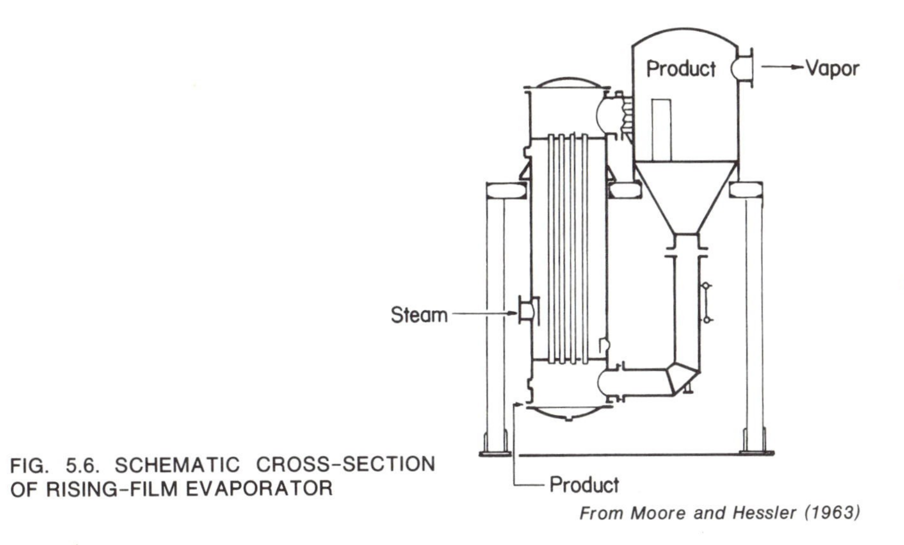


FIG. 5.6. SCHEMATIC CROSS-SECTION OF RISING-FILM EVAPORATOR

*From Moore and Hessler (1963)*

transfer coefficient on the product side. The product moves downward over the heat-transfer surface in these types of evaporators. A somewhat modified version of the agitated-film type is the falling-film evaporator, which is very similar to the agitated-film except for removal of the rotor mechanism. The product flows continuously downward in a film over the heat-transfer surface and evaporation occurs during a single pass over the heat-transfer surface.

The final type in the classification of Moore and Hessler (1963) is the rising-falling film concentrator shown in Fig. 5.8. In this type of evaporator, the product enters at the bottom of the evaporator body and is carried upward through the heating tubes in a manner similar to the rising-film evaporator. The product and vapor are separated at the top of the tubes, but the product is concentrated more as it moves downward again through the heating tubes to the outlet at the bottom of the evaporator body.

An additional type of evaporator developed specifically for milk and milk products is the plate evaporator. The product is heated in a plate heat exchanger where evaporation of the moisture occurs and the liquid and vapor are separated in a chamber following the heat exchanger.

## 5.3.1 Retention Time

A factor of concern in the evaporation of moisture from liquid food products is the exposure time of the product to the existing temperature conditions. As is the case with any heat-sensitive material, the time of exposure to the high-temperature conditions in the evaporator should be kept to a minimum. There is a distinct difference among the various evaporators discussed earlier in this chapter based on the exposure time,

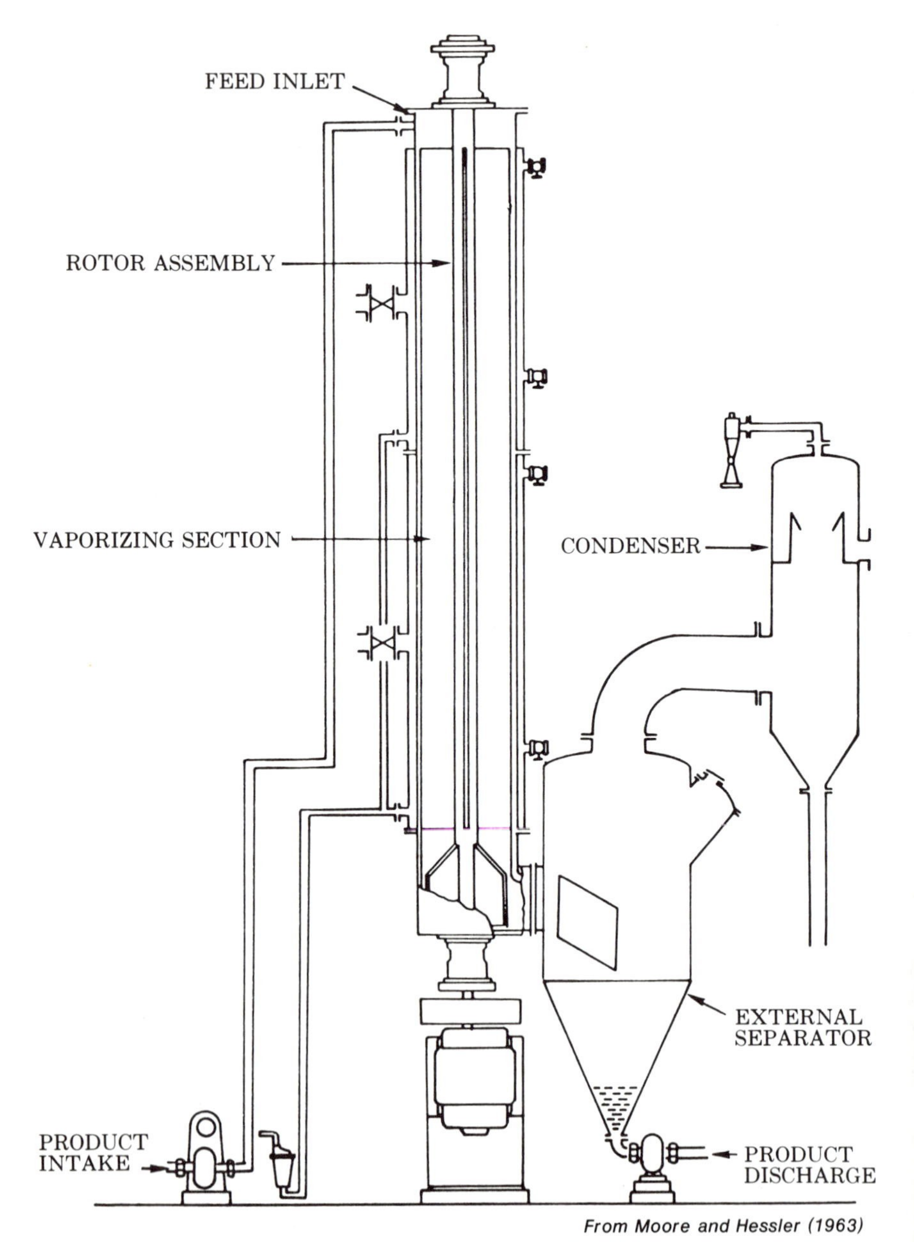


*From Moore and Hessler (1963)*

FIG. 5.7. SCHEMATIC CROSS-SECTION OF AGITATED FILM EVAPORATOR

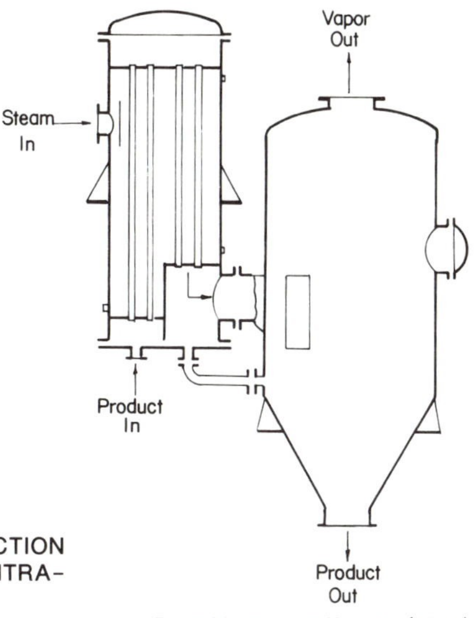


FIG. 5.8. SCHEMATIC CROSS-SECTION OF RISING-FALLING FILM CONCENTRATOR

*From Moore and Hessler (1963)*

which is directly related to the number of passes over the heat-transfer surface. Several of the evaporators discussed operate on the single-pass technique, while the remaining types require several passes of the product over the heat-transfer surface. In the case of multiple-pass evaporators, the product is introduced to the system in a continuous manner while the product which has been concentrated to the desired level is removed. This requires that the new unconcentrated product be introduced and mixed with the concentrated product on a continuous basis resulting in a relatively difficult requirement for describing retention time. Moore and Hessler (1963) used an approximation of the Taylor Series expansion [Danckwerts (1953)] to describe retention times as follows:

$$\% \text{ Removed} = [1 - \exp(-t/R)]\, 100 \tag{5.27}$$

where an assumption of ideal mixing of concentrated and unconcentrated product is required, and $R$ is equal to the ratio of evaporator volume to rate of product removal.

Equation (5.27) provides information on the time required for various percentages of the liquid to be removed from an evaporator body. This can be interpreted to indicate the exposure or retention time for various percentages of the product introduced into the evaporator. As would be expected, the times required for various percentage removals are significant, regardless of the value of $R$ in the expression. For the case of single-pass evaporators, the time will be considerably shorter. Moore and Hessler (1963) illustrated these differences by conducting experiments which used the pattern of a dye density at the evaporator discharge to evaluate the residence time of a finite amount of dye injected into the feed stream. The pattern of dye density (current, milliamperes) in three different multiple-pass evaporators is illustrated in Figure 5.9. The results indicate that retention characteristics of the Calandria, forced-circulation (*FC*), and rising-film (*VRC*) evaporator are very similar. Moore and Hessler (1963) showed that equation (5.27) can be used to predict retention time in multiple-pass evaporation.

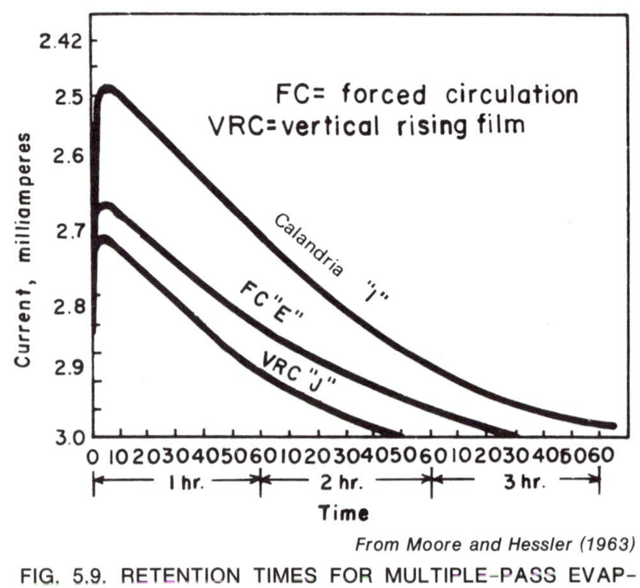


From Moore and Hessler (1963)

FIG. 5.9. RETENTION TIMES FOR MULTIPLE-PASS EVAPORATORS

The dye density pattern in three single-pass evaporators is illustrated in Figure 5.10. In general, the retention time in the single-pass systems is considerably shorter than in the multiple-pass equipment. The results in Figure 5.10 indicate small differences between a rising-falling film (*FRC*) system operated in two different ways and an agitated-film (Rotovak) system. Due to the rather significant differences between product retention for the two types of systems, considerable attention should be given to using single-pass evaporation systems with heat-sensitive liquid foods.

EXAMPLE 5.6 A multiple-pass continuous-type evaporator is being used to evaporate the moisture from 100 liters of product. The desired concentration allows the product to be removed at a rate of 10 liters/min. Compute the retention time for 10% of the product.

SOLUTION

(1) In equation (5.27)

$$R = \frac{100}{10} = 10 \text{ min}$$

(2) Using equation (5.27):

$90 = [1 - \exp(-t/R)]100$
$0.9 - 1 = -\exp(-t/R)$
$\ln(+0.1) = -t/10$
$t = 23.03$ min

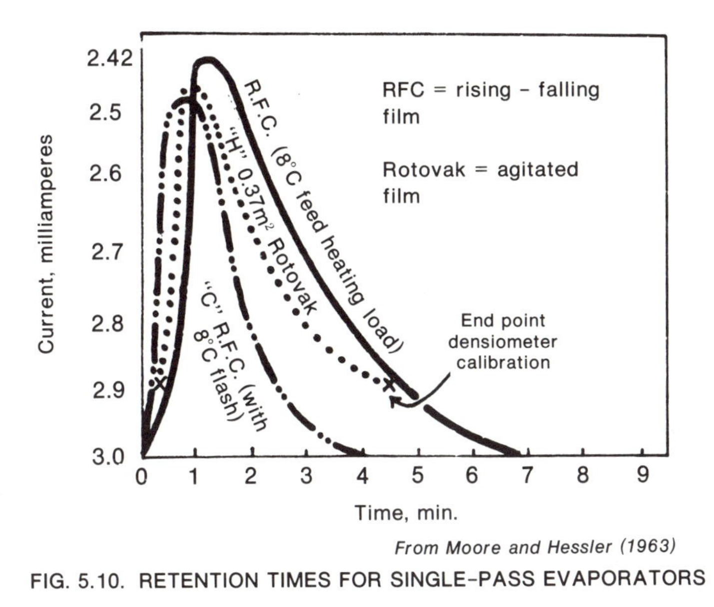


*From Moore and Hessler (1963)*

FIG. 5.10. RETENTION TIMES FOR SINGLE-PASS EVAPORATORS

## 5.3.2 Single-effect Systems

All the evaporators shown schematically in Figures 5.4 through 5.8 would be considered single-effect systems. Basically, this means that the product passes through one evaporator chamber and heat is supplied to one heat-transfer surface area. The design of such a system is relatively straightforward and can be accomplished by conducting mass and heat balances on the evaporator body. The mass balance results in the following two equations (see Figure 5.11):

$$F = V + P \tag{5.28}$$

$$X_F F = X_P P \tag{5.29}$$

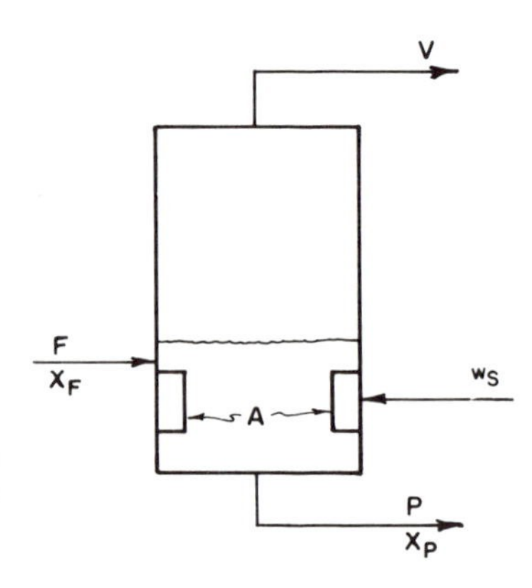


FIG. 5.11. SCHEMATIC ILLUSTRATION OF MATERIAL AND ENTHALPY BALANCE ON SINGLE-EFFECT EVAPORATOR

where equation (5.28) represents a total mass balance and equation (5.29) represents a mass balance on the product solids. An enthalpy balance conducted on the evaporation chamber shown in Figure 5.11 results in the following expression:

$$Fc_{pF}\,(T_F - 0) + w_s H_s = VH_v + Pc_{pP}\,(T_P - 0) + w_s H_C \qquad (5.30)$$

where a base temperature of 0°C is utilized to obtain the heat content values from standard steam tables. The last term on the right side of equation (5.30) introduces the factor which must be included to account for the heat leaving with the steam condensate. The heat added by cooling the condensate below the condensing temperature will be small and is neglected in most computations, as is heat loss from surfaces of the evaporator body. The expression which describes heat transfer from the heating medium to the product was introduced as equation (5.9). The following expression represents the heat which must be transferred to the product in order to accomplish the specified extent of evaporation:

$$q = w_s H_s - w_s H_C \qquad (5.31)$$

Equation (5.31) equates the total heat given up by the heating medium to the expression describing heat transfer and allows computation of the overall heat-transfer coefficient or the heat-transfer surface area if the quantity of heating medium required is known. A common expression used to describe the efficiency of the evaporation process is the steam economy presented by the following equation:

$$\text{Steam economy} = \frac{V}{w_s} \qquad (5.32)$$

which expresses the mass of water evaporated from the product per unit mass of steam utilized. A similar expression could be utilized if steam is not used as the heating medium. Equations (5.28) through (5.32) permit computation of most design parameters for a single-effect system and analysis of the evaporation process.

EXAMPLE 5.7 Orange juice with 11% total solids is being concentrated in a single-effect evaporator with a feed rate of 15,000 kg/hr at 20 C. The evaporator is being operated at sufficient vacuum to allow the product moisture to evaporate at 70 C while steam is being supplied at 198.5 kPa. The desired concentration of the final product is 50% total solids. Compute the steam requirements and steam economy for the process, when condensate is released at 70 C.

SOLUTION

(1) Using the mass balance equations (5.28) and (5.29):

$$15000 = V + P$$
$$0.11\ (15000) = 0.5\ P$$

Then:

$$P = 3300 \text{ kg/hr}$$
$$V = 15000 - 3300 = 11700 \text{ kg/hr}$$

(2) The use of equation (5.30) requires the input of several product properties as follows:

$c_{pF} = 3.822$ J/gK (from Table A.9) using a modification of equation (3.33)

$$c_{pF} = c_{pS}\ X_S + c_{pW}\ X_W$$

$$3.822 = c_{pS}\ (0.11) + 4.1865\ (0.89)$$

$$c_{pS} = 1.491 \text{ J/gK}$$

Then

$$c_{pP} = 1.491\ (0.5) + 4.1865\ (0.5)$$
$$= 2.839 \text{ J/gK}$$

$$\left.\begin{aligned} H_s &= 2706.3 \text{ kJ/kg} \\ H_v &= 2626.8 \text{ kJ/kg} \\ H_c &= 292.98 \text{ kJ/kg} \end{aligned}\right\} \quad \text{from Table A.6}$$

(3) By using equation (5.30) and a product boiling point of 70 C and assuming the condensate is related at saturated liquid conditions:

$$(15000)\ (3.89)\ (20 - 0) + w_s\ (2706.3) = (11700)\ (2626.8)$$
$$+ (3300)\ (2.839)\ (70 - 0) + w_s\ (292.98)$$

Then:

$$2706.3\ w_s - 292.98\ w_s = 30{,}733{,}560 + 655{,}809 - 1{,}167{,}000$$
$$2413.32\ w_s = 30{,}222{,}369$$
$$w_s = 12{,}523.15 \text{ kg/hr}$$

(4) The steam economy can be determined by using equation (5.32):

$$\text{Steam economy} = \frac{11700}{12{,}523.15} = 0.934 \text{ kg } H_20\text{/kg steam}$$

For single-effect continuous recirculating or multiple-pass systems, the changes in product during evaporation would normally result in changes in the overall heat-transfer coefficient and in the product temperature at which boiling occurs under the selected vacuum conditions. For design purposes, the parameters may be selected on the basis of the properties of the concentrated product leaving the evaporation system. This approach results in over-design of the equipment, but probably is least subject to error. In the design of continuous single-pass systems, the overall heat-transfer coefficient ($U$) and the boiling point of the product ($T_p$) will change as the product moves along the heat-transfer surface. As the product becomes more concentrated with distance along the heat-transfer surface, the overall heat-transfer coefficient will decrease, and for the same reason the boiling point elevation will increase. It should be possible to incorporate these design changes or changes in product properties into the design expressions.

EXAMPLE 5.8 Determine the surface area necessary for heat transfer in the evaporator described in Example 5.7. The overall heat-transfer coefficient is 1500 W/m² K.

SOLUTION

(1) Using equation (5.31):

$$q = 12{,}523.15\ (2706.3) - 12{,}523.15\ (292.98)$$

$$= 33{,}891{,}400 - 3{,}669{,}032 = 30{,}222{,}368 \text{ kJ/hr}$$

(2) Using equation (5.9):

$$30{,}222{,}368 = (1500)\ A\ (120 - 70)$$

$$A = \frac{(30{,}223{,}368 \text{ kJ/hr})(1000 \text{ J/kJ})}{(1500 \text{ W/m}^2\text{K})(50\text{K})(3600 \text{ s/hr})} = 111.94 \text{ m}^2$$

(3) The above results indicate that 112 m² of heat-transfer surface would be necessary to accomplish the desired concentration at the rate specified.

### 5.3.3 Multiple-effect Systems

The use of a typical single-effect system has the obvious disadvantage of releasing significant amounts of water vapor to the atmosphere surrounding this process. Most modifications to evaporation systems incorporate steps that make more efficient use of the heat release during condensation of this water vapor. The concepts of multiple-effect evaporation systems represent one approach to improving the efficiency of the process. A triple-effect evaporation system is shown schematically in Fig. 5.12. The system illustrated is a forward-feed system with both product feed and heating medium being introduced to the first effect. After partial concentration, the product moves to a second effect and the vapors produced by evaporation in the first effect are used as a heating medium in the second effect. After additional concentration in the

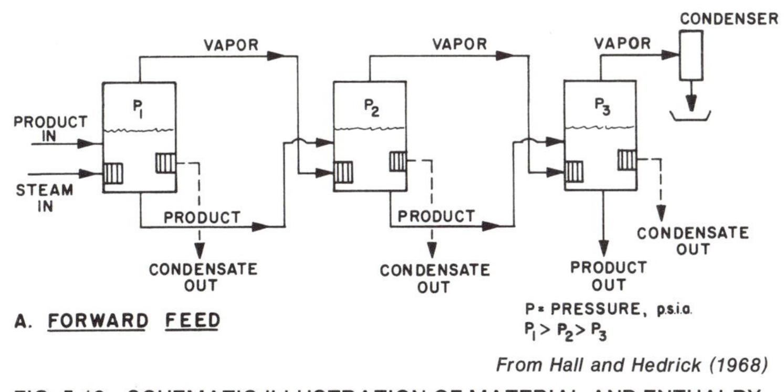


*From Hall and Hedrick (1968)*

FIG. 5.12. SCHEMATIC ILLUSTRATION OF MATERIAL AND ENTHALPY BALANCE ON FORWARD FEED TRIPLE-EFFECT EVAPORATOR

second effect, the product moves to a third effect and the vapors produced by evaporation in the second effect are utilized as the heating medium in the third effect. In summary, the total concentration desired is accomplished in three steps with the vapors produced in two of the effects being used as medium media in the following evaporation effect.

The design of a multiple-effect system becomes significantly more complex than outlined for a single-effect system. The normal recommended approach is to write mass and enthalpy balance equations for the entire system and for each effect.

The mass balance expressions for the forward-feed triple-effect system would be as follows:

$$F = V_1 + V_2 + V_3 + P \tag{5.33}$$

$$X_F F = X_P P \tag{5.34}$$

which describe changes in the product and in the product solids, respectively. An enthalpy balance can be written for each effect of the triple-effect system as follows:

$$Fc_{pF}\,(T_F - 0) + w_s H_s = V_1 H_v' + P' c_p' \,(T_1 - 0) + w_s H_C' \tag{5.35}$$

$$P' c_p' \,(T_1 - 0) + V_1 H_v' = V_2 H_v'' + P'' c_p'' \,(T_2 - 0) + V_1 H_C'' \tag{5.36}$$

$$P'' c_p'' \,(T_2 - 0) + V_2 H_v'' = V_3 H_v''' + P''' c_p''' \,(T_3 - 0) + V_2 H_C''' \tag{5.37}$$

Equations (5.35, 5.36, 5.37) have been written with 0 C as a base temperature and are not independent equations. The approach utilized, however, should allow for better understanding of the enthalpy balance which must be maintained on each effect of the system. For the case of triple-effect systems, the heat-transfer equations which describe the heat-transfer area must be written for each effect of the system as follows:

$$q_1 = U_1 A_1 (T_s - T_1) = w_s H_s - w_s H'_C \tag{5.38}$$

$$q_2 = U_2 A_2 (T_1 - T_2) = V_1 H'_v - V_1 H''_C \tag{5.39}$$

$$q_3 = U_3 A_3 (T_2 - T_3) = V_2 H''_v - V_2 H'''_C \tag{5.40}$$

In equations (5.38, 5.39, 5.40), the expressions for rate of heat transfer have been equated to the heat released by the heating medium in each effect of the triple-effect system. Equations (5.33) through (5.40) represent the necessary design equations for a triple evaporation system.

The objective of the design of a multiple-effect evaporation system is to evaluate the efficiency as expressed by steam economy and to compute the required heat-transfer surface area. The computation of steam economy as presented in equation (5.32) requires knowledge of the steam requirement or the heating-medium requirement for the system. This particular parameter must be computed by use of the mass and enthalpy balance equations. After computing the quantity of heating medium required, equations (5.38) through (5.40) can be utilized to compute the heat-transfer surface area after the heat-transfer coefficients for each effect are known or computed. Example computations will reveal that computing the steam requirement from equations (5.33) through (5.37) is difficult even when the feed capacity or rate is known and the desired concentration effect is established. The usual procedures for computations utilize assumptions, such as equal heat flux, equal heat-transfer area or equal temperature gradients in each effect of the system. The assumption of equal heat-transfer areas appears to be most logical, since each effect of a multiple-effect system would normally have similar construction. Other assumptions might include utilizing the boiling point of water at the evaporator pressure without accounting for boiling-point rise. In addition, changes in specific heat of product due to concentration may be ignored in the computation without significant error. The influence of product concentration on the overall heat-transfer coefficient should not be overlooked, and should be computed or estimated for each effect of a multiple-effect evaporator. Equations (5.25) and (5.26) can be used for this purpose if product property values are available.

EXAMPLE 5.9 Use a forward-feed triple-effect evaporator to accomplish the same product concentration for orange juice as presented in example 5.7. All product feeds, steam pressures and condensate temperatures are the same as in the previous example. The product boiling point of 70 C will be maintained in the third effect of the system. Compute the steam requirements and steam economy for this system under the conditions presented.

SOLUTION

(1) From the material balance equations (5.33) and (5.34):

$$15000 = V_1 + V_2 + V_3 + P$$

$$0.11\,(15000) = 0.5\,P$$

$$P = 3300 \text{ kg/hr}$$

$$\text{and } V_1 + V_2 + V_3 = 11700 \text{ kg/hr}$$

indicating the total amount of moisture evaporating.

(2) The overall heat transfer coefficients will decrease as the product becomes concentrated and have been estimated as follows:

$$U_1 = 1420 \text{ W/m}^2 \text{ K}$$
$$U_2 = 1050 \text{ W/m}^2 \text{ K}$$
$$U_3 = 650 \text{ W/m}^2 \text{ K}$$

(3) Since steam is being supplied at 198.5 kPa, the steam temperature is 120 C and the total temperature gradient will be 120 – 70 = 50 C and:

$$\Delta T_1 + \Delta T_2 + \Delta T_3 = 50.$$

(4) Using the assumption that $A_1 = A_2 = A_3$, then

$$\frac{q_1}{U_1(T_s - T_1)} = \frac{q_2}{U_2(T_1 - T_2)} = \frac{q_3}{U_3(T_2 - T_3)}$$

or:

$$\frac{(w_s H_s - w_s H_c')(1000 \text{ J/kJ})}{(1420)(T_s - T_1)(3600 \text{ s/hr})} = \frac{(V_1 H_v' - V_1 H_c'')(1000)}{(1050)(T_1 - T_2)(3600)}$$

$$= \frac{(V_2 H_v'' - V_2 H_c'')(1000)}{650\,(T_2 - T_3)(3600)}$$

(5) Using equation (5.35):

$$(15000)(3.8)(20 - 0) + w_s\,(2706.3) = V_1 H_v' + P' C_p'\,(T_1 - 0) + w_s\,H_c'$$

(6) Using equation (5.36):

$$P' C_p'\,(T_1 - 0) + V_1\,H_v' = V_2\,H_v'' + P'' C_p''\,(T_2 - 0) + V_1\,H_c''$$

(7) Using equation (5.37):

$$P'' C_p''\,(T_2 - 0) + V_2\,H_v'' = V_3\,(2626.8) + 3300\,(2.839)\,(70 - 0) + V_2\,(292.98)$$

(8) As an assumption for the first computation:

$$\Delta T = \Delta T_2 = \Delta T_3 = 16.7 \text{ C}$$

then:

$T_s$ = 120 C; $H_s$ = 2706.3; $H_c'$ = 292.98
$T_1$ = 103.3C; $H_v'$ = 2681.2; $H_c''$ = 292.98
$T_2$ = 86.5C; $H_v''$ = 2654.4; $H_c'''$ = 292.98
$T_3$ = 70C; $H_v'''$ = 2626.8

(9) Using step (4)

$$\frac{w_s(2706.3-292.98)(1000)}{1420(16.7)(3600)} = \frac{V_1(2681.2-292.98)(1000)}{1050(16.7)(3600)}$$

$$= \frac{V_2(2654.4-292.98)(1000)}{650(16.7)(3600)}$$

or: $V_1 = 0.7472\ w_s$
$V_2 = 0.4678\ w_s$

(10) Using step (5)

$$1{,}140{,}000 + 2706.3\ w_s = V_1\ (2681.2) + P'\ (3.54)(103.3) + w_s\ (292.98).$$

where $P' = 15000 - V_1$

then:

$$2706.3\ w_s - 292.98\ w_s = 2681.2\ (0.7472\ w_s) + 365.682\ (15000 - V_1) - 1{,}140{,}000$$

$$2413.32\ w_s - 2003.39\ w_s = 5{,}485{,}230 - 365.682\ (0.7472\ w_s) - 1{,}140{,}000$$

$$409.93\ w_s + 273.24\ w_s = 4{,}345{,}230$$

$$w_s = 6360.4\ \text{kg/hr}$$

and

$$V_1 = 4752.5\ \text{kg/hr}$$

$$V_2 = 2975.4\ \text{kg/hr}$$

$$V_3 = 3972.1\ \text{kg/hr}$$

(11) By using these values in the equation presented in step (7), with $P'' = 15000 - V_1 - V_2 = 7272.1$ kg/hr:

$$(7272.1)(3.19)(86.3) + (2975.4)(2654.4) = (3972.1)$$

$$(2626.8) + (3300)(2.839)(70) + (2975.4)\ 292.98$$

$$2{,}001{,}987{,}314 + 7897901.76 = 10433912.28 + 655809 + 871732.69$$

$$9{,}899{,}889{,}074 \neq 11{,}961{,}453.97$$

(12) The computations will be repeated using adjusted temperature gradients:

$$\Delta T_1 = 16.3\text{C};\ \Delta T_2 = 16.7\text{C};\ \Delta T_3 = 17\text{C}$$

then:

$$T_s = 120\text{C};\ H_s = 2706.3:\ H_c' = 292.98$$

$$T_1 = 103.7\text{C};\ H_v' = 2681.7;\ H_c'' = 292.98$$

$$T_2 = 87\text{C};\ H_v''' = 2655.3;\ H_c'' = 292.98$$

$$T_3 = 70\text{C};\ H_v''' = 2626.8$$

(13) Using the new property and temperature values in step (4):

$$\frac{w_s(2706.3-292.98)(1000)}{(1420)(16.3)(3600)}=\frac{V_1(2681.7-292.98)(1000)}{(1050)(16.7)(3600)}$$

$$=\frac{V_2(2655.3-292.98)(1000)}{(650)(17)(3600)}$$

then: $V_1 = 0.7654\ w_s$
$V_2 = 0.4877\ w_s$

(14) By repeating step (5):

$$1{,}140{,}000 + 2706.3\ w_s = V_1\ (2681.7) + P'\ (3.54)(103) + w_s\ (292.98)$$

where: $P' = 15000 - V_1$

then:

$$2706.3\ w_s - 292.98\ w_s = 2681.7\ (0.7654\ w_s) + 364.62\ (15000 - V_1) - 1{,}140{,}000$$

$$2413.43\ w_s - 2052.57\ w_s = 5{,}469{,}300 - 364.62\ (0.7654\ w_s) - 1{,}140{,}000$$

$$360.75\ w_s + 279.08\ w_s = 4{,}329{,}300$$

$$w_s = 6766.33\ \text{kg/hr}$$

then:

$$V_1 = 5178.95\ \text{kg/hr}$$

$$V_2 = 3299.94$$

$$V_3 = 3221.11$$

(15) These results can be checked by repeating step (7) with $P'' = 15000 - V_1 - V_2 = 6521.11$ kg/hr

then:

$$(6521.11)(3.19)(87) + (3299.94)(2655.3) = (3221.11)(2626.8) + 3300\ (2.839)(70) + (3299.94)\ (292.98)$$

$$1{,}809{,}803.66 + 8{,}762{,}330.68 = 8{,}461{,}211.75 + 655{,}809 + 966{,}816.42$$

$$10{,}572{,}134.34 = 10{,}083{,}837.17$$

which is acceptable agreement.

(16) Using the results obtained, the steam economy is:

$$\frac{11700\ \text{kg vapor/hr}}{6766.33\ \text{kg steam/hr}} = 1.73\ \text{kg vapor/kg steam}$$

In addition to the forward-feed multiple-effect evaporator shown in Fig. 5.12, backward-feed and parallel-feed evaporators may be utilized. Backward-feed and parallel-feed triple-effect evaporators are shown schematically in Figs. 5.13 and 5.14, respectively. In the backward-feed system, the product flows from the third effect toward the first effect as illustrated in Fig. 5.13. As in the forward-feed system, the vapors formed the first effect are used as a heating medium for the second effect, and the vapors from the second effect are used as a heating medium for the third effect. This requires that the product be pumped from the third effect to the second and from the second to the first effect due to the decreasing pressures as the vapors move from the first through the third effects. This means that the product entering the evaporator at a given feed temperature is exposed to low temperatures while the concentrated product in the first effect is exposed to the largest temperature gradient. These conditions are opposite those of the forward-feed system, and may result in more heat damage of the high-viscosity concentrated product due to the higher temperature gradient in the first effect.

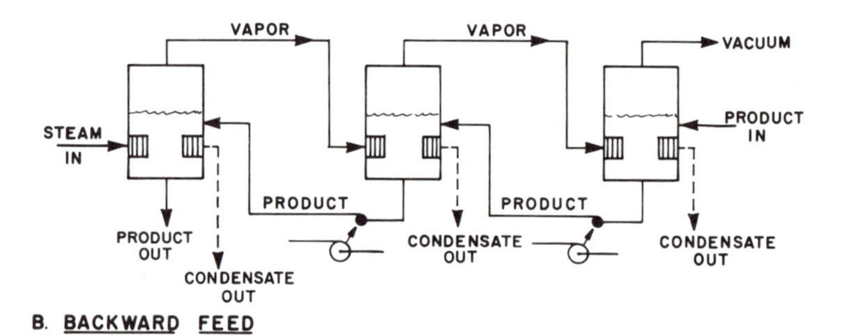


*From Hall and Hedrick (1968)*

FIG. 5.13. SCHEMATIC DIAGRAM OF BACKWARD FEED TRIPLE-EFFECT EVAPORATOR

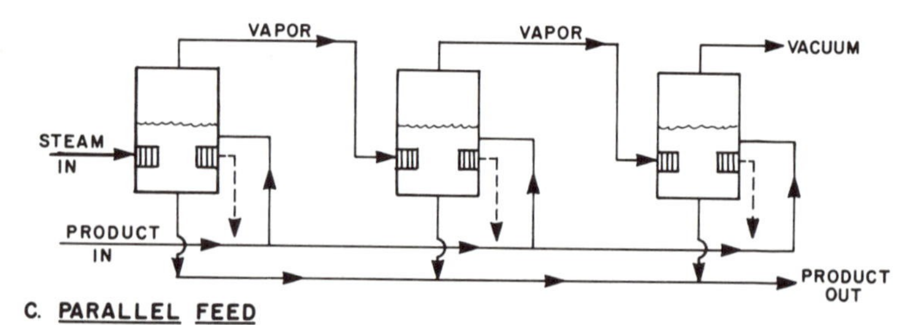


*From Hall and Hedrick (1968)*

FIG. 5.14. SCHEMATIC DIAGRAM OF PARALLEL FEED TRIPLE-EFFECT EVAPORATOR

For the case of a parallel-feed system, the heating medium is supplied to the first effect and the vapors produced in the first and second effects are utilized in the second and third effects, respectively. The feed is supplied separately and individually to each of the three effects and the concentrated product is removed from each of the three effects individually. This approach results in the total concentrating effect being the combined influence of all effects of the evaporator. The product which is concentrated in the first effect is exposed to a relatively large temperature gradient throughout the concentration process, while the product concentrated in the third effect is exposed to a much lower temperature gradient. The resulting product, which is a mixture of the product being removed from the three effects, probably has a composition similar to the product from the forward- or backward-feed systems.

As indicated earlier, one of the main advantages of the multiple-effect systems is to increase the efficiency or steam economy. This factor is illustrated in Fig. 5.15, where the steam economy is plotted versus the initial feed temperature. For normal feed temperatures, steam economies are between 2 and 2.5. This represents a doubling of the steam economy when compared to single-effect systems. A rather interesting influence of initial temperature is illustrated by Fig. 5.15 and the interrelationship to the types of triple-effect evaporator is illustrated. At low initial tem-

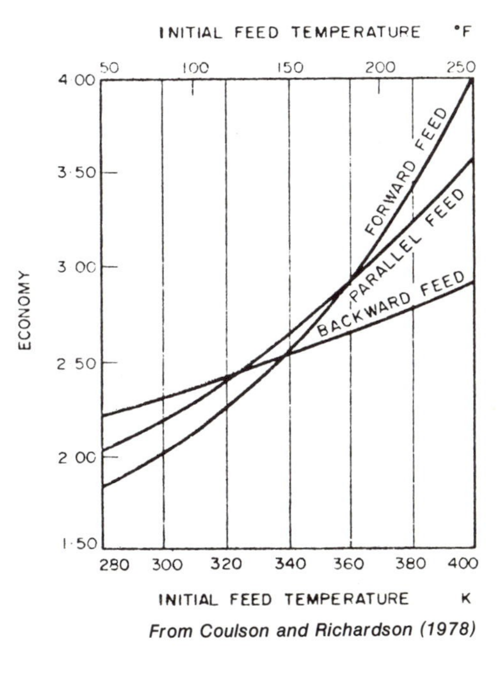


FIG. 5.15. ECONOMY OF TRIPLE EFFECT EVAPORATORS

*From Coulson and Richardson (1978)*

peratures the backward-feed system has the highest steam economy, while at higher feed temperatures the forward-feed evaporator has higher steam economy. An explanation of this interrelationship results from examining the heat content of the product entering the evaporator as compared to the temperature at which evaporation is occurring in that particular effect of the system.

EXAMPLE 5.10 Compute the surface area required for the triple-effect evaporator designed in Example 5.9.

SOLUTION

(1) Using the information presented in Example 5.9 along with the results of computations:

$$q_1 = U_1 A_1 \Delta T_1 = w_s H_s - w_s H_c'$$

then:

$$(1420)A_1\,(16.3) = (6766.33)\,(2706.3) - (6766.33)\,(292.98)$$

$$= 16{,}329{,}319.52$$

$$A_1 = \frac{(16{,}329{,}319.52)\,(1000)}{(1420)\,(16.3)\,(3600)} = 195.97\ \text{m}^2$$

(2) Since the heat transfer surface area is the same in all three effects:

$$A_1 = A_2 = A_3 = 196\ \text{m}^2$$

and the total heat transfer surface area will be $A_1 + A_2 + A_3 =$ 588 m$^2$

# 5.4 IMPROVING EVAPORATION EFFICIENCY

Although the use of multiple-effect evaporation has been the most popular approach to improving efficiency of evaporation systems, other approaches have become more visible. These alternative approaches may be as simple as preheating of product feed or as complex as the addition of a condensate flash tank. The more typical modifications or additions involve thermal recompression or mechanical recompression. Although these systems may be used with either single or multiple-effect evaporation, our discussion will be limited to illustration of their influence on efficiency of single-effect evaporation.

## 5.4.1 Thermal Recompression Systems

A thermal recompression system incorporates a steam ejector or thermocompressor into the evaporation system in order to reduce the steam requirements (Figure 5.16). This reduction is achieved by using a steam jet designed to mix low-pressure vapors leaving the evaporator with high-pressure steam. This process results in a low-pressure steam to

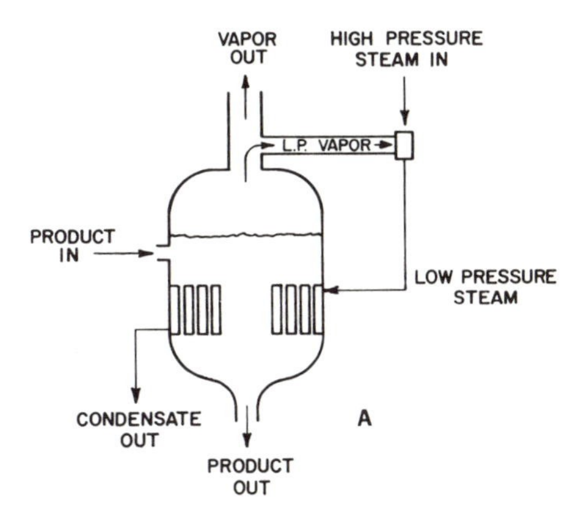


FIG. 5.16. THERMORECOMPRESSION EVAPORATOR UTILIZING STEAM JET

*From Hall and Hedrick (1968)*

be used as a heating medium for the evaporation system. In addition, the thermal recompression system provides an opportunity to utilize product vapors as a heating medium without using multiple-effect systems. In fact, the addition of thermal recompression provides an efficiency improvement equivalent to the addition of one effect in a multiple-effect system.

The addition of a steam ejector or thermocompressor to an evaporation system is based on several design considerations: (a) existing operating conditions, (b) the pressure of the motive steam, high-pressure steam source and (c) the influence of the thermocompressor on operating conditions and heat-transfer rates. A basic design consideration is the capacity ratio: the ratio of suction-vapor flow rate to motive-steam flow rate. Although most parts of the evaporation system design involve material and heat balances as previously described, the design must incorporate specific characteristics of the steam ejector as illustrated in Figure 5.17. These curves allow the determination of the capacity ratio for the thermocompressor from knowledge of the motive steam pressure and the compression ratio. This ratio of the suction pressure (product evaporation pressure) to the discharge pressure (heating-medium pressure in evaporator) cannot exceed 1.8 in order to use a steam ejector. From the capacity ratio, the flow rates of motive steam and product vapors used as the heating medium are computed.

EXAMPLE 5.11 A single-effect evaporation system with thermal recompression is being used to concentrate skim milk from an initial total solids content of 9.5% to 25% total solids. The product feed temperature is 5°C and evaporation is occurring at 57.83 kPa. The heating medium for the evaporator is 100°C while the steam supply is at 790.9 kPa. If the product feed rate is 8250 kg/hr, compute the improvement in efficiency achieved by the addition of the thermal recompression system.

SOLUTION

(1) Using the mass balance equations (5.28) and (5.29):

$$8250 = V + P$$

$$0.095\ (8250) = 0.25\ P$$

then:

$$P = 3135 \text{ kg/hr}$$

$$V = 8250 - 3135 = 5115 \text{ kg/hr}$$

(2) The use of equation (5.30) requires values for product-specific heat, before and after concentration. Using equation (3.32):

$$c_p = 4.1865\ (0.905) + 1.549\ (0.035) + 1.4234\ (0.051) + 2.093\ (0.001) + 0.8373\ (0.008)$$

$$c_p = 3.924 \text{ kJ/kgK for skim milk}$$

Using the same equation:

$$c_p = 4.1865\ (0.75) + 1.549\ (0.0921) + 1.4234\ (0.1342) + 2.093\ (0.0026) + 0.8373\ (0.0211)$$

(3) Additional property data for steam and vapor are obtained from Table A.6.:

$$H_s = 2676.1 \text{ kJ/kg for steam at } 100^\circ\text{C}$$

$$H_v = 2651.9 \text{ kJ/kg for vapor at 57.83 kPa } (85^\circ\text{C})$$

$$H_c = 419.04 \text{ kJ/kg for condensate at 100 C.}$$

(4) Using equation (5.30):

$$(8250)(3.924)(5 - 0) + w_s\ (2676.1) = (5115)(2651.9) + (3135)(3.4967)(85 - 0) + w_s\ (419.04)$$

then:

$$2676.1\ w_s - 419.04\ w_s = 13{,}564{,}468.5 + 931{,}783.13 - 161{,}865$$

$$2257.06\ w_s = 14334386.63$$

$$w_s = 6351 \text{ kg/hr}$$

(5) The steam requirement of 6351 kg/hr represents the amount for a single-effect evaporator without thermal recompression and indicates a steam economy of 0.805 kg water/kg steam.

(6) The use of the thermal recompression system requires:

$$\text{Compression Ratio} = \frac{101.35}{57.83} = 1.75$$

(7) Using Figure 5.17 with a discharge pressure of 101.35 kPa and a suction pressure of 57.83, the Capacity Ratio = 1.2.

(8) Since the Capacity Ratio represents the ratio of vapor flow to motive steam flow and the sum of these must equal 6351 kg/hr:

motive steam flow = 2886.8 kg/hr

vapor flow = 3464.2 kg/hr

(9) The motive steam flow represents the steam requirement for the evaporator with the thermal recompressor; the new steam economy is

$$\frac{5115 \text{ kg water/hr}}{2886.8 \text{ kg steam/hr}} = 1.77 \text{ kg water/kg steam}$$

(10) As indicated by the steam economy, the efficiency of the evaporator has been doubled by using the thermal recompression system. This improvement appears to be equivalent to a triple-effect evaporator.

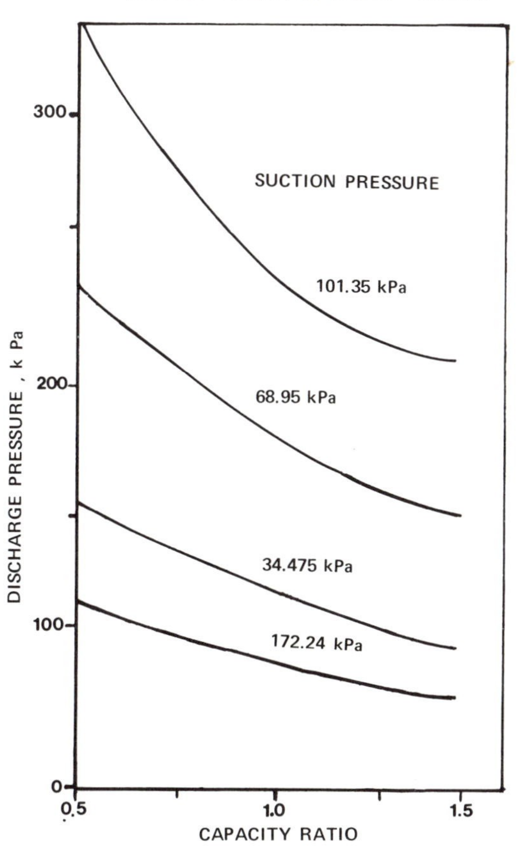


FIG. 5.17. CAPACITY RATIO CURVES FOR FIXED MULTIPLE NOZZLE THERMOCOMPRESSOR

In summary, it seems evident that thermal recompression provides a viable alternative for reducing steam requirements for food product concentration. The applications of this system are limited when "boiling-point-rise" is high, surface heat-transfer coefficients decrease significantly with increasing product concentration, the evaporator being used uses high-pressure steam, a low-pressure steam source acts as a heating medium or the steam condensate is returned to the boiler for make-up.

### 5.4.2 Mechanical Recompression Systems

By introducing mechanical recompression into the evaporation system, significant reductions in steam utilization can be achieved. These savings are achieved by installing a compressor to recycle all of the product vapors to be used as heating medium (Figure 5.18). The system in Figure 5.18 illustrates that some make-up steam might be required after the system is operating and the illustration describes the use of a heat exchanger for preheating product feed by using heat from the condensate leaving the heating coils of the evaporator.

The design considerations associated with mechanical recompression include computation of steam requirements and sizing of the compressor. The calculation of compression work requires the inlet and outlet enthalpy values for the compressor. This information can be obtained from

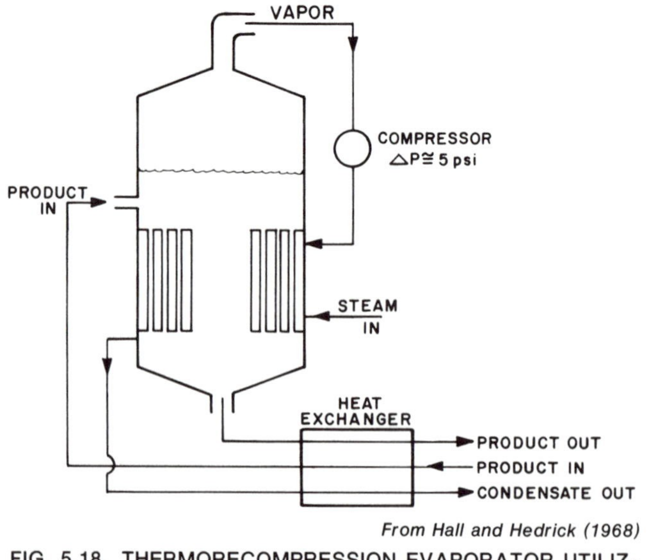


*From Hall and Hedrick (1968)*

FIG. 5.18. THERMORECOMPRESSION EVAPORATOR UTILIZING MECHANICAL COMPRESSION

steam tables (Appendix Table A.6) by recognizing that compression is theoretically a constant entropy process. By introducing factors to account for inefficiency in the compressor, estimates of the desired values can be obtained. The vapors are superheated during compression and desuperheaters are introduced to eliminate the influence of this factor on heat transfer in the evaporator.

EXAMPLE 5.12 Compute the efficiency improvement achieved by the use of a mechanical recompressor for conditions described in Example 5.11. Estimate the size of compressor required for the system.

SOLUTION

(1) Based on computations for Example 5.11, use of a single-effect system without efficiency measures has the following parameters:

$$P = 3135 \text{ kg/hr}$$

$$V = 5115 \text{ kg/hr}$$

$$w_s = 6351 \text{ kg/hr}$$

(2) The properties of vapor before and after compression can be obtained from Table A.6.:

$H_s$ = 2676.1 kJ/kg for steam at 100°C (101.35 kPa)

$H_v$ = 2651.9 kJ/kg for vapor at 57.83 kPa (85°C)

$H_c$ = 419.04 kJ/kg for condensate at 100°C

$S_v$ = 7.5445 kJ/kg for vapor at 57.83 kPa

The enthalpy after compression ($H_s'$) is obtained from tables at $S_v$ = 7.5445 kJ/kg and pressure of 101.35 kPa. By interpolation, an estimated value of $H_s'$ = 2749 kJ/kg is obtained. The amount of superheat is evident at the vapor temperature of approximately 136 C.

The change in enthalpy:

$$H_s' - H_v = 2749 - 2651.9 = 97.1 \text{ kJ/kg}$$

(3) By introducing a compression efficiency of 78%, the energy requirement for compression becomes:

$$\frac{97.1}{0.78} = 124.5 \text{ kJ/kg}$$

(4) The work of compression is the product of the energy for compression and vapor flow:

$$124.5 \text{ kJ/kg} \times 5115 \text{ kg/hr} = 636{,}817.5 \text{ kJ/kg}.$$

(5) The enthalpy balance for an evaporator with mechanical recompression is a modification of equation (5.30) to incorporate the energy for compression:

$$Fc_{pf}\,(T_F - 0) + w_s H_s + (H_s' - H_v) = Pc_{pp}\,(T_p - 0) + w_s H_c$$

then:

$$(8250)(3.924)(5 - 0) + w_s(2676.1) + 636{,}817.5 = (3135)(3.4967)(85 - 0) + w_s(419.04)$$

or:

$$2676.1\ w_s - 419.04\ w_s = 931{,}783.13 - 636{,}817.5 - 161{,}865$$

$$2257.06\ w_s = 133100.63$$

$$w_s = 58.97 \text{ kg/hr}$$

This result indicates that a small amount of make-up steam is required for the process. When expressed as steam economy:

$$\frac{5115 \text{ kg water/hr}}{58.97 \text{ kg/hr}} = 86.74 \text{ kg water/kg steam}$$

the efficiency improvement is dramatic. This calculation does not account for electrical energy required to drive the compressor.

In some cases, the calculation of steam requirement can be negative, indicating that not all of the compressed vapors are needed. In these situations, the excess vapors are removed before compression.

(6) The size of the compressor can be expressed in terms of volumetric flow rate. From Table A.6, the specific volume of the inlet vapor is 2.828 $m^3$/kg. Using this value:

$$5115 \text{ kg/hr} \times 2.828 \text{ m}^3\text{/kg} = 14465 \text{ m}^3\text{/hr}$$

or: 241 $m^3$/min vapor flow is obtained.

This volumetric flow rate indicates the capacity of the compressor required for the system.

In addition to volumetric flow rate several other factors must be considered in selection of a compressor for a mechanical recompression application. These factors include compression ratios, variability of evaporator flow rates and operating conditions, vapor temperature as well as expected installation and maintenance costs. It seems evident that mechanical recompression offers significant potential for evaporator efficiency improvement.

### 5.4.3 Low-temperature Evaporator

The evaporation of highly heat-sensitive materials must be accomplished at even lower temperatures than normally possible in the evaporators discussed up to this point. The low-temperature evaporator, utilizing a refrigeration cycle, is illustrated in Figure 5.19. This system provides conditions which allow evaporation of moisture from the product at temperatures between 25 and 38 C. Basically the system consists of a simple refrigeration cycle and appropriate heat exchangers to allow exposure of the product and condensate vapors to the appropriate conditions. The evaporator portion of the refrigeration system is utilized to condense the vapors which are being removed from the

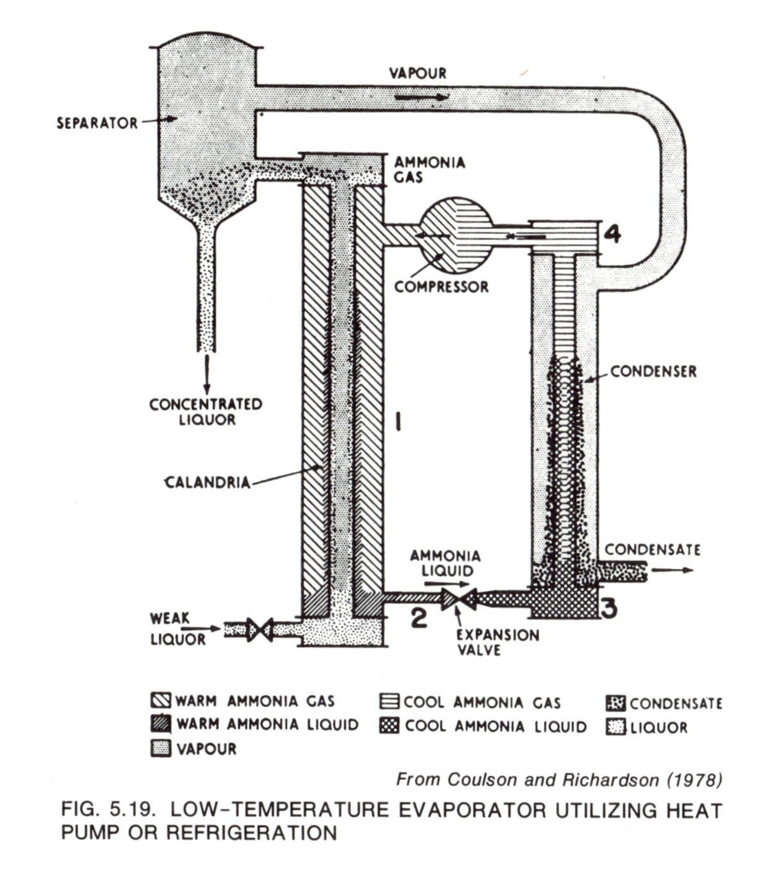


*From Coulson and Richardson (1978)*

FIG. 5.19. LOW-TEMPERATURE EVAPORATOR UTILIZING HEAT PUMP OR REFRIGERATION

product evaporator. The product evaporator is surrounded by the condenser portion of the refrigeration cycle. This refrigeration condenser provides the required heat to remove the vapors from the product. In most cases this evaporator would be a single-effect system so that product vapors will be removed from the evaporator portion to a separator, as illustrated in the evaporator portion of the refrigeration cycle. The refrigerant vapors are produced by heat exchange from the product vapors coming from the product evaporator, which results in condensation of the product vapors and production of the refrigerant vapors. The advantages of the low-temperature evaporation system are difficult to evaluate, but it is unlikely that such a system would be utilized unless the product being concentrated is highly heat-sensitive.

## PROBLEMS

5.1. *Heat transfer during evaporation.* Orange juice is being concentrated in a natural convection evaporator operating at 65C. The product is heated inside vertical tubes with 0.035 m inside diameter and 2 m length with the log mean velocity equal to 1.5 m/s. The tube walls have 0.05 cm thickness and are constructed with stainless steel. If the heating medium is steam at 169 kPa, compute the overall heat transfer coefficient. Properties of orange juice include $c_p$ = 3.5 kJ/kgK, $k$ = 0.7 W/mK, $\rho$ = 980 kg/m$^3$ and $\mu = 4 \times 10^{-4}$ kg/ms.

5.2. *Concentration in a single-effect evaporator.* Apple juice is being concentrated from an initial composition of 12.8% total solids to a final total solids content of 50.2% in a single-effect evaporator. Compute the feed rate and steam requirements for a product output of 2500 kg/hr. The steam supply is available at 232 kPa and the product feed is at 25C while the evaporator is being operated at 25 kPa.

5.3. *Concentration in a triple-effect evaporator.* A triple-effect evaporation system is being used to concentrate a sugar solution from 10% to 30% total solids. Each effect in the system is the same with 500 m$^2$ of heat-transfer surface area, and the overall heat-transfer coefficient in the first effect is 1000 W/m$^2$K. The steam supply is at 198.5 kPa and the pressure in the first effect is 38.50 kPa, while the condensate from the first effect is released at saturated liquid conditions. If the product feed is at 15C and the solution is concentrated to 18% total solids in the first effect, compute the operating feed rate.

5.4. *Concentration in a mechanical recompression system.* A mechanical recompression system is being used to concentrate a fruit juice extract from 15% to 25% total solids. The vapors from the evaporator, operating at 57.83 kPa, are pumped by the compressor with a volumetric flow rate of 250 m$^3$/min and the pressure is increased to 101.35 kPa. If the product feed temperature is 10C, compute the steam economy. Assume the condensate is released at saturated liquid conditions.

## COMPREHENSIVE PROBLEM V

*Design Calculations for Multiple-Effect Evaporation Systems*

### Objectives:

1. To become acquainted with procedures involved in design of multiple-effect evaporators for liquid foods.
2. To evaluate the influence of product feed temperature on steam economy and heat-transfer surface area for multiple-effect evaporators.

3. To investigate the relationship of product flow through a multiple-effect evaporation system to product feed rate as indicated by magnitudes of steam economy and heat-transfer surface area.

**Procedures**

A. The system being designed is for concentration of fruit juice which has an initial total solids content of 8 percent. Evaporation will be accomplished under vacuum corresponding to an absolute pressure of 70.14 kPa in the third effect. Saturated steam is available at 617.8 kPa absolute pressure. The product is being concentrated to 45 percent total solids. The specific heat of the feed is 3.77 kJ/kg C while the concentrated product has a specific heat of 3.35 kJ/kg C. The overall heat-transfer coefficients are: (a) First effect, $U_1$ = 1800, (b) Second effect, $U_2$ = 1300, and (c) Third effect, $U_3$ = 1000 W/m$^2$C.

The variations in the system to analyzed include:

a. Product feed temperatures of 25, 55, and 85C
b. Product feed rates of 0.75, 2.00 and 3.5 kg/s
c. Product flow patterns of forward, backward, and parallel flow.
d. The factors to be evaluated include concentrated product flow rate, steam consumption rate, steam economy and heat-transfer area.

B. Develop a computer program to accommodate the required analysis.
C. Discuss the following:
   1. Relationships between feed temperature and steam economy.
   2. Influence of product flow pattern on steam economy and heat-transfer surface area.
   3. Relationships between feed rate and steam economy.
   4. Influence of feed temperature and rate on heat-transfer surface area.

## NOMENCLATURE

$A$ = area, m$^2$.
$C$ = constant in equations (5.1), (5.2) and (5.7).
$C'$ = constant in equation (5.3).
$C_R$ = constant in equation (5.7).
$C_n$ = constant in equation (5.16)
$C_L$ = constant in equation (5.17)
$C_{sf}$ = constant in equation (5.18)
$c_p$ = specific heat; $c_p'$ = first effect, $c_p''$ = second effect, etc., kJ/kgK.
$D$ = diameter, m.
$F$ = feed, amount or rate, kg or kg/s.
$g$ = acceleration due to gravity, m/s$^2$.
$g_c$ = constant in equations (5.16) and (5.18).
$H$ = enthalpy or heat content; $H'$ = first effect, $H''$ = second effect,

etc., kJ/kg.
$h$ = convective heat transfer coefficient, $W/m^2K$.
$k$ = thermal conductivity, W/mK.
$L$ = length, m.
$L_v$ = latent heat of vaporization for water, kJ/kg.
$L'_v$ = latent heat of vaporization for product, kJ/kg.
$M$ = rate of condensate formation, kg/s.
$N_{Nu}$ = Nusselt number.
$P$ = product, amount or rate; $P'$ = first effect, $P''$ = second effect, etc., kg or kg/s.
$N_{Pr}$ = Prandtl number.
$p$ = vapor pressure of water, Pa.
$p'$ = vapor pressure of product, Pa.
$q$ = heat transfer rate, kJ/s
$R$ = retention constant in equation (5.25)
$R_g$ = gas constant, J/mole K.
$N_{Re}$ = Reynolds number.
$S$ = entropy, kJ/kg.
$T$ = temperature, C.
$T_A$ = absolute temperature, K.
$\Delta T_B$ = boiling point elevation, °C or °K.
$t$ = time, hr or min or sec.
$U$ = overall heat transfer coefficient, $W/m^2K$.
$u$ = velocity, m/s.
$u_m$ = log mean liquid-vapor velocity using velocities at inlet and outlet of tube, m/s.
$V$ = vapor, amount or rate, kg or kg/s.
$W$ = molecular weight.
$w_s$ = amount of steam, kg.
$X$ = mole fraction or mass fraction.
$X_{tt}$ = parameter defined by equation (5.25).
$x$ = thickness, m.
$y$ = mass fraction of vapor.
$\beta$ = angle between surface and bubble during boiling.
$\lambda'$ = constant defined in equation (5.21).
$\lambda_v$ = latent heat of vaporization per unit mole, kJ/mole.
$\mu$ = viscosity, kg/ms.
$\sigma$ = surface tension, N/m.
$\rho$ = density, $kg/m^3$.

## Subscripts

$A$ = solvent component of product (water).
$B$ = solute component of product solids.

$b$ = boiling process.
$C$ = condensate.
$c$ = convection.
$F$ = product feed.
$f$ = fluid or liquid state.
$m$ = average or mean.
$o$ = pure state.
$p$ = product.
$S$ = solid phase.
$s$ = heating medium (steam).
$v$ = vapor phase.
$W$ = water phase.
$w$ = wall condition.
1,2,3 = effect in multiple-effect system.

## BIBLIOGRAPHY

BLATT, T.S. and ADT, R.R. 1964. An experimental investigation of boiling heat transfer and pressure drop characteristics of Freon II and Freon 113 refrigerants. Am. Inst. Chem. Engrs. J. *10,* No. 3, 369.

BROMLEY, L.A. 1950. Heat transfer in stable film boiling. Chem. Eng. Progr. *46,* 221.

BROWN, G.G. 1950. Unit Operations. John Wiley & Sons, New York.

CHARM, S.E. 1978. The Fundamentals of Food Engineering, 3rd Edition. AVI Publishing Co., Westport, Conn.

COULSON, J.M. and MEHTA, R.R. 1953. Heat transfer coefficients in a climbing film evaporator. Inst. Chem. Engrs. *31,* 208.

COULSON, J.M. and RICHARDSON, J.F. 1978. *Chemical Engineering,* Vol. II, 3rd Edition. Pergamon Press, New York.

DANCKWERTS, P.V. 1953. Continuous flow systems. Chem. Eng. Sci. *2,* No. 1.

DUKLER, A.E. 1960. Fluid mechanics and heat transfer in vertical falling-film systems. Chem. Eng. Progr. Symp. Ser. *30.56,* 1–10.

ERDA Technology Applications Manual. 1977. Upgrading existing evaporators to reduce energy consumption. Technical Information Center. P.O. Box 62, Oak Ridge, Tennessee 37830.

FARBER, E.A. and SCORAH, R.L. 1948. Heat transfer to water boiling under pressure. Trans. Am. Soc. Mech. Engrs. *70,* 369.

HALL, C.W. and HEDRICK, T.I. 1968. Drying Milk and Milk Products, 2nd Edition. AVI Publishing Co., Westport, Conn.

KRIETH, F. 1965. Principles of Heat Transfer. International Textbook Co., Scranton, Pa.

KROLL, J.E. and McCUTCHEN, J.W. 1968. Heat transfer in an LTV falling film evaporator: A theoretical and experimental analysis. J. Heat Transfer *90C,* No. 2, 201.

LOCKHART, R.W. and MARTINELLI, R.C. 1948. Proposed correlation of data for two-phase, two-component flow in pipes. Chem. Engr. Prog. *45*:39–48.

McCABE, W.L. and SMITH, J.C. 1976. Unit Operations of Chemical Engineering. McGraw-Hill Book Co., New York.

MOORE, J.G. and HESSLER, W.E. 1963. Evaporation of heat-sensitive materials. Chem. Eng. Progr. *59,* No. 2, 87–92.

PIRET, E.L. and ISBIN, H.S. 1954. Natural circulation evaporation. Chem. Eng. Progr. *50,* No. 6, 305.

ROHSENOW, W.M. 1952. A method of correlating heat transfer data for surface boiling of liquids. Trans. Am. Soc. Mech. Engrs. *74,* 969.

SINEK, J.R. and YOUNG, E.H. 1962. Heat transfer in falling-film long-tube evaporators. Chem. Eng. Progr. *58,* No. 12, 74.

# 6

# Food Dehydration

In general, food dehydration is interpreted as the removal of moisture from a food product. In order to distinguish this process from evaporation as described in the previous chapter, additional specifications related to characteristics of the final product are usually required. There are many objectives of dehydration as applied to food products, probably the most evident being to preserve the product during prolonged storage. The dehydration process meets this objective by reducing the moisture content of the product to levels which are adequate to limit microbial growth or other reactions. In addition, reduction of the moisture content results in preservation of quality characteristics such as flavor and nutritive value. Another objective of dehydration is the significant reduction in product volume, which promotes efficiency in both transportation and storage of the important components of the food product. The somewhat less evident objective in dehydration is to provide or manufacture a product which is convenient to use—a factor which may not have reached its full potential as yet.

The removal of moisture to produce most dehydrated foods is accomplished by thermal dehydration, a process that utilizes heat to remove the moisture held in the product. The design of systems that will accomplish this is not possible without a thorough understanding of the complex changes that occur in the product during dehydration. Only by gaining this understanding will more efficient drying methods be developed, in addition to processes that will maintain the optimum quality of the product. The dehydration process incorporates many of the unit operations or processes which have been discussed previously. First of all, heat transfer is involved in moving the heat from the heating medium to the point at which evaporation occurs. Following evaporation of the moisture, the vapors produced must be transported through the product structure to the surrounding medium. This process may involve fluid flow, and in some cases liquid must be transported through the structure during the dehydration process. Due to the complex and varying structures of food products these processes are accomplished differently in almost any food product considered. It is possible, therefore, to present only the basic principles of processes which may occur during dehydration and the process which must be applied to the specific product.

Analysis of the unit operation of drying offers a considerable challenge to an engineer since many of the dryers in use today were invented for a specific product or application. In most cases, the design of these dryers was based on little theory. In addition, both heat and mass transfer play an equally important role in the drying process, adding to the complexities. Recent advances in computer simulation have been helpful in the analysis of drying systems. In this chapter attempts will be made to describe the more commonly used dryers in the food industry. Mathematical treatment useful in the analysis and design of drying systems will be presented.

Several approaches have been used in classifying drying processes. Porter *et al.* (1973) has used nineteen types of dryers that may be used to handle eight types of materials such as for liquids, slurries, pastes and sludges, free-flowing powders, granular, crystalline or fibrous solids, special forms, continuous sheets and discontinuous sheets.

A more recent classification is done by Dittman (1977) which involves two general classes and five subclasses. This classification is shown in Figure 6.1. The two general classes are for adiabatic processes and non-adiabatic processes. The adiabatic dryers involve a drying gas that provides the heat of vaporization and later carries the vapor away from the product. The solid surface will undergo a decrease in temperature to the adiabatic saturation temperature of the pure liquid. This temperature will remain constant until the solid has no free water surface. The non-adiabatic dryers involve heat flows into the solid from sources other than the drying gas, e.g., radiation or conduction by contact with a surface. The temperature of the solid surfaces exposed to the heat source increases. The moisture vapors are removed either by vacuum or sweeping the drying product with a purge gas.

Further distinction among adiabatic dryers is made on whether the drying gas passes through the material or across its surface. For non-adiabatic dryers the distinction is based on whether heat is applied through a heat transfer area or direct radiation and whether moisture is carried by vacuum or purge gas.

Thus, it is clear that these processes involve a variety of mechanisms for heat transfer to the product and within the product as well as movement of moisture from the product. Prior to examining selected examples of drying systems it is important to understand the basic principles of dehydration.

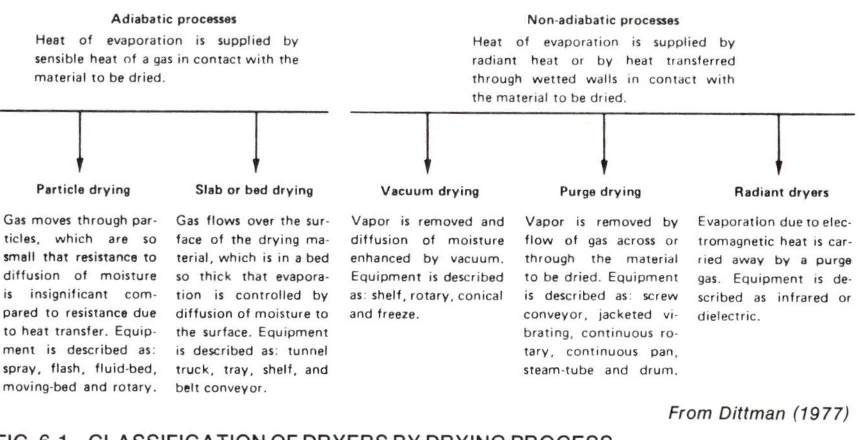


FIG. 6.1. CLASSIFICATION OF DRYERS BY DRYING PROCESS

# 6.1 BASIC PRINCIPLES OF DEHYDRATION

One of the basic considerations in the dehydration of food products is the role of water in the product. Rockland (1969), while discussing the chemical, physical, and thermodynamic properties of bound water, indicated that three types of bound water may exist in food products, namely: (a) water molecules which are bound to ionic groups such as carboxyl and amino groups, (b) water molecules which are hydrogen-bonded to hydroxyl and amide groups, and (c) unbound free water found in interstitial pores in which capillary forces and soluble constituents cause lowering of vapor pressure. The water is removed with a certain degree of difficulty during the dehydration process, depending on the classification within which it falls. The above types of water are listed in order of decreasing difficulty of moisture removal. In addition, they are listed in reverse order in which moisture would be removed as the moisture content of the product is decreased. This means that the free water would be evaporated and removed initially, followed by the water molecules that are hydrogen bonded and finally, the water which is bound to ionic groups. The influence of this information relates to the energy requirements for moisture removal in that each of the types of water listed will probably require different amounts of energy for removal. Since the energy requirements for moisture removal will vary with the product as well as type of water, the design of the dehydration equipment must account for as many factors as possible. The efficiency of the equipment designed will be closely related to the extent to which moisture removal is desired. In addition, the different mechanisms of water bonds with the solid influence of the food quality characteristics during storage. These relationships will be discussed later in section 6.2.3.

### 6.1.1 Psychrometrics

The study of the thermodynamic relationships between water vapor and air is referred to as psychrometrics. These relationships are very basic and important to food dehydration because of the role of the forces involved in removing moisture from the product.

The terminology used in psychrometrics is somewhat unique and must be presented before an adequate explanation of the processes can be provided.

The partial pressure ($p$) of a component can be defined as the pressure that the component would exert if it was completely isolated. Both dry air and water vapor will exert partial pressures in air-vapor mixtures.

Relative humidity ($H_R$) is the ratio of the amount of vapor in a gaseous mixture to the maximum amount of vapor which the mixture can hold. In addition, relative humidity is the ratio of the partial vapor pressure ($p_v$) to the saturation vapor pressure ($p_s$) for a vapor-air mixture.

The absolute humidity ($H_A$) is the ratio of the mass of water vapor per unit mass of dry air. The expression which relates absolute humidity to vapor pressure and relative humidity can be derived directly from the ideal gas law as follows:

$$pV = nR_gT_A = \frac{m}{M}R_gT_A \tag{6.1}$$

which states that the product of the pressure ($p$) and volume ($V$) of an ideal gas must be equal to the product of the number of molecules ($n$), the gas constant ($R_g$) and absolute temperature ($T_A$). By assuming that the ideal gas law will describe both dry air and water vapor, the following expression can be obtained for a constant volume and constant temperature:

$$\frac{m_v}{m_a} = \frac{p_vM_v}{p_aM_a} = H_A \tag{6.2}$$

By introducing the values for the molecular weights of water vapor, 18, and air, 29, into equation (6.2), the following expression is obtained:

$$H_A = \frac{18}{29}\frac{p_v}{P - p_v} = 0.622\frac{p_v}{P - p_v} \tag{6.3}$$

where the partial pressure is equated to the difference between total pressure ($P$) and partial pressure of vapor ($p_v$). In addition, when the definition of relative humidity is introducted into equation (6.3), the following expression is obtained:

$$H_A = 0.622\frac{H_Rp_s}{P - H_Rp_s} \tag{6.4}$$

which provides the desire relationship between absolute humidity, relative humidity, saturation vapor pressure and total pressure. Equation (6.4), along with additional equations which express the heat content of air-water vapor mixtures, can be utilized to construct a psychrometric chart such as presented in Appendix Figure A.5.

The psychrometric chart is a plot of absolute humidity along the vertical axis versus the dry-bulb temperature on the horizontal axis. The curved upper axis represents the saturation point and gives the wet-bulb temperatures that can be utilized in the computation of relative humidity. The curves which appear parallel to the saturation curve represent lines of constant relative humidity. Another factor of considerable importance in psychrometrics is the heat content or enthalpy at saturation. The only other value that can be obtained from the chart shown is the volume of the mixture per kilogram of dry air.

There are several processes which can be described by observing changes on the psychrometric chart. These include sensible heating or cooling, described as a process occurring along constant absolute humidity lines or parallel to the dry-bulb temperature axis. Evaporative cooling or chemical dehydration processes occur at a constant wet-bulb temperature. Cooling and dehumidifying or heating and humidifying processes are somewhat more difficult to describe on the psychrometric chart, but they can be evaluated by appropriate computation of final initial conditions of the air. The mixing of air with different moisture conditions can also be evaluated by utilizing the chart. The final conditions of the mixture must fall on a line drawn between the points describing the conditions of the two air volumes being mixed. The location along the line describing the mixture depends on the proportion of the two volumes being mixed. Probably the most important process which occurs during most drying operations is the evaporative cooling which occurs during constant-rate drying. This process is recognized by the addition of moisture to the drying air at the constant wet-bulb temperature.

### 6.1.2 Rate-of-Drying Curve

The rate-of-drying curve is obtained from data on moisture contents obtained by exposing a moist sample to a stream of air. The sample is usually suspended in a cabinet or a duct. A stream of air of constant temperature, humidity, velocity and direction of flow across the drying surface is used to dry the sample. The weight of the sample is recorded continuously as a function of time. These data allow one to calculate the moisture content dry basis, $w$. Next, free moisture content, $w$-$w_e$, is computed where $w_e$ is the equilibrium moisture content. A plot of free moisture content vs. time is shown in Fig. 6.2. The drying curve shown in this figure includes various periods of drying. First, the moisture is removed by evaporation from the saturated surface, next the area of saturated surface gradually decreases, followed by water evaporation in the interior parts of the sample.

The free moisture content vs. time curve can be expressed as a rate curve as shown in Fig. 6.3. The rate curve clearly shows the presence of a warming-up period AB, a constant-rate period BC, and a falling-rate period CE. The moisture content at point C, where the rate changes from constant rate to falling rate, is termed critical moisture content. The critical moisture content depends on several factors which are characteristic of the product being dried. The falling rate period is made up of two parts, first falling-rate period CD and second falling-rate period DE.

**6.1.2a Constant-rate period dehydration.**—During the constant-rate drying period, the rate of moisture removal from the product is limited only by the rate of evaporation from water surfaces on or within the pro-

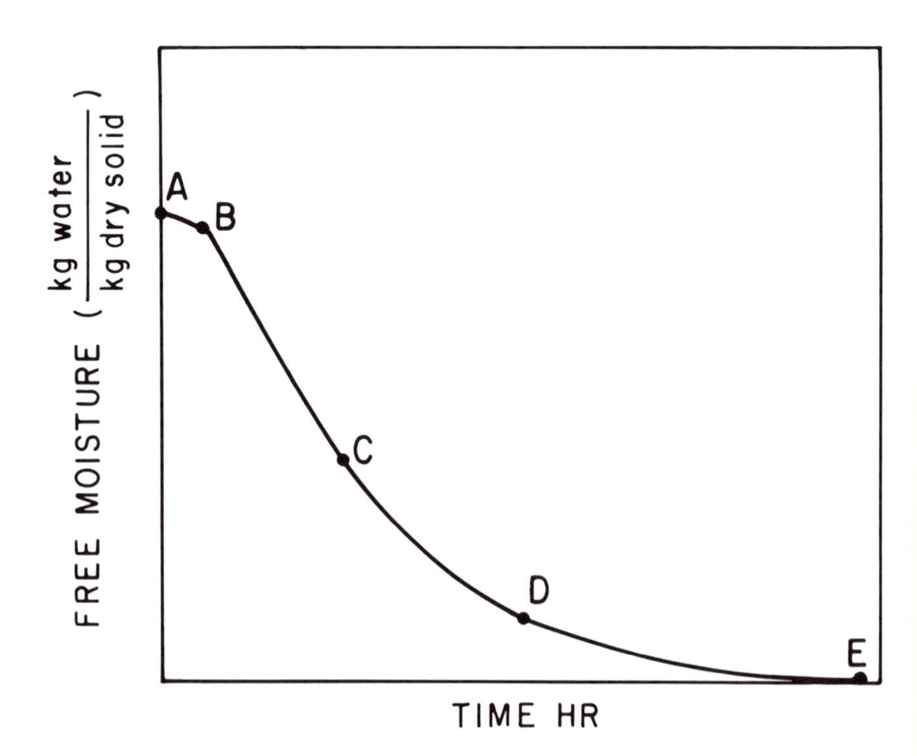


FIG. 6.2. LOSS OF FREE MOISTURE WITH TIME FOR A SOLID

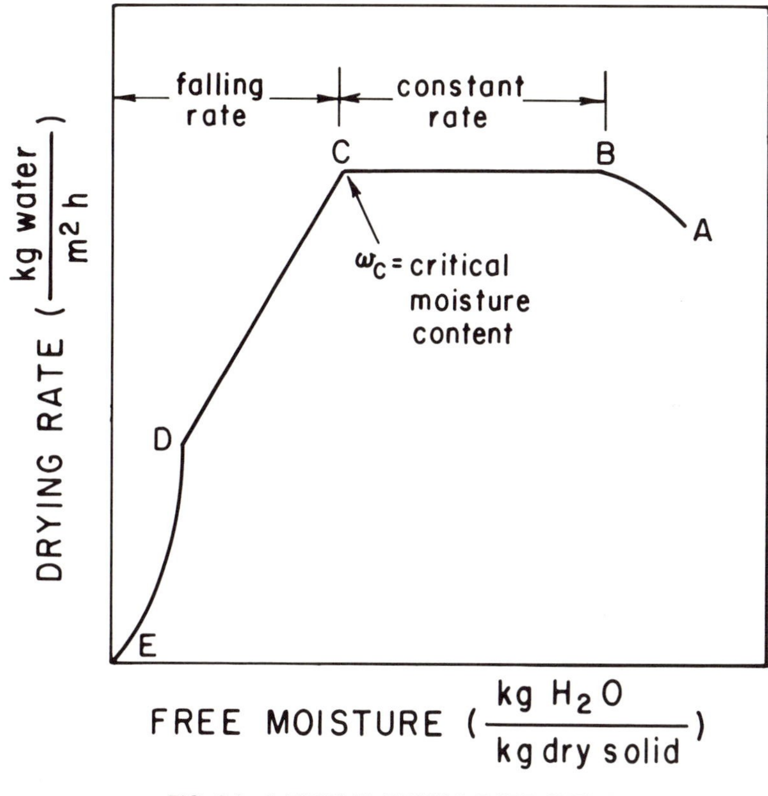

FIG. 6.3. A TYPICAL DRYING-RATE CURVE

duct. This drying rate will continue as long as the migration of moisture to the surfaces at which evaporation is occurring is more rapid than the evaporation taking place at the surface.

The rate at which evaporation occurs at a water surface is dependent on two factors illustrated in the following equation:

$$\frac{dw}{dt} = \frac{hA(T_a - T_w)}{L} = k_m A(H_w - H_A) \qquad (6.5)$$

where $h$ is a heat-transfer coefficient describing conditions which exist at the surface, $k_m$ is the mass-transfer coefficient describing moisture transfer to the surrounding air, $T_a$ is air dry bulb temperature, $T_w$ is the wet-bulb temperature, and $H_w$ is absolute humidity at wet-bulb conditions.

Based on equation (6.5), evaporation rate may be limited either by low heat-transfer rate from the air to the water surface or by low mass-transfer rate for moisture moving from the water surface to the air. The heat-transfer coefficient in equation (6.5) will most frequently be a convective heat-transfer coefficient and can be predicted by expressions presented in Chapter 3. In some situations, this heat-transfer coefficient may be radiation-dependent and will require other expressions presented in Chapter 3.

The mass-transfer coefficient ($k_m$) is not easily measured for conditions which exist during constant-rate drying. A commonly used approximation involves the use of the Lewis number defined as:

$$N_{Le} = \frac{h}{k_m\, c_p} \tag{6.6}$$

which normally applies to conditions existing when heat and mass transfer are influenced by flow conditions in the same manner. These conditions may exist during constant-rate drying, and the Lewis number is approximately unity. It follows that $k_m \simeq 0.8\, h$ since $c_p = 1.21$ kJ/kg C for air.

**6.1.2b Falling-rate period dehydration.**—After reaching a critical moisture content, the drying process proceeds at a decreasing rate. In most cases, this is brought about by the decreased rate at which evaporation occurs. After reaching the critical moisture content, the drying rate may decrease linearly with decreasing moisture content for the remaining portion of the process. In some products, there may be more than one falling-rate period, and a second or possibly a third period of decreasing drying rate with decreasing moisture content will be evident.

In the first falling-rate period the saturated surface area decreases as the moisture movement within the solid can no longer supply enough moisture. The drying rate decreases as unsaturated surface area increases. The factors that influence the drying rate include those that effect the moisture movement away from the solid in addition to the rate of internal moisture movement. When the entire surface area reaches the unsaturated state the internal moisture movement becomes the controlling factor.

Gorling (1958) has discussed the various mechanisms which may result in moisture migration within a product. These include (a) liquid movement by capillary forces, (b) diffusion of liquids, (c) surface diffusion, and (d) water-vapor diffusion.

## 6.2 ESTIMATION OF DRYING TIME

Most calculations involving the design and analysis of dryers require estimating the length of time needed to dry a product from some initial moisture content $w_1$ to a final moisture content $w_2$. In this section several approaches will be discussed to estimate drying time for both constant-rate and falling-rate periods. It should be emphasized that the knowledge of the complex drying mechanisms in food products is still imperfect. Thus, theoretically derived expressions are useful only in obtaining rough estimates. Experimental trials to predict drying times are most appropriate where possible.

### 6.2.1 Experimental Approaches to Predict Drying Time

**6.2.1a Constant Rate Period.**—A batch of moist material is dried under conditions as similar as possible to the expected conditions in a commercial dryer. These conditions include using relatively same surface area, temperature, relative humidity and velocity of air. The weight change with time is continuously recorded. From these data, a curve exhibiting change of moisture content with time is plotted as shown previously in Fig. 6.2. This curve can be used directly to estimate time of drying within the constant rate period.

The experimental drying curve can be transformed into a drying rate curve as shown earlier in Fig. 6.3. The drying rate in constant rate period, $N_c$, can be expressed as:

$$N_c = -\frac{m}{A}\frac{dw}{dt} \tag{6.7}$$

The above equation can be integrated to obtain time of drying:

$$\int_0^{t_c} dt = \frac{m}{A}\int_{w_2}^{w_1} \frac{dw}{N_c} \tag{6.8}$$

where both $w_1$ and $w_2$ are within the constant-rate drying period. Thus, the drying time constant-rate period is:

$$t_c = \frac{m}{AN_c}(w_1 - w_2) \tag{6.9}$$

**6.2.1b Falling-Rate Period.**—The experimentally obtained drying rate curve can be again used to estimate time of drying. From equation (6.7), for falling-rate period:

$$t_F = \frac{m}{A}\int_{w_2}^{w_1} \frac{dw}{N_F} \tag{6.10}$$

where initial moisture $w_1$ is below the critical moisture content.

From the experimental data, a plot of $\frac{1}{N_F}$ vs. $w$ is drawn. The area under the curve is determined to calculate the rate of drying.

EXAMPLE 6.1 In drying a formulated food the following drying-rate data were obtained during the falling-rate period.

| $w$ | $N_F$ | $w$ | $N_F$ |
|---|---|---|---|
| 0.6 | 2.25 | 0.45 | 1.1 |
| 0.55 | 1.8 | 0.4 | 0.8 |
| 0.5 | 1.4 | 0.38 | 0.6 |

If the equilibrium moisture content $w_e$ = 0.35 kg water/kg dry solid, total mass of product $m$ = 10 kg and top surface area = 1 m$^2$, calculate the time required to dry the product from 0.6 kg water/kg dry solid to 0.38 kg water/kg dry solid.

SOLUTION

From the given information the following table can be constructed:

| $w$-$w_e$ | $N_F$ | $1/N_F$ | $w$-$w_e$ | $N_F$ | $1/N_F$ |
|---|---|---|---|---|---|
| 0.25 | 2.25 | 0.44 | 0.10 | 1.1 | 0.91 |
| 0.20 | 1.8 | 0.56 | 0.05 | 0.8 | 1.25 |
| 0.15 | 1.4 | 0.71 | 0.03 | 0.6 | 1.67 |

From equation (6.10):

$$t_F = \frac{m}{A} \int_{w_2}^{w_1} \frac{dw}{N_F}$$

$\int_{w_2}^{w_1} \frac{dw}{N_F}$ can be evaluated by graphical integration from a plot of $1/N_F$ and $w$ or using a desk-top computer with a packaged subroutine on numerical integration:

$$\int_{0.25}^{0.03} \frac{dw}{N_F} = 0.177$$

$$t_F = \frac{10}{1} \times 0.177$$

$$t_F = 1.77 \text{ hour}$$

## 6.2.2 Theoretical Expressions Useful to Predict Drying Time:

**6.2.2a Constant-rate Period.**—Equation (6.5) can be modified for applications in drying calculations for constant-rate period. For a case of evaporation from a tray of moist material, assuming no changes in the volume,

$$\frac{dw}{dt} = \frac{h}{\rho_s L y}(T_a - T_w) \tag{6.11}$$

where $\rho_s$ is the bulk density dry material, kg/m$^3$; $y$ is the thickness of bed, m; $h$ is total heat-transfer coefficient, W/m$^2$ C; and $L$ is latent heat of vaporization, kJ/kg.

For a through-circulation type of drying, equation (6.5) can be modified to give:

$$\frac{dw}{dt} = \frac{ha}{\rho_s L}(T_a - T_w) \tag{6.12}$$

where $a$ is heat-transfer area $m^2/m^3$ of bed. The value of '$a$' can be estimated from the following expressions.

For a packed bed of spherical particulates of diameter $d_s$:

$$a = \frac{6(1-e)}{d_s} \tag{6.13}$$

For a packed bed of cylindrical particulates of diameter $d_c$ and length $l$

$$a = \frac{4(1-e)(l + 0.5\, d_c)}{d_c} \tag{6.14}$$

**6.2.2b Diffusion-controlled Falling-rate Period.**—A governing equation that expresses liquid diffusion in a solid can be written as:

$$\frac{\partial C}{\partial t} = D\left[\frac{\partial^2 C}{\partial r^2} + \frac{j}{r}\frac{\partial C}{\partial r}\right] \tag{6.15}$$

where $j$ is equal to 0 for an infinite slab, 1 for an infinite cylinder and 2 for a sphere.

This governing equation for a sphere can be solved with the following boundary and initial conditions:

$$\frac{\partial C}{\partial r} = 0 \qquad r = 0 \qquad t \geq 0$$

$$C = C_e \qquad r = R_1 \qquad t > 0$$

$$C = C_o \qquad 0 \leq r \leq R_1 \qquad t = 0$$

The first boundary condition specifies a finite concentration at the center of the sphere. The second boundary condition means that the surface is at equilibrium moisture content and the initial concentration is uniform throughout the object according to the initial condition.

For spherical geometry, equation (6.15) with $j = 2$ can be solved using separation of variables to obtain the following relationship.

$$\frac{C - C_e}{C_o - C_e} = \frac{2R}{\pi r}\sum_{n=1}^{\infty}\frac{(-1)^n}{n}\sin\left(\frac{n\pi r}{R}\right)\exp\left(\frac{-Dn^2\pi^2 t}{R^2}\right) \tag{6.16}$$

The above equation can be integrated over radius to get moisture removal as a function of time.

$$\frac{w_o - w}{w_o - w_e} = 1 - \frac{6}{\pi^2}\sum_{n=1}^{\infty}\frac{1}{n^2}\exp\left(\frac{-Dn^2\pi^2 t}{R^2}\right) \tag{6.17}$$

For long drying times the second and higher terms of the series can be neglected; thus,

$$\frac{w_o - w}{w_o - w_e} = 1 - \frac{6}{\pi^2} \exp\left(\frac{-D\pi^2 t}{R^2}\right) \tag{6.18}$$

Equation (6.18) can be solved for time of drying:

$$t = \frac{R^2}{D^2\pi^2} \ln\left(\frac{6}{\pi^2} \frac{(w_o - w_e)}{(w - w_e)}\right) \tag{6.19}$$

Similar equation for a slab is:

$$t = \frac{4\delta^2}{\pi^2 D} \ln\left(\frac{8}{\pi^2} \frac{(w_o - w_e)}{(w - w_e)}\right) \tag{6.20}$$

and for an infinite cylinder of radius $r$,

$$t = \frac{r^2}{5.78\,D} \ln\left(\frac{0.692\,(w_o - w_e)}{(w - w_e)}\right) \tag{6.21}$$

Most of the earlier investigations on drying have assumed that diffusion is the primary mechanism involved in movement of moisture to the surface for evaporation. They have illustrated, in most cases, that the diffusion coefficient ($D$) will be dependent on moisture content. Jason (1958) and Fish (1958) have shown, however, that for fish and for starch jells the diffusion coefficient is essentially constant until the moisture content (dry basis) has decreased to 15%. At moisture contents below 15%, a second coefficient is required to describe the moisture removal. The products represent just two examples, and it is very likely that other products will respond somewhat differently. It must be emphasized that diffusion coefficients ($D$) for vapor are not available for most food products.

**6.2.2c Additional Expressions to Predict Falling-rate Period.**—An equation similar to equation (6.5) can be used to express falling-rate period (Porter *et al.*, 1973). Thus,

$$\frac{dw}{dt} = -K_1(w - w_e) \tag{6.22}$$

where $K_1$ is a function of the constant rate

$$K_1 = -\frac{(dw/dt)_c}{w_c - w_e} \tag{6.23}$$

For evaporation from a moist material in a tray, from equation (6.11) substituting the value for $\left(\frac{dw}{dt}\right)_c$

$$K_1 = \frac{-h(T_a - T_w)}{\rho_s Ly(w_c - w_e)} \tag{6.24}$$

Thus the falling rate can be given as:

$$\left(\frac{dw}{dt}\right)_F = -\frac{h(T_a - T_w)(w - w_e)}{\rho_s Ly(w_c - w_e)} \tag{6.25}$$

The above equation can be integrated to determine the falling-rate period:

$$t_F = \frac{\rho_s Ly(w_c - w_e)}{h(T_a - T_w)} ln \frac{w_c - w_e}{w - w_e} \tag{6.26}$$

Similarly for a through circulation drying:

$$t_F = \frac{\rho_s L(w_c - w_e)}{ha(T_a - T_w)} ln \frac{w_c - w_e}{w - w_e} \tag{6.27}$$

**6.2.2d Use of Heat and Mass Balances in the Analysis of Continuous Dryers.**—Consider a flow diagram, as shown in Fig. 6.4, where product and air move counter-currently to each other. A mass balance on moisture gives:

$$GH_2 + m_t w_1 = GH_1 + m_t w_2 \tag{6.28}$$

By conducting a heat balance with a datum temperature of 0 C and neglecting the heat of wetting, the enthalpy of air can be expressed as:

$$H_G = c_s (T_G - T_o) + HL_o \tag{6.29}$$

where $L_o$ is the latent heat of water at $T_o$ (0 C) = 2501 kJ/kg and $c_s$ is humid heat given by $c_s$ = 1.005 + 1.88H. The enthalpy of wet solid is:

$$H_m = c_{pS} (T_s - T_o) + w\, c_{pA} (T_s - T_o) \tag{6.30}$$

Thus the energy balance on the dryer gives:

$$GH_{G2} + m H_{m1} = GH_{G1} + m H_{m2} + q \tag{6.31}$$

where $q$ is heat loss from the dryer.

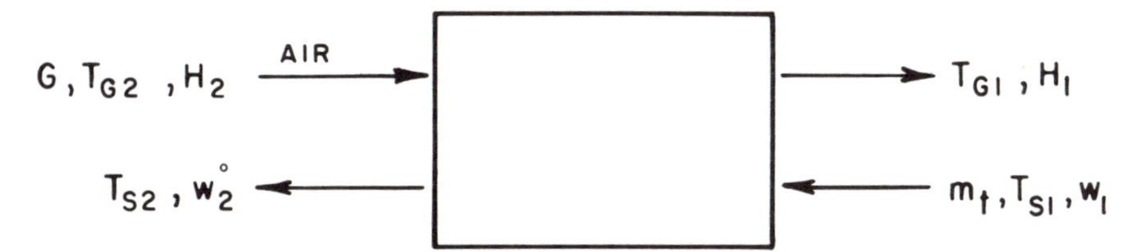


FIG. 6.4. A COUNTER-CURRENT CONTINUOUS DRYER

These expressions are useful in determining the air flow requirements and analysis of the product and air conditions at the exit from the dryer.

EXAMPLE 6.2 A continuous dryer is being operated counter-currently to dry a product containing 0.2 kg water/kg dry solid to final moisture content of 0.05 kg water/kg dry solid. The feed rate is 500 kg dry solid/h. The solid product enters the dryer at 20 C and exits at 70 C. The heat capacity of the dry solid is 2.0 kJ/kg C. The drying air enters at 95 C with 0.005 kg water/kg dry air. The moist air exits at 45 C. If the specific heat of the product is assumed constant and there is no heat loss from the dryer, calculate the air flow rate and the exit humidity ratio.

SOLUTION

From equation (6.28), the mass balance is:

$$GH_2 + m_t w_1 = GH_1 + m_t w_2$$

$$G\,(.005) + 500\,(.2) = GH_1 + 500\,(.05)$$

From equation (6.29), for $L_v$ = 2501 kJ/kg from steam table:

$$H_{G1} = (1.005 + 1.88H_1)(45-0) + H_1\,(2501)$$

$$H_{G1} = 45.225 + 2585.6\,H_1$$

for the entering air:

$$H_{G2} = [1.005 + 1.88\,(.005)\,]\,(95-0) + .005 \times 2501$$

$$H_{G2} = 108.87 \text{ kJ/kg dry air}$$

From equation (6.30), for entering solid:

$$H_{m1} = 2.0\,(20-0) + 0.2\,(4.187)\,(20-0)$$

$$H_{m1} = 56.75 \text{ kJ/kg dry solid}$$

For exit solids, using 0.05 in equation (6.30)

$$H_{m2} = 2.0\,(70-0) + .05\,(4.187)(70-0)$$

$$H_{m2} = 154.65 \text{ kJ/kg dry solid}$$

Substituting in equation (6.31), the energy balance

$$108.87\,G + 500 \times 56.75 = G\,(45.225 - 2585.6\,H_1) + 500 \times 154.65 + 0$$

Solving the energy balance and the mass balance simultaneously,

Air flow rate = 4789.2 kg dry air/h
Exit humidity ratio = 0.0207 kg water/kg dry air

### 6.2.3 Equilibrium Moisture Content and Water Activity

The equilibrium moisture content of a material is defined as the moisture content that exists when the material is at vapor pressure equilibrium with its surrounding. In the dehydration process, this represents the moisture content of the product which is approached at completion of the process. The ratio of the equilibrium vapor pressure to the saturation vapor pressure is known as the equilibrium relative humidity or water activity, and values will correspond to the appropriate values of equilibrium moisture content for the product.

The magnitude of the equilibrium moisture content is established by the structure of the material and the maner in which water is held or bound by the product. Most biological materials, including food products, will have moisture molecules adsorbed on the interior walls of the porous capillaries. The adsorbed water may be in the form of single molecular layers at low equilibrium relative humidities or of multi-layers of adsorbed molecules at higher equilibrium relative humidities. Because of this type of bound moisture in food products, most products will exhibit a Type-2 or S-shaped isotherm according to the classification of Brunauer *et al.* (1938). A typical isotherm for a freeze-dried food product is illustrated in Figure 6.5. In addition, Figure 6.5 illustrates the hysteresis effect between adsorption and desorption isotherms. Most theories explain the hysteresis effect as being created by irreversible changes which occur during the adsorption of moisture.

Probably one of the first and simplest equations to describe equilibrium isotherms was developed by Langmuir (1918). This equation was based on the assumption that the ratio of the actual number of adsorbed molecules to the difference between the maximum number and the actual number was proportional to the vapor pressure. Based on this assumption, the Langmuir equation can be written as:

$$w_e = \frac{(w_e)_{max}\, bp}{1 + bp} \tag{6.32}$$

where $b$ represents an adsorption coefficient. The primary weakness in the Langmuir equation is that the expression will not describe adsorption isotherms when more than one monomolecular layer of moisture is adsorbed.

Utilizing the assumption that van der Waals forces account for the adsorption of water molecules on product surfaces, Brunauer *et al.* (1938) developed a well-known expression to describe the S-shaped isotherms. The resulting expression, which has become known as the BET equation, can be written as:

$$\frac{p_v}{w(p_s - p_v)} = \frac{1}{w_m\, S} + \left(\frac{S-1}{w_m\, S}\right)\frac{p_v}{p_s} \tag{6.33}$$

where $w_m$ and $S$ represent constants in the expression. The constant ($w_m$) is assumed to represent the moisture content of the product when one monomolecular layer of moisture has been adsorbed. The assumptions made in the derivation of the BET equation are as follows: (a) more than one layer of water molecules may be adsorbed on the surface; (b) the energy of adsorption for molecules in all layers except the monomolecular

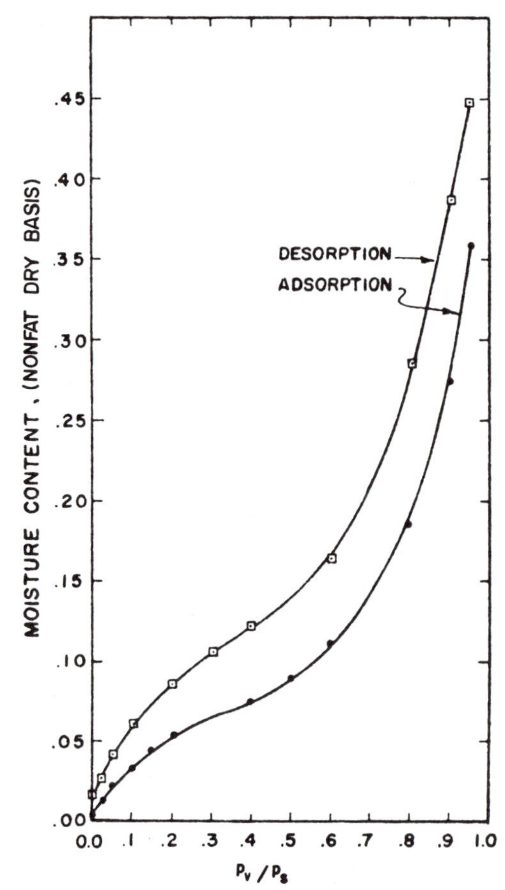


FIG. 6.5. EQUILIBRIUM MOISTURE CONTENT ISOTHERM FOR A FREEZE-DRIED FOOD PRODUCT ILLUSTRATING HYSTERESIS

moisture layer is equal to the heat of condensation for water; and (c) the energy adsorption for the monomolecular layer is the same for all molecules existing in the layer. The BET equation usually can be expected to describe equilibrium moisture content data up to an equilibrium relative humidity of 40%.

EXAMPLE 6.3 The following moisture adsorption equilibrium data were obtained for ground, precooked, freeze-dried beef at 10 C:

| $p_v/p_s$ | Moisture Content (kg water/kg dry solids) | $p_v/p_s$ | Moisture Content (kg water/kg dry solids) |
|---|---|---|---|
| 0.0 | 0.0 | 0.10 | 0.077 |
| 0.1 | 0.011 | 0.15 | 0.093 |
| 0.02 | 0.019 | 0.20 | 0.106 |
| 0.03 | 0.027 | 0.30 | 0.121 |
| 0.04 | 0.041 | 0.40 | 0.137 |

Use the BET equation to evaluate the monomolecular layer moisture content ($w_m$) and the energy constant ($S$).

SOLUTION

(1) Since equation (6.33) is written in the form of an equation of a straight line, the data given can be plotted by computing the values of $p_v/w(p_s - p_v)$ and plotting versus $p_v/p_s$ as illustrated in Fig. 6.6.

(2) From Fig. 6.6 the intercept of the vertical axis represents $1/w_m S$ and the slope should equal $(S - 1)/w_m S$.

(3) In Fig. 6.6 the intercept equals 0.7, therefore: $1/w_m S = 0.7$ and the slope is 8.7; $(S - 1)/w_m S = 8.7$.

(4) The two previously given equalities provide two equations with two unknowns to be solved.

(a) Since $1/w_m = 0.7S$; then $0.7S(S - 1)/S = 8.7$

$$\text{or:} \quad 0.7S = 8.7 + 0.7 = 9.4$$
$$S = 13.4$$

(b) Then: $w_m = 1/0.7\ (13.4) = 0.1065$

(5) The monomolecular layer moisture content for the situation given is 0.1065 kg water/kg nonfat dry solids. The energy constant, $S = 13.4$.

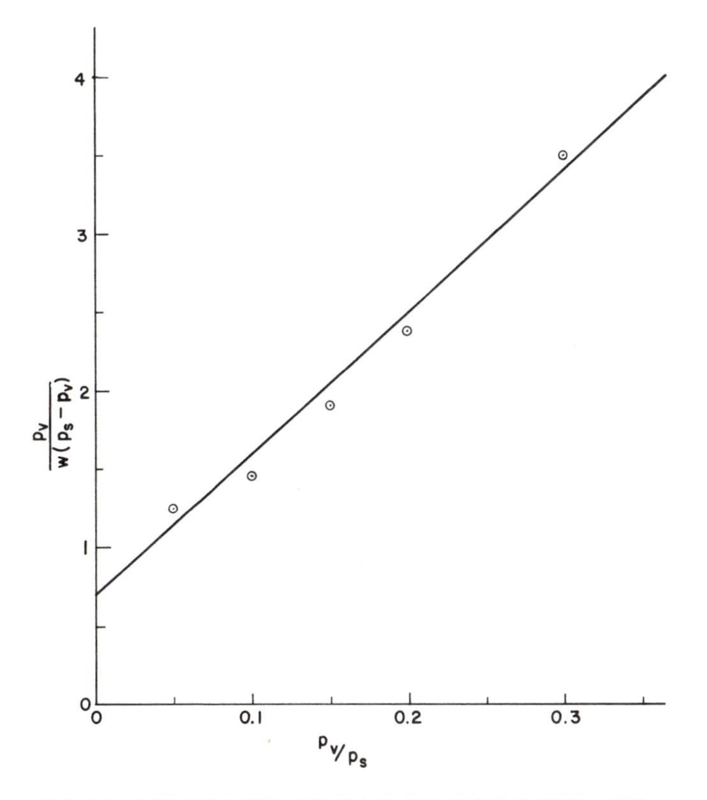


FIG. 6.6. BET ISOTHERM USED TO COMPUTE CONSTANTS

Considerable emphasis is placed on the accurate measurement of equilibrium moisture contents of dehydrated foods. The importance of water activity in food storage has been studied by researchers (see Labuza, 1968 and Labuza, 1980) and reviewed recently by Rockland and Nishi (1980). Several reactions and growth of microorganisms occur selectively within certain ranges of water activities. A diagram of the influence of water activity on chemical, enzymatic and microbial changes is shown in Fig. 6.7. There appears to be the presence of two water activity optima related to the stability of certain heterogenous food systems. The figure also includes major types of water-binding regions mentioned in section 6.1. Between water activity of 0 and 0.25 the region dominated by water bound by ionic groups such as $NH_3^+$ associated with proteins and $COO^-$ groups associated with proteins, pectins and other polyuronic acids. The region between water activity of 0.25 and 0.75 is influenced by covalantly bound water, such as amide groups in proteins and OH groups in proteins and carbohydrate polymers such as starch, pectin, cellulose and hemicellulose. Between water activity values of 0.75 to 1.0, there are multilayers of water on proteins and carbohydrate polymers. In addition, there is water with low vapor pressure due to presence of dissolved solutes such as sugars and free amino acids.

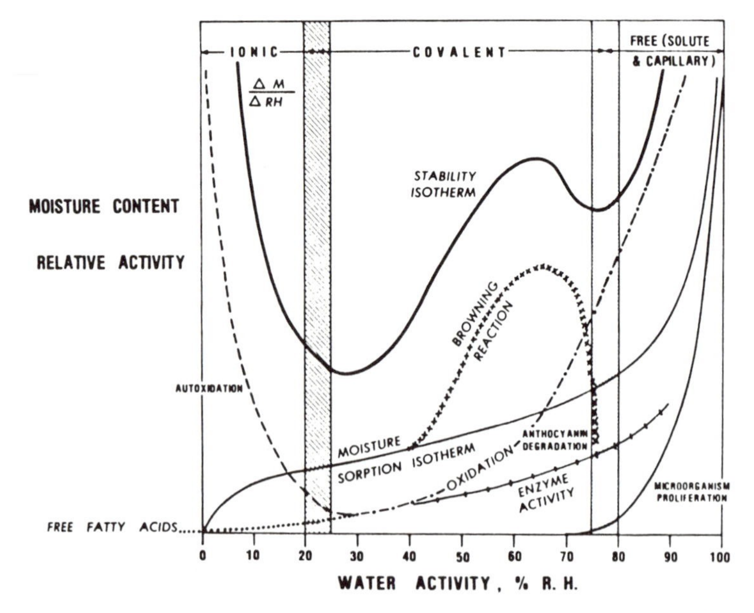


*From Rockland and Nishi (1980)*

FIG. 6.7. DIAGRAMATIC REPRESENTATION OF THE INFLUENCE OF WATER ACTIVITY ON CHEMICAL, ENZYMATIC, AND MICROBIOLOGICAL CHANGES AND ON OVERALL VELOCITY AND MOISTURE SORPTION PROPERTIES OF FOOD PRODUCTS

## 6.3. FIXED-TRAY DEHYDRATION

According to the discussion by Forrest (1968), several of the more important dehydration methods for food are in the fixed-tray classification. In general, all the methods to be discussed under this classification are used for products which have a solid structure before the dehydration process begins. The product in this form can be placed on a surface or tray and can be exposed to heated air in various ways. In most cases, the primary design parameter to be computed is the drying time, although such factors as the volumes of drying air required and the temperature distribution within the product cannot be overlooked.

### 6.3.1 Cabinet Drying

Food products dried in a cabinet dryer are placed on trays and moved into a drying compartment where the product is exposed to the drying air. This arrangement is illustrated schematically in Fig. 6.8. As is evident from the illustration, after the air is heated it passes through a stack of trays and over the product which is exposed in each of the trays before returning to the heating section. In the illustration, the air is passing over the product in a parallel fashion, but other designs may have the air passing vertically through the product trays. One of the major problems in this type of dehydration system is related to obtaining uniform drying of the product at various locations on the drying trays. This is caused by the lack of uniformity of air flow across the product and of temperature and absolute humidity of the air entering the area of drying. Forrest (1968) indicated that velocities in the 2.5 m/s to 5 m/s range may be utilized in these types of dryers, and added that there is no definite advantage to the higher velocities. The second problem is related to the more rapid dehydration of product near the point where air enters the drying area, since product downstream will be exposed to higher-humidity air and will dry less rapidly. Solutions to this problem include rotation of the trays or reversal of the air flow direction so that different product is exposed initially at different times during the drying period.

The computations of drying time in a cabinet dryer can be done using expressions presented previously in section 6.2.

### 6.3.2 Tunnel Drying

The second type of fixed-tray dryer in the classification by Forrest (1968) is the tunnel dryer, shown schematically in Fig. 6.9. In this type of dryer, the product segments are placed on trays which, in turn, are placed in stacks, and several stacks of product trays are then placed in an air

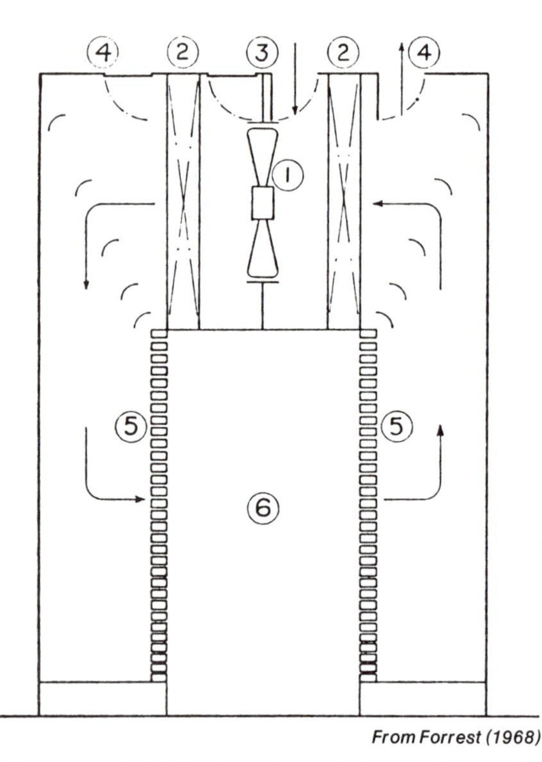


From Forrest (1968)

FIG. 6.8. SCHEMATIC ILLUSTRATION OF A CABINET DRIER: 1, CIRCULATING FAN, FULLY REVERSIBLE; 2, HEATER BATTERIES; 3, VENTED AIR INLET PORTS; 4, VENTED AIR EXHAUST PORTS; 5, ADJUSTABLE LOUVRE WALLS; 6, TRUCK SPACE

tunnel. As illustrated in Fig. 6.9, several approaches may be used in the introduction and removal of product to and from the tunnel. These variations include counter-current flow of air with respect to the movement of product through the tunnel, parallel flow of air and product through the tunnel, and counter-current flow of air with product moving directly through the tunnel and with portions of the air being recirculated. There are advantages and disadvantages to each approach, as discussed by Forrest (1968) and Charm (1978). One of the problems in the tunnel dryer, similar to the cabinet dryer, is the non-uniform drying of product at different locations in the tunnel. By obtaining a uniform distribution of air velocities through the tunnel, this disadvantage can be minimized.

The primary computation involved in design of the tunnel dryer is the computation of drying time; this is accomplished in a manner similar to that described for the cabinet dryer. One additional feature of the computation for design purposes is the determination of the length of tunnel necessary to accomplish the desired dehydration. The specific details of this computation are discussed by Charm (1978).

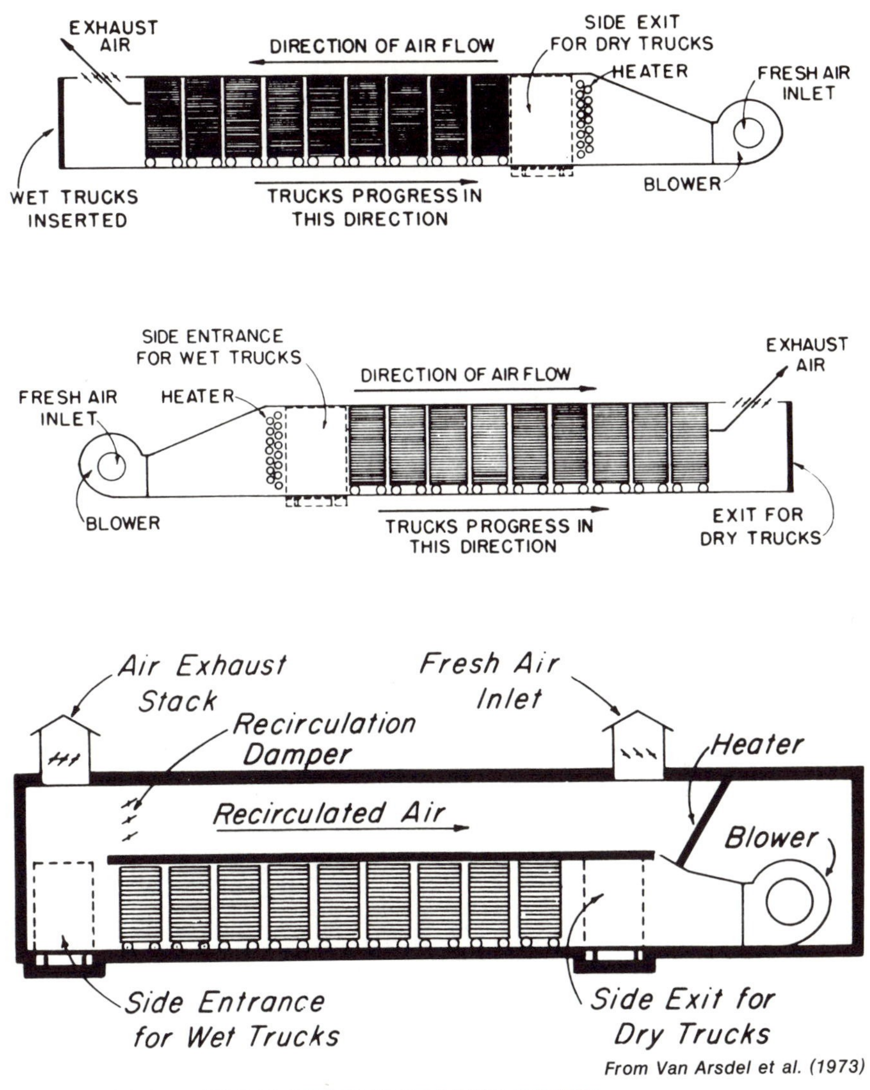


*From Van Arsdel et al. (1973)*

FIG. 6.9. SCHEMATIC ILLUSTRATION OF TUNNEL DRYERS

## 6.4. MOVING-BED DEHYDRATION

The second type of dehydration method in the classification by Forrest (1968) includes moving-bed dryers. Probably the main feature which distinguishes this from the fixed-tray type of dehydration is that movement of the product creates agitation within the pieces or particles making up the product and enhances the dehydration process.

### 6.4.1 Conveyor Drying

By placing a layer of the product material on a moving conveyor which is perforated, the dehydration process can be accomplished with a conveyor dryer. The air being used for dehydration passes through the perforations in the conveyor in an upward or downward direction. The direction of the air through the product is usually determined by the characteristics of the product, but in some cases there are advantages to reversing the direction at different times during the process, so that it will proceed efficiently. As the conveyor moves through the enclosed area of the dryer and the moisture content of the product decreases, the properties of the air utilized at various locations along the conveyor can be varied to meet the demands and desired exposure conditions of the product. The conveyor dryer does have the limitation that dehydration cannot be economically accomplished to moisture contents below 10%. The usual procedure is to transfer the product to a secondary dryer when the product moisture content is at 27% or less.

One of the best examples of a conveyor dryer is the foam-mat procedure developed and described by Morgan *et al.* (1961). This type of dryer can be used for various heat-sensitive liquids which have been converted to a stable foam and placed on a perforated conveyor. As illustrated schematically in Fig. 6.10, air is blown through the perforated conveyor and the product in an upward direction to accomplish the dehydration. The foam layer is usually about $3 \times 10^{-3}$m in thickness, and investigations have shown that velocity of air and humidity do not appreciably affect drying time.

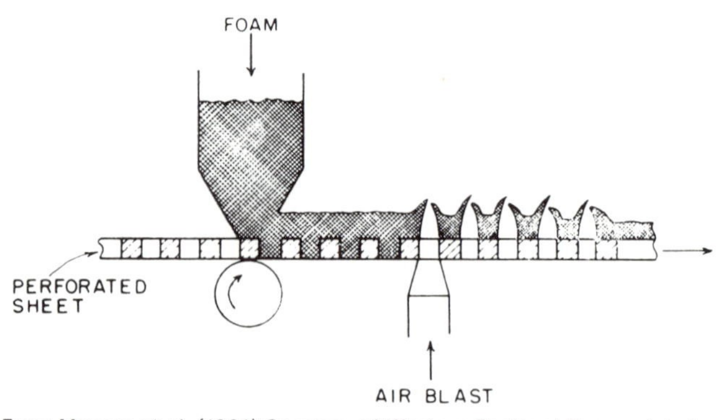


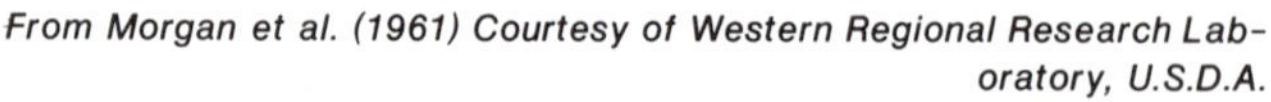
*From Morgan et al. (1961) Courtesy of Western Regional Research Laboratory, U.S.D.A.*

FIG. 6.10. SCHEMATIC DIAGRAM OF FOAM-MAT DRIER

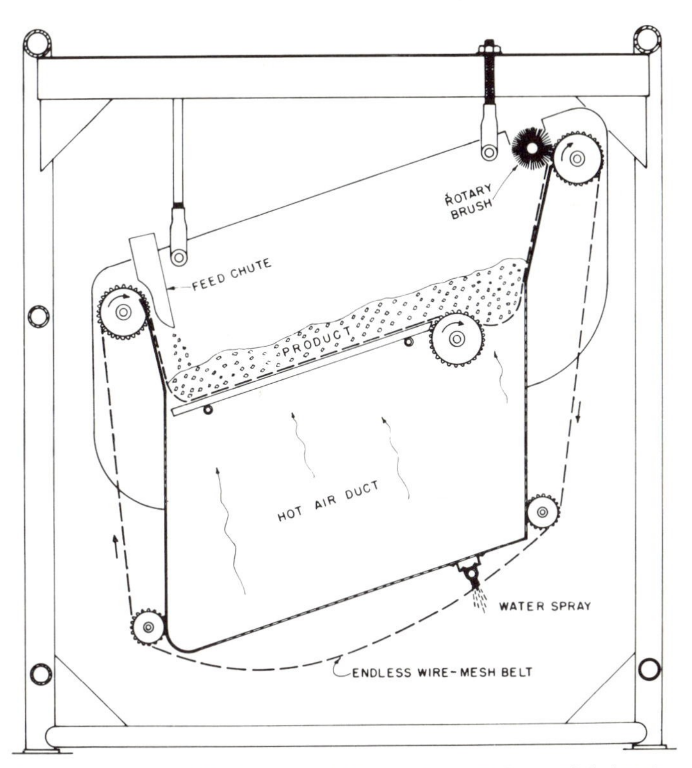


*From Lowe et al. (1955) Courtesy of Western Regional Research Laboratory, U.S.D.A.*

FIG. 6.11. SCHEMATIC CROSS-SECTION OF BELT TROUGH DRYER

### 6.4.2 Belt Drying

The belt dryer shown schematically in Fig. 6.11 provides additional agitation of the products during drying. Additional agitation results in a more uniform drying of all food-product particles. The belt trough dryer was described in detail by Lowe *et al.* (1955).

## 6.5. AIR-SUSPENDED PRODUCT

The third classification of Forrest (1968) includes dehydration methods which suspend the food product in the drying air. The most common type of dryer, and one which is probably the most important for fluid food products, is the spray dryer. This type will be discussed in detail under the air-suspended product classification.

### 6.5.1 Spray Drying

The dehydration process best adapted to food products which have high initial moisture contents and are in a liquid state initially is spray drying. The process was first applied to drying of milk shortly after 1900, and was applied to eggs and coffee in the 1930's. It has been adapted to many other products which are liquid initially and are heat-sensitive. The unique features of spray drying include a rapid drying cycle, a short holding or retention of the product in the drying chamber, and a final product which is ready for packaging as it leaves the dryer. The fact that the retention time may be as low as 3 to 10 seconds and the product particle may never be at a higher temperature than the wet-bulb temperature of the air used for drying is of considerable significance. The latter situation allows the use of rather high temperatures at the inlet to the dehydration chamber without causing damage to the product.

Although spray dryers may have many designs, Patsavas (1963) has classified the types under four general classifications. The first of these is the counter-current type illustrated in Fig. 6.12. In the counter-current spray dryer, the liquid is atomized near the top of the drying chamber and falls downward, while the air is introduced near the bottom of the drying chamber and moves upward through the liquid droplets. The dried product leaves the bottom of the chamber, while the air is removed near the top of the drying chamber. The inlet air, which is at a relatively high temperature, is brought into direct contact with the product which is dried or nearly dried.The latter situation is the primary disadvantage of the counter-current type of dryer—that of reduced product quality due to the influence of heat on the dry product. In addition, the air flow rate must be relatively low in order to avoid entrainment of large quantities of the product when the air is removed at the top of the drying chamber.

The co-current spray dryer illustrated Fig. 6.13 mixes the inlet air with newly formed droplets at the atomizer. After the initial mixing, the product and air move in the same direction as dehydration proceeds. The product and most of the air leave the drying chamber at the lower exit and move to a separation system. This arrangement is probably best for heat-sensitive food products, since the liquid product contacts the high-temperature inlet air and the dehydrated product is in contact with the air after the temperature has been reduced significantly.

Masters (1976) has illustrated that co-current flow may occur in three patterns in addition to the downward rotary flow in Fig. 6.13. These patterns include: (a) downward parallel flow, (b) upward flow with air introduced at bottom of drying chamber near point of atomization and (c) horizontal flow with both drying air and product droplets moving horizontally through the drying chamber.

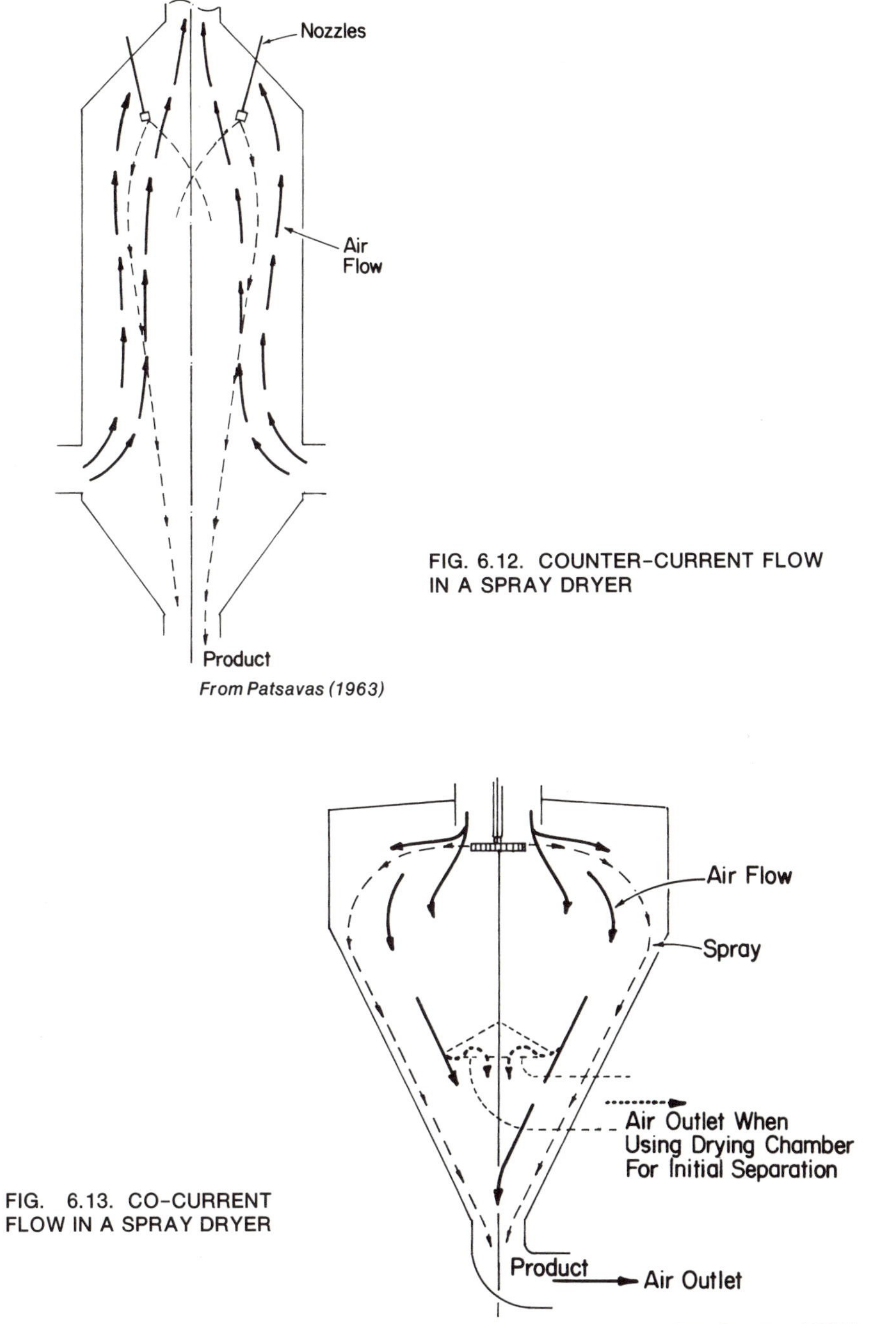


FIG. 6.12. COUNTER-CURRENT FLOW IN A SPRAY DRYER

FIG. 6.13. CO-CURRENT FLOW IN A SPRAY DRYER

The flow of product and air through a mixed-flow type of spray dryer is illustrated in Fig. 6.14. As is evident, the product is introduced by an atomizer at some location near the center of the drying chamber; the air introduced at the top moves toward the bottom of the drying chamber, where it contacts the product before moving upward again toward the air outlet. The product leaves through an exit near the lower part of the drying chamber. High inlet air temperatures may create some reduction in product quality with this particular flow arrangement, but the system does have a high evaporating capacity per unit volume. Masters (1976) refers to the arrangement in Fig. 6.14 as cyclonic and indicates that an alternate mixed flow pattern can be created by introducing the atomized product near the bottom of the drying chamber.

The parallel-flow arrangement of a spray dryer is illustrated in Fig. 6.15. With this arrangement the product and air flow in rather uniform lines from the top to the bottom of the narrow drying chamber. Product and air leave the drying chamber together and move to a separate component of the system for separation. The feature of this dryer which differs from the co-current type is the high air velocity utilized permitting high inlet air temperatures. This air velocity will normally be in the range of 2 to 3 m/s.

In general, spray drying must be considered a two-step process. The first step is atomization or formation of liquid droplets, while the second step is the actual drying or evaporation of the moisture to produce the solid product particles. The importance of recognizing these two steps will become more evident as the discussion proceeds.

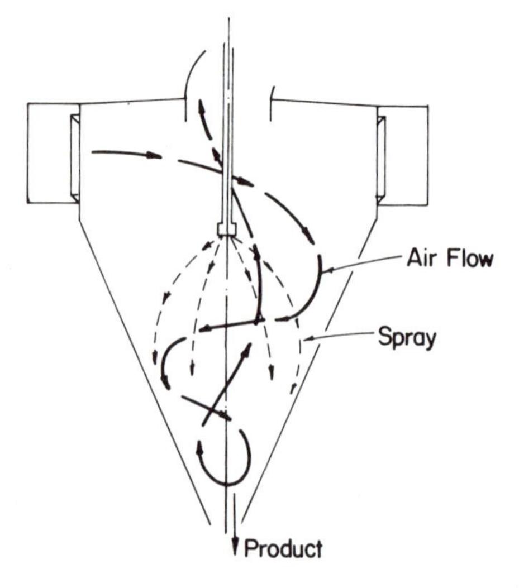


*From Patsavas (1963)*

FIG. 6.14. SPRAY DRYER WITH MIXED FLOW PATTERN

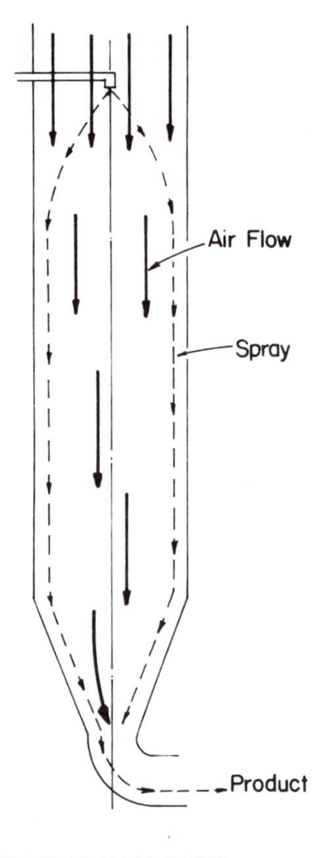


*From Patsavas (1963)*

FIG. 6.15. SPRAY DRYER WITH PARALLEL FLOW

One of the primary functions of the atomization is to generate small droplets, which create a large surface area for moisture evaporation. In addition, the atomizer acts as as a metering device to control the flow rate of product into the dryer. The atomizer distributes the liquid into the air stream in a relatively uniform manner and should produce droplets of the proper size to result in the desired product particle size after the process is complete. There are numerous types of atomizers available for use in spray dryers. Tate (1965) classified these atomizers under four different headings: (a) centrifugal pressure nozzles, (b) fan-spray nozzles, (c) two-fluid atomizers, and (d) rotary atomizers.

Pressure nozzles, either centrifugal or fan-spray types, utilize pressure to force the liquid through an orifice and form a liquid sheet. This sheet of liquid then breaks up into the droplets required for spray drying. Centrifugal-pressure nozzles usually result in the formation of conical sheets of liquid. The liquid sheet is formed by forcing the liquid to flow through a narrow divergent angular orifice of the type illustrated in Fig. 6.16. This type is usually referred to as a swirl-spray nozzle. If an additional jet of liquid is introduced in the center of a swirl chamber, a full-cone spray nozzle results, as illustrated in Fig. 6.17.

A fan-spray nozzle is utilized to form liquid droplets by impinging a stream of liquid on an orifice which has been designed to form a liquid sheet in a plane perpendicular to the plane of the liquid stream. As shown in Fig. 6.18, a flat sheet of liquid is formed and breaks up into liquid droplets. The fan-spray nozzle operates best at high pressures and with a wide spray angle. A somewhat similar atomizer is referred to as an impinging jet nozzle. The latter type has not been used in spray-drying operations.

Two-fluid atomizers utilize high-velocity gas streams to impinge on and break up low-velocity liquid streams to form the divided liquid droplets. The actual impingement of the liquid in the gas streams may occur inside the atomizer body, as illustrated in Fig. 6.19. In general, these types of atomizers tend to use considerable power and may be uneconomical at high capacities. Fine droplets can be produced at low flow rates and with high-viscosity products.

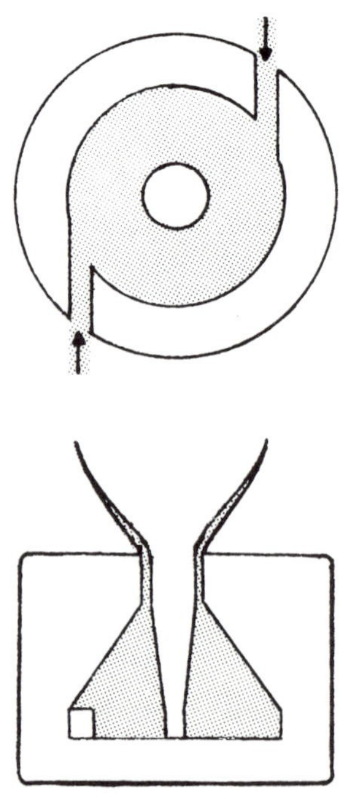

*From Dombrowski and Munday (1968)*

FIG. 6.16. CROSS-SECTIONAL DIAGRAM OF SWIRL SPRAY NOZZLE

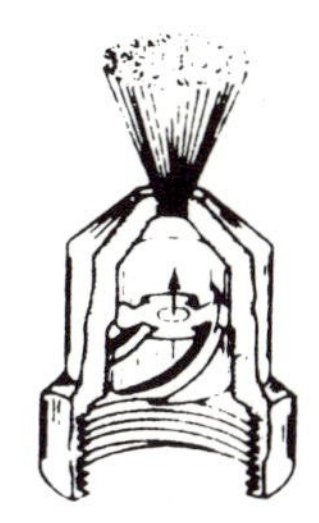

FIG. 6.17. SCHEMATIC ILLUSTRATION OF FULL CONE SPRAY

*From Dombrowski and Munday (1968)*

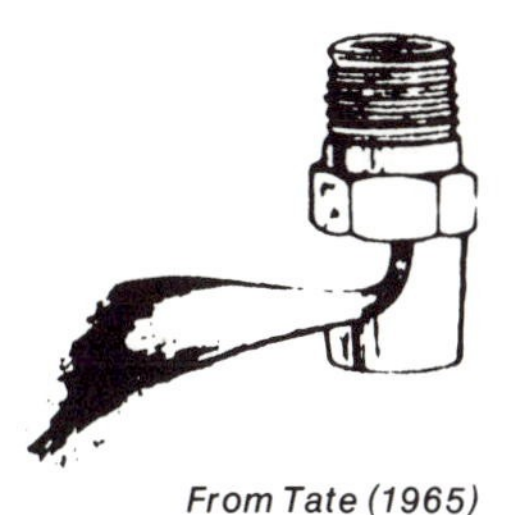

FIG. 6.18. FORMATION OF FAN SPRAY SHEETS

*From Tate (1965)*

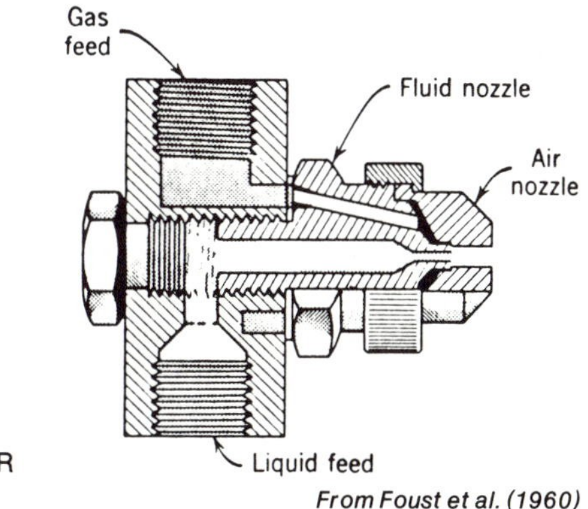


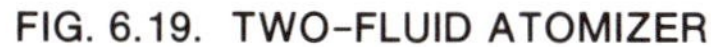

FIG. 6.19. TWO-FLUID ATOMIZER

*From Foust et al. (1960)*

Rotary atomizers utilize centrifugal force to form a liquid sheet which breaks up into the desired droplets. The liquid is fed onto a rotating surface and moves across the surface to form a thin sheet at the periphery, as illustrated in Fig. 6.20. Since the force utilized in forming the liquid sheet is directly dependent on the speed of rotation, this type of atomizer is very versatile and can be utilized for a wide range of feed rates and liquid properties. The actual droplet size which results is influenced by both disc speed and feed rate. Considerably more information is given on the mechanics of flow through all types of atomizers in the review of Dombrowski and Munday (1968) and by Masters (1976).

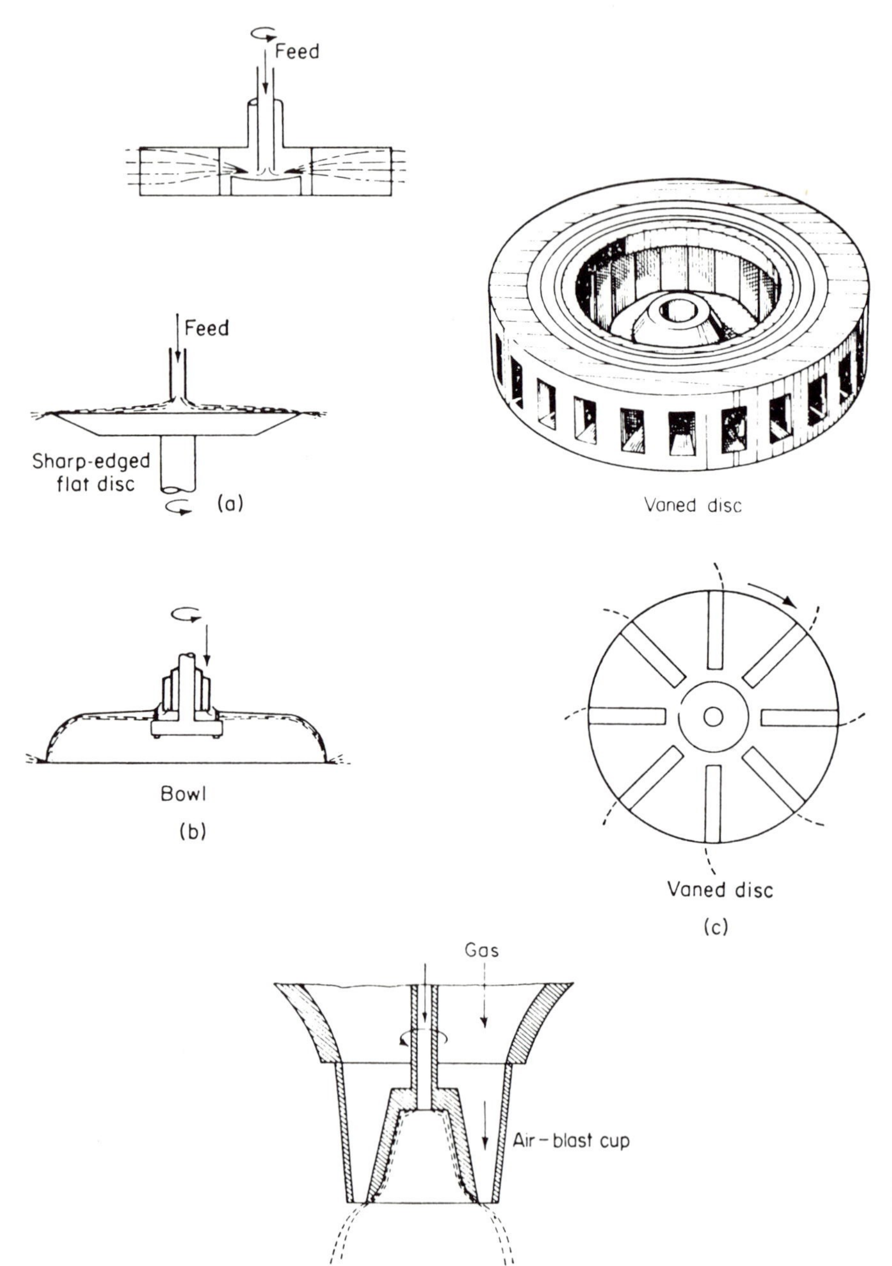


*From Dombrowski and Munday (1968)*

FIG. 6.20. SCHEMATIC ILLUSTRATION OF ROTARY ATOMIZER

Dombrowski and Munday (1968) presented various expressions for energy requirements of atomizers. The energy required for the pressure nozzle is given by equation (6.34):

$$E = 7Q\Delta P \times 10^{-4} \tag{6.34}$$

where $Q$ is the liquid flow rate in gal/min and $\Delta P$ is the pressure in $lb_f/in^2$. For a two-fluid or air blast atomizer the following expression will apply:

$$E = 0.136GT_A\{0.5\,M_c^2 + 2.5\,[1 - P^{0.286}]\} \tag{6.35}$$

where $G$ is the air mass flow rate in $lb_m$/sec and $T_A$ is the temperature in R. Computations for both types of atomizers as shown by equations (6.34) and (6.35) illustrate that the energy applied is used primarily for kinetic energy imparted to the liquid or gas, and only a minor portion is utilized for breaking the liquid into droplets. For rotary atomizers the following expression applies:

$$E = 42.5G_L(d_L\,N')^2 \times 10^{-12} \tag{6.36}$$

where $G_L$ is the liquid flow rate in $lb_m$/hr, $N'$ is the rotation speed of the atomizer in revolutions per minute, and $d_L$ is the diameter of the atomizer in ft. In all three equations, the resulting computation gives energy in horsepower (HP).

These relationships for power requirements to produce droplets indicate that a very small amount of the energy is used for liquid breakup. Most of the energy is transferred to the liquid in the form of kinetic energy.

The fundamental theory which allows the formation of liquid droplets by any type of atomizer is based on the theory that a free cone of liquid is unstable when its length is greater than its circumference and that, for a highly viscous liquid, the wavelength of any disturbance will grow most rapidly when the amplitude is 4 times greater than the diameter. Physically, the formation of a droplet is accomplished by increasing the surface area of a liquid sheet until it becomes unstable and disintegrates. The actual manner in which the disintegration occurs is influenced by nozzle design and the surrounding atmosphere, as discussed in detail by Dombrowski and Munday (1968) and Masters (1976). The properties of the liquid influence droplet formation considerably; increased viscosity makes droplet formation more difficult. The influence of surface tension is dependent on the system utilized, whereas density has very little effect on droplet formation. Again, Dombrowski and Munday (1968) and Masters (1976) have provided excellent reviews of available information.

There are three ways to express the mean diameter of a droplet aerosol. Although there are several equations, the three expressions to be presented are most frequently used. The first equation is for the count or number mean diameter defined as:

$$\text{CMD} = \frac{\Sigma N_i d_i}{N} \tag{6.37}$$

which is the mean size of droplet or particle diameter based on the total number of droplets. The second type of mean diameter, usually referred to as the mass mean diameter, is defined as:

$$\text{MMD} = \left[\frac{\Sigma N_i d_i^3}{N}\right]^{1/3} \tag{6.38}$$

and represents the mean droplet or particle diameter computed on the basis of the droplet mass. The Sauter or volume-surface mean diameter is defined as:

$$\text{SMD} = \frac{\Sigma N_i d_i^3}{\Sigma N_i d_i^2} \tag{6.39}$$

The SMD should be the diameter having the same surface area to volume as the entire distribution of the spray. In addition, the SMD is commonly used to describe droplet size distributions.

In general, the prediction of mean droplet diameters produced by various atomizers is made by using empirical relationships based on experimental investigations. Only in the case of fan-spray nozzles have theoretical investigations produced usable results and expressions. Ford and Furmidge (1967) and Dombrowski and Munday (1968) have described the process of droplet formation from fan-spray nozzles. In general, it is felt that two mechanisms are involved. The first is the formation of perforations in the liquid sheet, which expand to form ligaments and then droplets. The second mechanism involves the development of unstable waves at right angles to the direction of flow. These waves grow in amplitude until a break in the sheet occurs and results in droplet formation. Ford and Furmidge (1967) have shown that the mechanism involved is dependent on the flow characteristics in the nozzle. At low Reynolds number, only ligaments may form, whereas at high Reynolds number, the second mechanism mentioned results in droplet formation. Between the two extremes, there is a region of unstable conditions which is difficult to describe in terms of droplet formation. Dombrowski and Munday (1968) presented the following expression for a fan-spray nozzle based on the theoretical development and appropriate constant to account for experimental verification:

$$\text{SMD} = \frac{B_1}{C_Q}\left[\frac{\text{FN}\gamma\rho_L}{\sin\theta\,\Delta P}\right]^{1/3}\frac{1}{\rho_g{}^{1/6}} \tag{6.40}$$

The use of equation (6.40) requires that the units of surface tension ($\gamma$) are dyne/cm, density ($\rho$) is expressed as $lb_m/ft^3$ and pressure ($\Delta P$) is $lb_f/in^2$. The constant ($B_1$) varies from 26.3 to 43 based on experimental data used for verification. The discharge coefficient ($C_Q$) can be determined from experimental data presented in Fig. 6.21 and the flow number (FN) is a constant representing the flow rate divided by the square root of pressure differential through the spray nozzle. The equation (6.40) has been converted to SI units by Coulson and Richardson (1978) as:

$$SMD = \frac{0.000156}{C_Q}\left[\frac{(FN)\gamma\rho_L}{\sin\theta\,\Delta P}\right]^{1/3}\rho_g^{-1/6} \qquad (6.41)$$

with $\Delta P$ in kPa

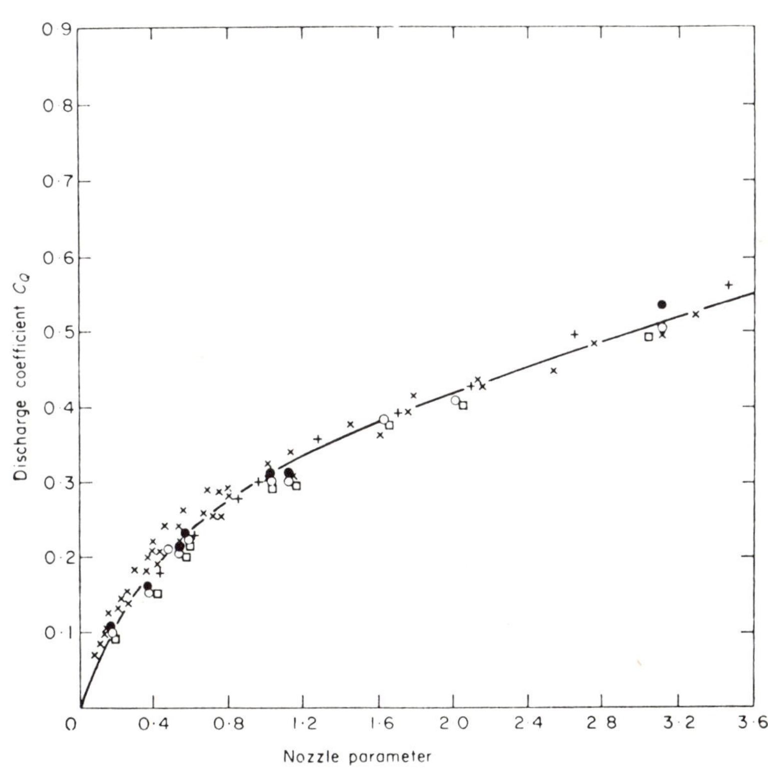


*From Dombrowski and Munday (1968)*

FIG. 6.21. DISCHARGE COEFFICIENTS FOR A FAN-SPRAY NOZZLE

EXAMPLE 6.4 A fan-spray nozzle is being used to generate a spray of tomato juice in a spray dryer. The surface tension of the product is 0.05 N/m and the density is 960 kg/m$^3$. The Flow Number (FN) for the nozzle has been established at 0.8 with a spray angle of 120° and pressure of 700 kPa. The nozzle parameter for the system has been computed to be 3.0. Determine the mean diameter of atomized droplets.

SOLUTION:

(1) The use of equation (6.41) requires a value for the discharge coefficient ($C_D$) and a value for density of air in order to compute the Sauter mean diameter (SMD) for the droplets from the nozzle.

(2) The discharge coefficient can be determined from Fig. 6.21, using the given nozzle parameter of 3.0, to be 0.52. A density of 1.04 kg/m$^3$ for air would apply at 65 C.

(3) Using the information given and equation (6.41):

$$\text{SMD} = \frac{0.000156}{0.52}\left[\frac{(0.8)(0.05\ \text{N/m})(960\ \text{kg/m}^3)}{\sin(120) \times 700\ \text{kPa}}\right]^{1/3}(1.04\ \text{kg/m}^3)^{-1/6}$$

$$= (3 \times 10^{-4})\,(0.0633)^{1/3}\,(0.9935)$$

$$= 1.1878 \times 10^{-4}\ \text{m}$$

$$\text{SMD} = 119\ \text{micron}$$

Masters (1976) presented several empirical relationships to predict spray droplet size for centrifugal pressure nozzles. One of the more useable equations was developed by Tate and Marshall (1953):

$$\text{SMD} = 286(d_o + 0.17)\exp\left[\frac{13}{u_v} - 0.0094\,u_t\right] \tag{6.42}$$

where $d_o$ is orifice diameter (in.) for nozzle, $u_v$ is axial velocity in ft/sec and $u_t$ is tangential velocity in ft/sec. Coulson and Richardson (1978) presented an expression for swirl-plate centrifugal pressure nozzle as:

$$\text{SMD} = 0.0134\,\frac{(\text{FN})^{0.209}(\mu/\rho)^{0.215}}{\Delta P^{0.348}} \tag{6.43}$$

where the Flow Number (FN) can be computed from:

$$\text{FN} = \frac{\text{volumetric flow}}{(\text{pressure})^{1/2}} \tag{6.44}$$

where volumetric flow is expressed by $2.23 \times 10^6$ (m$^3$/sec) and pressure is kPa.

EXAMPLE 6.5 Compute the droplet size generated by a centrifugal pressure nozzle (swirl-plate) for conditions presented in Example 6.4. Additional information needed includes a volumetric flow rate for the product of 0.035 m$^3$/hr and a liquid viscosity of $1.1 \times 10^{-3}$ kg/ms.

SOLUTION:

(1) In order to use equation (6.43), the Flow Number must be computed from equation (6.44):

$$FN = \frac{2.23 \times 10^6 (0.035 \text{ m}^3/\text{hr})}{(700 \text{ kPa})^{1/2} (3600 \text{ sec/hr})}$$

$$FN = 0.82$$

(2) Using equation (6.43)

$$SMD = 0.0134 \frac{(0.82)^{0.209} (1.1 \times 10^{-3} \text{ kg/ms}/960 \text{ kg/m}^3)^{0.215}}{(700 \text{ kPa})^{0.348}}$$

$$= 0.0134 \frac{(0.959)(0.0528)}{9.7745}$$

$$= 6.94 \times 10^{-5} \text{ m}$$

$$SMD = 69.4 \text{ micron}$$

Several empirical relationships for predicting droplet size in sprays from two-fluid converging nozzles are presented by Masters (1976). The expression developed by Nukiyama and Tanasawa (1937) is:

$$SMD = \frac{1410}{u_a}\left[\frac{\gamma}{\rho_L}\right]^{0.5} + 191\left[\frac{\mu_L}{(\gamma\rho_L)^{0.5}}\right]^{0.45}\left[\frac{Q_L}{Q_a} \times 10^3\right]^{1.5} \quad (6.45)$$

with $u_a$ = air velocity (ft/sec), $\gamma$ = surface tension (dyne/cm), $\rho_L$ = liquid density ($lb_m/ft^3$), $\mu_L$ = liquid viscosity ($cP$), $Q_L$ = liquid food rate ($ft^3$/min) and $Q_a$ = air feed rate ($ft^3$/min). The Nukiyama-Tanasawa equation is limited to nozzles with internal mixing and for feed rates up to 1 $lb_m$/min with small nozzle diameter.

Masters (1976) has summarized the expressions available for computation of droplet size and size distribution from rotary atomizers. The Herring-Marshall equation (1955) is an example of expressions available:

$$VMD = \frac{K\, G_L^{\,0.24}}{(N'd)^{0.83}(n'h')^{0.12}} \quad (6.46)$$

where $G_L$ = mass flow rate for liquid ($lb_m$/min), $N'$ = rotation speed (rpm), $d$ = wheel diameter (in.), $n'$ = number of vanes of atomizer wheel and $h'$ = height of vanes (in.). For this equation, the mean droplet size is volume mean diameter (micron) and the constant ($K$) is a function of spray dryer size. A value of $92.5 \times 10^4$ is generally accepted as average. The distribution of droplet size can be predicted by procedures developed by Herring and Marshall (1955). The procedures are presented in detail by Masters (1976).

EXAMPLE 6.6 Compute the volume mean diameter for droplets generated by a rotary atomizer from tomato juice described in Example 6.4. A liquid feed rate of 1000 kg/hr is used with a 8 cm diameter wheel rotating at 10,000 rev/min. The wheel has 180 vanes with 0.75 cm height.

SOLUTION:

(1) In order to use equation (6.46), modifications in the equation must be developed in order to accommodate SI units. The following revised equation can be used:

$$\text{VMD} = \frac{K\, G_L^{0.24}}{(N'd)^{0.83}\,(n'h')^{0.12}}$$

where the constant $K$ becomes $2.71 \times 10^6$, the feed rate is kg/min and the wheel diameter and vane height are in cm.

(2) Using the modified form of equation (6.46)

$$\text{VMD} = \frac{(2.71 \times 10^6)\,(1000/60)^{0.24}}{(10{,}000 \times 8)^{0.83}(180 \times 0.75)^{0.12}}$$

$$= \frac{(2.71 \times 10^6)\,(1.96)}{(11{,}737)(1.8)}$$

$$= 251 \text{ micron.}$$

The removal of moisture from droplets during spray drying to obtain dry food solids is accomplished by simultaneous heat and mass transfer. Spray drying differs from other dehydration methods in that a large percentage of the moisture is removed during a constant-rate drying period when the moisture content is very high. During the constant-rate period, evaporation of moisture from the droplet takes place in the same manner as for a pure water droplet. The simultaneous heat and mass transfer occurs in the following way: (a) heat for evaporation is transferred by conduction and convection from the heated air to the droplet surface, and (b) the vapor produced by moisture evaporation is transferred by diffusion and convection from the droplet surface to the heated air. The rate at which this process occurs is a function of several factors, including air temperature, humidity or vapor pressure, transport properties of the air, droplet diameter, droplet temperature, relative velocity, and the nature of the solids in the liquid droplet.

Frossling (1938, 1940) presented the partial differential equations which describe the heat and mass transfer occurring around a liquid droplet during evaporation. The analytical solution to these equations is difficult and presentation is not justified, but they do lead to a definition of appropriate dimensionless numbers which describe the heat and mass transfer during droplet evaporation. Marshall (1954) showed that these dimensionless groups indicate the analogy between heat and mass transfer for the spray-drying process.

The only solution which can be obtained directly from the basic partial differential equations is the limiting condition of $N_{Re} = 0$. This solution leads to the following equation:

$$\frac{hd}{k} = \frac{k_m M_m d\rho_L}{D\rho_g} = 2 \qquad (6.47)$$

where the dimensionless numbers represent the Nusselt number ($N_{Nu}$) and a form of the Sherwood number ($N_{Sh}$), respectively. Use of equation (6.47) illustrates that the rate of surface-area change in a pure liquid droplet is constant during evaporation.

For information on the rate of heat and mass transfer during the more realistic case of turbulence around the liquid droplet, only experimental evidence is available. Frossling (1938) obtained the following expressions:

$$N_{Nu} = 2 + K_1 N_{Re}{}^{1/2} N_{Pr}{}^{1/3} \tag{6.48}$$

$$N_{Sh} = 2 + K_2 N_{Re}{}^{1/2} N_{Sc}{}^{1/3} \tag{6.49}$$

where the constants $K_1$ and $K_2$ are equal to each other and to a value of 0.6 based on experimental data obtained by Ranz and Marshall (1952), and a value of 0.52 based on experimental data obtained by Frossling (1940). The basis for these empirical expressions is given in Fig. 6.22, where experimental data have been used to establish a single curve representing the correlation.

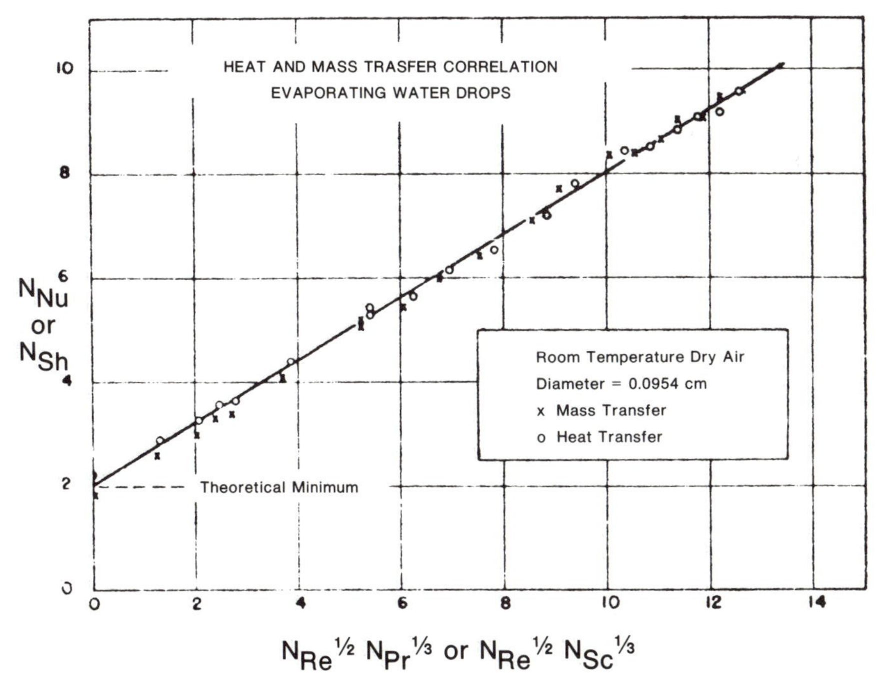


*From Ranz and Marshall (1952)*

FIG. 6.22. CORRELATION ILLUSTRATING HEAT AND MASS TRANSFER FROM LIQUID DROPLETS

From equations (6.48) and (6.49), it is evident that one of the critical factors is the relative velocity between the droplet and the drying air. This velocity, which will change throughout the drying process, can only be predicted based on the motion of a sphere in air. The expression relating the force to the velocity of a droplet is as follows:

$$F = C_d u_r^2 \rho_g A/2 \qquad (6.50)$$

where:

$$C_d = 24\, \mu/u_r \rho_g d \qquad (6.51)$$

The drag coefficient ($C_d$) is a function of Reynolds number, as indicated in equation (6.51). The latter equation applies for Reynolds numbers less than 1, the situation expected during spray drying. The influence of Reynolds number on the drag coefficient for spheres over a wide range of Reynolds numbers is illustrated in Fig. 6.23. It is unlikely that the droplets are exposed to conditions described by Reynolds numbers greater than $10^3$ even when the droplets are first introduced into the drying chamber from the atomizer. In addition, it is unlikely that a spherical shape exists at this point and detailed analysis would require accounting for shapes other than spherical. For conditions when the droplet is either accelerating or decelerating, the drag coefficient may be as much as 20% higher. These conditions create distortion of the liquid droplets from true spherical shape.

Since the initial moisture removal from a liquid droplet during spray drying occurs at a constant rate, the rate of moisture evaporation can be described adequately by treating the droplet as pure liquid. Utilizing this assumption, a heat and mass balance conducted on the liquid droplet leads to the following expressions:

$$\frac{dw}{dt} = k_m A(p_w - p_a) = \frac{2\pi D \rho_L d}{\rho_g}(p_w - p_a) \qquad (6.52)$$

and

$$\frac{dw}{dt} = \frac{hA(T_a - T_w)}{L} = \frac{2\pi k_g d}{L}(T_a - T_w) \qquad (6.53)$$

which describe the rate of moisture evaporation where either expression could be utilized to define the time of droplet disappearance in the case of a pure liquid. Equation (6.52) requires a minor revision by introducing the ideal gas law to obtain the expression presented by Charm (1978):

$$\frac{dw}{dt} = \frac{2\pi DMd}{R_g T_A}(p_w - p_a) \qquad (6.54)$$

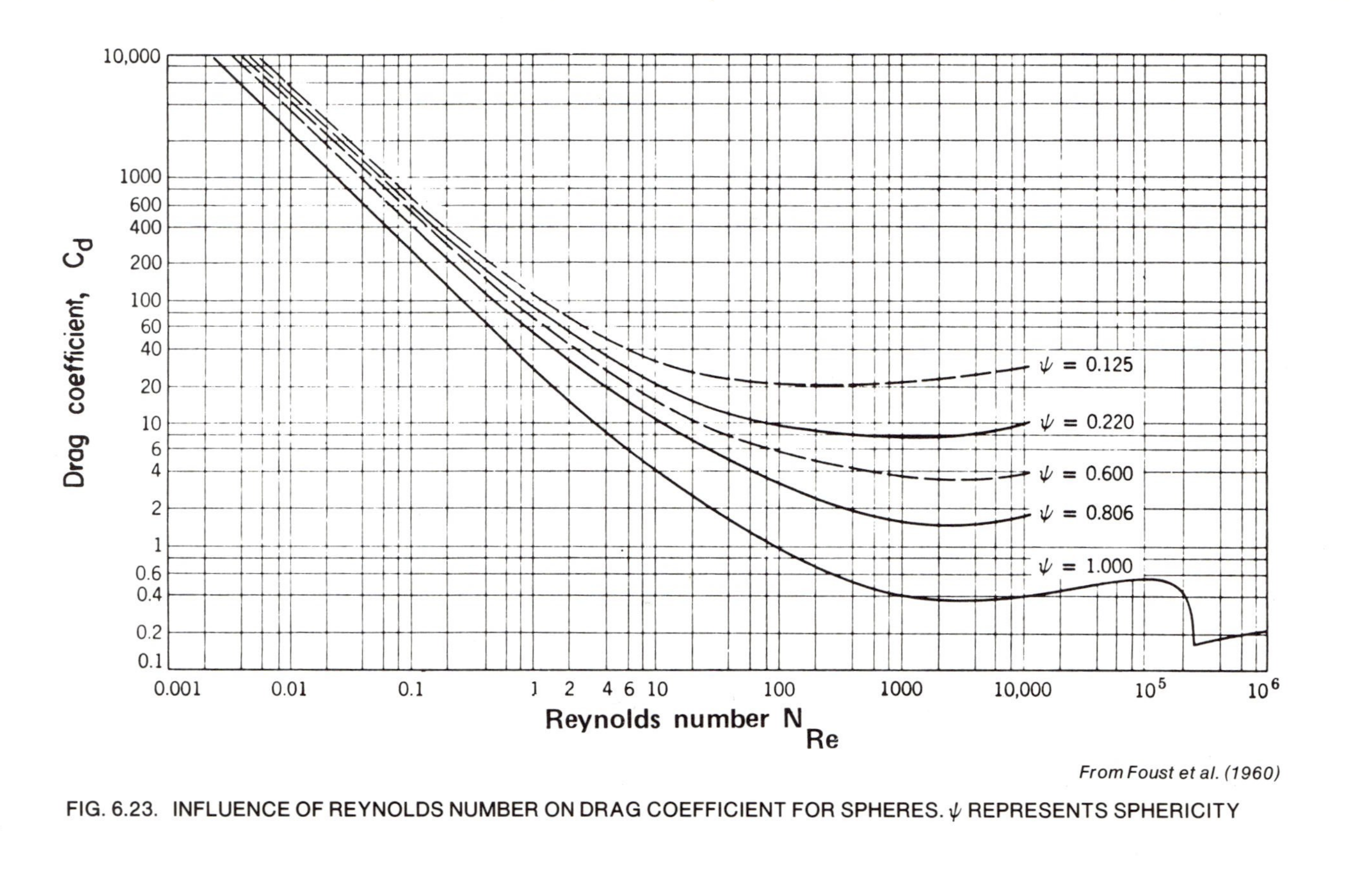


*From Foust et al. (1960)*

FIG. 6.23. INFLUENCE OF REYNOLDS NUMBER ON DRAG COEFFICIENT FOR SPHERES. ψ REPRESENTS SPHERICITY

where $M$ is the molecular weight of the diffusing vapor. The actual form of equation (6.53) was obtained by acknowledging that the heat-transfer coefficient could be expressed as a function of film thermal conductivity and particle diameter as follows:

$$h = \frac{2k_f}{d} \tag{6.55}$$

Equation (6.53) can be utilized to obtain an expression for the drying time in the constant rate period by integration in the following manner:

$$t_c = \frac{\rho_L L}{2\Delta T} \int_{d_0}^{0} \frac{dd}{h} = \frac{\rho_L L d_o^2}{8k_g(T_a - T_w)} \tag{6.56}$$

where the $o$ at the upper limit of integration indicates that the droplet completely disappears, as it would in the case of a pure liquid. Charm (1978) introduces a variation of this equation in which the particle size at the end of the constant-rate period is assumed and obtained the following expression:

$$t_c = \frac{L}{8k_g(T_a - T_w)} [\rho_1 d_1^2 - \rho_2 d_2^2] \tag{6.57}$$

where the subscripts (1 and 2) represent the initial and final conditions of the constant-rate drying period. Since equation (6.55) was utilized in arriving at the expressions given as equations (6.56) and (6.57), they apply only for low Reynolds number or stationary droplets.

EXAMPLE 6.7 Compare the constant-rate drying time for 20-micron droplets of skim milk when computed with and without the assumption of finite particle size at the end of the drying period. The air for drying is 115 C with 2% RH. For the case of finite particle size after drying, the particle has a 10-micron diameter at the completion of the constant-rate period. The latent heat of vaporization is 2177 kJ/kg.

SOLUTION:

(1) The use of equation (6.56) requires the following input parameters (estimated for skim milk):

$$\rho_L = 1035 \text{ kg/m}^3 \qquad T_a = 115 \text{ C}$$

$$k_g = 0.035 \text{ W/mK} \qquad T_w = 41 \text{ C}$$

(2) Using equation (6.56):

$$t_c = \frac{(1035 \text{ kg/m}^3)(2177 \text{ kJ/kg})(20 \times 10^{-6}\text{m})^2(1000 \text{ J/kJ})}{8(0.035 \text{ W/mK})(115 - 41)(3600 \text{ sec/hr})}$$

$$t_c = 1.208 \times 10^{-5} \text{ hr} = 0.0435 \text{ s}$$

(3) The use of equation (6.57) requires knowledge of additional parameter:

(a) Particle size after constant-rate drying will be:

$$d_2 = 10 \times 10^{-6} \text{ m}$$

(b) In order to determine the change in density due to drying, the following computations are required:

initial volume = $4/3\ \pi(10 \times 10^{-6})^3 = 4.19 \times 10^{-15}\text{m}^3$

final volume = $4/3\ \pi(5 \times 10^{-6})^3 = 5.24 \times 10^{-16}\text{m}^3$

(c) Volume change = $3.665 \times 10^{-15}$ m$^3$

(d) Weight loss (with water at 961 kg/m$^3$):

$= 3.665 \times 10^{-15} \text{ m}^3 \times 961 \text{ kg/m}^3$

$= 3.52 \times 10^{-12}$ kg water

(e) Density change will be function of the mass of water loss and the volume change:

$$\rho_2 = \frac{\text{initial mass of droplet} - \text{water mass lost}}{\text{initial droplet volume} - \text{volume change}}$$

$$= \frac{(4.34 \times 10^{-12} - 3.52 \times 10^{-12})\text{ kg}}{(4.19 \times 10^{-15} - 3.665 \times 10^{-15})\text{m}^3}$$

$$\rho_2 = 1561.9 \text{ kg/m}^3$$

(4) Using equation (6.57):

$$t_c = \frac{(2177 \text{ kJ/kg})(1000 \text{ J/kJ})}{8(0.035 \text{ W/mK})(115 - 41)(3600 \text{ sec/hr})}$$

$$[(1035 \text{ kg/m}^3)(20 \times 10^{-6}\text{m})^2$$

$$- (1561.9 \text{ kg/m}^3)(10 \times 10^{-6}\text{m})^2]$$

$$= (29.185)(2.5781 \times 10^{-7})$$

$$= 7.52 \times 10^{-6} \text{ hr} = 0.027 \text{ s}$$

Marshall (1954) developed an empirical expression for drying time in the constant-rate period by introducing an expression for the convective heat-transfer coefficient which applies in the range of Reynolds numbers greater than 20 and utilizes the terminal velocity of the droplet. The expression obtained was:

$$t_c = \frac{\rho_L L d_0^2}{8k_g(T_a - T_w)} \left[1 - \frac{2\beta_0}{d_0^2}\right] \int_{100}^{d_0} \frac{d^{2.08}\, dd}{(1 + \beta_0 d^{1.08})} \tag{6.58}$$

where:

$$\beta_0 = 0.3 \left[1.53 \frac{g\rho_a(\rho_L - \rho_a)^{0.71}}{\mu^2}\right]^{1/2} \left[\frac{c_p \mu}{k_g}\right]^{1/3} \tag{6.59}$$

where the lower limit of the integration in equation (6.58) is 100 to indicate that droplets must have 100-micron diameter for the terminal velocity to be significant in the computation. The terms outside the brackets in equation (6.58) are the same as those for low Reynolds number given in equation (6.56). The term in brackets represents a velocity direction factor or a direction term necessary to account for higher Reynolds number.

At some point during the spray-drying process, the droplet reaches some constant size, as illustrated by Fig. 6.24. In addition, the droplet temperature begins to increase more rapidly as indicated by the decrease in temperature difference ($\Delta T$) between air and particle in Fig. 6.24. These results obtained by Ranz and Marshall (1952) seem to establish the end of the constant-rate drying period and the beginning of the falling-rate drying period for spray drying. The time for completing the first phase of drying can no longer be based on the changes in droplet diameter but becomes a function of moisture content of the constant-diameter particle. The moisture content of the droplet or particle at the end of the constant-rate period is usually referred to as the critical moisture content.

A heat balance on the suspended particle during the falling-rate period is given by the following expression:

$$\rho_p V_p L \frac{dw}{dt} = hA\,(T_a - T_w) + \rho_p\, V_p c_p\, \frac{dT}{dt} \tag{6.60}$$

which can be used to describe the rate of moisture change in the particle as illustrated by the following expression:

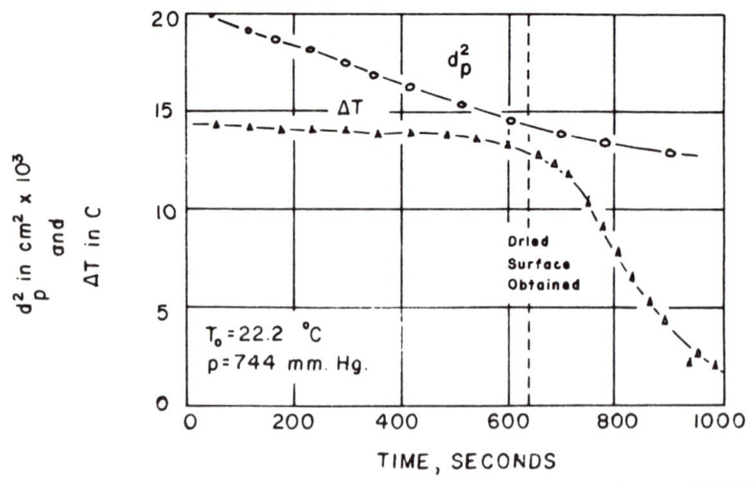


From Ranz and Marshall (1952)

FIG. 6.24. VARIATIONS IN DROPLET SIZE AND TEMPERATURE WITH TIME DURING SPRAY DRYING

$$\frac{dw}{dt} = \frac{hA(T_a - T_w)}{\rho_p V_p L} + \frac{c_p}{L}\frac{dT}{dt} \tag{6.61}$$

According to results obtained by Ranz and Marshall (1952), the last term of equation (6.61) accounts for only 3% of the moisture transfer and can be dropped from consideration without causing significant error. The remaining portion of equation (6.61) can be integrated to obtain the following expression for drying time during the falling-rate period:

$$t_F = \frac{\rho_P d_c L(w_c - w_e)}{6h\Delta T_{ave}} \tag{6.62}$$

where $\Delta T_{ave}$ represents the time average temperature difference between particle and air during the falling-rate drying period. An expression of the total drying time for a droplet in static air ($N_{Re} \approx 0$) would be the following:

$$t = \frac{\rho_L L d_o^2}{8k_g(T_a - T_w)} + \frac{\rho_p d_c L(w_c - w_e)}{6h\,\Delta T_{ave}} \tag{6.63}$$

which represents the sum of equations (6.56) and (6.62).

EXAMPLE 6.8 Estimate the time required to complete the falling-rate drying period for conditions presented in Example 6.7.

SOLUTION

(1) The falling-rate drying time can be estimated from equation (6.62) after moisture contents at the beginning and end of the period are known.

(2) Using an initial total solids content of 10% for the skim milk, the mass of water per unit volume is 0.9 (1035 kg/m$^3$) = 931.5 kg water/m$^3$ and the solids content would be 103.5 kg solids/m$^3$.

(3) Based on an initial droplet volume of $4.19 \times 10^{-15}$ m$^3$ (from Example 6.8), the droplet contains $4.337 \times 10^{-13}$ kg solids and $3.9 \times 10^{-12}$ kg $H_20$

(4) Since the droplet evaporates $3.52 \times 10^{-12}$ kg water during the constant-rate period, the water in the droplet at the beginning of the falling-rate period becomes

$$3.9 \times 10^{-12} - 3.52 \times 10^{-12} = 3.8 \times 10^{-13} \text{ kg } H_2O$$

and the moisture content is

$$w_c = \frac{3.8 \times 10^{-13} \text{ kg } H_2O}{4.337 \times 10^{-13} \text{ solids}} = 8.76 \text{ kg } H_2O/\text{kg solids}$$

(5) The final moisture content of the product will be 0.05 kg $H_2O$/kg product or 0.0526 kg $H_2O$/kg solids.

(6) Since the particle density does not change during the falling-rate drying period, the density will be 1561.9 kg/m$^3$.

(7) The average temperature difference during the falling-rate period can be computed assuming a linear change during drying and equal temperature of product and air at the end of the drying period:

$$\Delta T_{ave} = \frac{(115 - 41) + 0}{2} = \frac{74}{2} = 37\,C$$

(8) Equation (6.62) can be modified by recognizing

$$h = \frac{2k_f}{d} \quad \text{for } N_{Re} < 20$$

$$t_F = \frac{\rho_p d_c^2 L (w_c - w_e)}{12\, k_f\, T_{ave}}$$

(9) The time for falling-rate period drying becomes:

$$t_F = \frac{(1561.9\,kg/m^3)(10 \times 10^{-6} m)^2 (2177\,kJ/kg)(1000\,J/kJ)(0.876 - 0.0526)}{12(0.035\,W/mK)(37\,K)(3600\,sec/hr)}$$

$t_F$ = 5.005 × $10^{-6}$ hr = 0.018 sec.

Based on equation (6.63), drying times will vary with droplet diameter and with the size of the droplet or particle when it reaches the critical moisture content. In an actual spray-drying situation, the spray produced by the atomizer will have the droplet size distribution which causes a corresponding distribution of drying times. The drying time for this situation will be controlled by the time required for the larger droplets, while the smaller droplets will be dried in a shorter length of time. In addition, product solids contained in the smaller particles will be exposed to higher temperatures before drying of the larger droplets is completed. Results obtained by Probert (1946) indicated that sprays with small mean droplet sizes and narrow size distributions will evaporate in a shorter time than sprays with larger mean droplet sizes and larger size distributions.

Charlesworth and Marshall (1960) conducted a rather detailed study of the physical changes in an atomized droplet during spray drying. Utilizing sophisticated photographic techniques, these reseachers were able to visually observe the decrease in droplet size during evaporation of moisture in the constant-rate drying period and the actual formation of solid-particle crystals near the bottom of the droplet. This phase was followed by what the researchers called a surface crust, which corresponded to the end of the constant-rate drying period. The changes which occurred in the particle after the formation of the initial surface crust were dependent on the characteristics of the solids in the droplet. The final shape of the dry particle depended on several factors, including (a) the type of solids, (b) whether the air temperature was above or below the boiling point of the liquid in the droplets, and (c) the characteristics of the crust which formed at the particle surface. Charlesworth and Marshall (1960) were able to conclude from their studies that the moisture content of any solid at the time of crust formation was independent of initial droplet size, initial solids concentration, and the drying conditions.

An additional factor which influences the design of the spray-drying chamber is the air-flow pattern within the drying chamber. This air-flow pattern influences the direction of droplet or particle movement and the residence time of the droplet and particle in the drying chamber. Unfortunately, very little research has been done which has led to acceptable design expressions. Edeling (1950) conducted research on the droplet projectory when exposed to spiral flow. His development assumed that three forces were acting on the droplet or particle, as illustrated in Fig. 6.25, namely, air drag, centrifugal force, and the force due to gravity. Edeling (1950) presented the following expression for prediction of the tangential velocity of the droplet:

$$u_t = (u_t)_0 \sqrt{\frac{r}{R_0}} \tag{6.64}$$

where $(u_t)_o$ represents the tangential velocity at the drying-tower wall. The second expression presented a ratio of the radial velocity to the terminal velocity as follows:

$$\frac{u_R}{u_t} = \frac{gR_0}{(u_t)_0^2} \tag{6.65}$$

Both equations (6.64) and (6.65) require knowledge of the tangential velocity of the particle at the spray-tower wall. This is the most obvious limitation on the use of these equations, although other assumptions are required for their use in design computations. (See Fig. 6.29, page 314.)

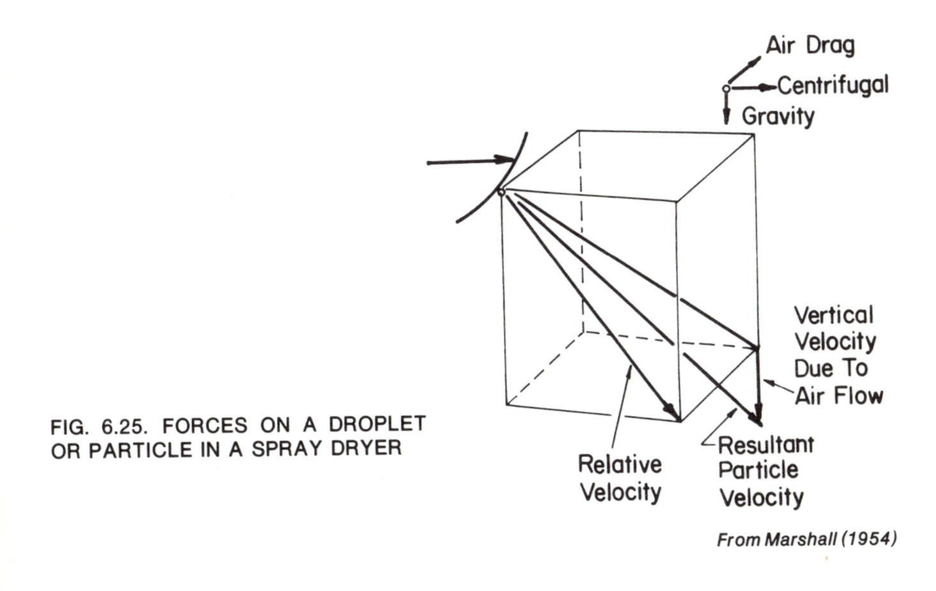


FIG. 6.25. FORCES ON A DROPLET OR PARTICLE IN A SPRAY DRYER

EXAMPLE 6.9 A co-current spray dryer is being designed for drying a product with 20% total solids. Ambient air at 30 C and 70% RH is being heated to 110 C before entering the dryer. The product droplets generated by the atomizer range in diameter from 40 to 70 microns and the moisture content of the particles is 0.5 kg water/kg product at the beginning of the falling-rate drying period. The product leaves the dryer in equilibrium with the air at the exit temperature of the dryer (55 C) at a moisture content of 0.04 kg water/kg product. Assume the latent heat of vaporization is 2325 kJ/kg. Estimate the size of the drying chamber required to dry the product when the feed rate is 1000 kg/hr at 20 C.

SOLUTION

(1) Two factors of importance in estimating the size of the drying chamber are the volume of air required for drying and the time required to achieve the desired amount of drying.

(2) The first step in the computation is a mass and energy balance:

for product:

$$1000 = m_p + m_w^o$$

for product solids:

$$1000\ (0.2) = m_p\ (0.96)$$

(3) Solution of the mass balance indicates:

$$m_p = \frac{1000\,(0.2)}{0.96} = 208.3 \text{ kg/hr (product leaving the dryer)}$$

and:

$$m_w^o = 1000 - 208.3 = 791.7 \text{ kg/hr (water leaving the dryer).}$$

(4) The energy balance on the drying chamber would be:

$$(1000 \text{ kg/hr})(3.68 \text{ kJ/kgK})(20 - 0) + m_a^i\,(c_{pa}^i)(110 - 0) =$$

$$m_w^o\,(2325 \text{ kJ/kg}) + m_a^o\,(c_p^o)(55 - 0) + m_p\,(1.775 \text{ kJ/kgK})(55 - 0)$$

where:

$c_p^i$ = 3.68 kJ/kgK = specific heat of product entering as estimated from water content.

$c_p^o$ = 1.775 kJ/kgK = specific heat of product out as estimated from moisture content.

$c_{pa}^i = 1.005 + 1.88\ (0.0188) = 1.0403$ kJ/kgK

and the energy balance uses 0 C as a reference for energy content.

(5) The energy balance equation can be solved by recognizing that $m_a^i = m_a^o$ on a dry air basis and estimating

$c_{pa}^o = 1.005 + 1.88\ (0.04) = 1.0802$ kJ/kgK for air leaving the dryer at 55°C and 38% RH.

then:

$$73600 + m_a^i\,(1.0403 \text{ kJ/kgK})(110 - 0) = (791.7)(2325) +$$
$$m_a^i\,(1.0802 \text{ kJ/kgK})(55 - 0) + (208.3)(1.775)(55 - 0)$$

or

$$73600 + 114.4\, m_a^i = 1840702.5 + 59.4\, m_a^i + 20335.3$$
$$55.0\, m_a^i = 1787437.8$$
$$m_a^i = 32498.9 \text{ kg/hr dry air}$$

(6) Using the absolute humidity ($H_A$) of 0.0188 kg water/kg dry air at the dryer inlet:

$$m_w^i = 32498.9(0.0188) = 611 \text{ kg water/hr}$$

then:

$$m_a^i = 33109.9 \text{ kg air/hr entering dryer}$$

and:

$$m_a^o = 33109.9 + 791.7 = 33901.6 \text{ kg air/hr leaving the dryer.}$$

(7) The actual absolute humidity of air leaving the dryer would be:

$$H_A = \frac{706 + 791.7}{32090.5} = 0.0432 \text{ kg water/kg dry air}$$

which is slightly higher than the estimated value in step (5).

(8) Using the psychrometric chart (Fig. A.5), a specific volume of 1.12 $m^3$/kg dry air is obtained for the inlet air. Based on this value:

$$\text{Volumetric flow rate} = (32498.9)(1.12) = 36399 \text{ m}^3\text{/hr}$$

or

$$607 \text{ m}^3\text{/min}$$

(9) The total drying time for the product droplets can be estimated using equation (6.63); using the following parameter values:

$\rho_L$ = 1075 kg/$m^3$ (estimated for liquid product)
$d_o$ = 70 micron (using largest droplet size for longest drying time).
$L$ = 2325 kJ/kg
$k_g$ = 0.032 W/mK (for air at 110 C)
$T_a$ = 110 C
$T_w$ = 41 C (from psychrometric chart)
$\rho_p$ = 1440 kg/$m^3$ (assumed)
$d_c$ = unknown
$w_c$ = 0.5 kg water/kg product = 1.0 kg water/kg solids.
$w_e$ = 0.04 kg water/kg product = 0.0417 kg water/kg solids.

$$\Delta T_{ave} = \frac{(110 - 41) + 0}{2} = 34.5 \text{ C}$$

(10) In order to compute the droplet diameter at the end of the constant-rate drying period, the change in water content can be used:

(a) Initially, the amount of water in a droplet (70 micron) will be:

(Volume)(Density)(Moisture Content)

so:

$$(4/3)\, \pi\, (35 \times 10^{-6})^3 (1075)(0.8) = 1.5445 \times 10^{-10} \text{ kg water}$$

(b) Since the moisture content at the beginning of the falling-rate drying period is 0.5 kg water/kg product, then 3/8 of the water mass has evaporated and:

$$(5/8)\,(1.5445 \times 10^{-10}) = (4/3)\,\pi\, r_c^3\,(1440)(0.5)$$

$$r_c^3 = \frac{(5/8)\,(1.5445 \times 10^{-10})}{(4/3)\,\pi\,(1440)(0.5)} = 3.2 \times 10^{-14}$$

$$r_c = 3.175 \times 10^{-5}\ \text{m}$$

$$r_c = 31.75\ \text{micron}$$

(11) Using equation (6.63) with $h = 2k_g/d_c$:

$$t = \frac{(1075\ \text{kg/m}^3)(2325\ \text{kJ/kg})(70 \times 10^{-6}\ \text{m})^2(1000\ \text{J/kJ})}{8\,(0.032\ \text{W/mK})(110 - 41)(3600\ \text{sec/hr})}$$

$$+ \frac{(1440\ \text{kg/m}^3)(63.5 \times 10^{-6}\ \text{m})^2(2325\ \text{kJ/kg})(1 - 0.0417)(1000\ \text{J/kJ})}{12\,(0.032\ \text{WmK})(34.5\ \text{K})(3600\ \text{sec/hr})}$$

$$t = 1.926 \times 10^{-4} + 2.713 \times 10^{-4} = 4.64 \times 10^{-4}\ \text{hr}$$

$$t = 1.67\ \text{s}$$

(12) The answer in step (11) indicates that the largest droplet generated by the atomizer should have a minimum residence time in the drying chamber of 1.67 sec.

(13) Since the length and diameter of the chamber are not independent dimensions, a 2 meter diameter will be assumed.

(14) Using the volumetric air flow rate of 607 m$^3$/min, the mean air speed can be computed:

$$\bar{u} = \frac{607}{\pi(1)^2} = 193.2\ \text{m/min} = 3.22\ \text{m/sec}$$

(15) In order to assure sufficient residence time, the length of the chamber would be:

$$\text{Length} = 3.22\ \text{m/sec} \times 1.67\ \text{sec} = 5.38\ \text{meters.}$$

(16) Chamber dimensions of 2 meter diameter and 5.38 meter length would be initial values of the design. Factors such as air flow patterns and deviations from single droplet drying would cause variations to be considered in a more detailed analysis.

Marshall (1954) concluded that the design of a spray dryer depends on several variables including the following: (a) the features and type of product, (b) the atomization methods, (c) the product solids content and properties of the product, (d) the drying temperature and the corresponding influence of product temperature, (e) the procedure used for handling the product, and (f) the method used for product recovery. A detailed analysis of these factors is presented by Masters (1976).

### 6.5.2 Pneumatic Drying

If the product to be dried has granular, flaky or powdery characteristics before drying, a pneumatic dryer of the type illustrated in Fig. 6.26 may be utilized for the drying process. In this type of dryer, heated air carries the product through the drying zone and into a separating section. Utilizing this approach, very intimate contact between the product and the drying air is attained, and if the exposure time is kept to a minimum, product quality may be maintained. In some applications, retention time is sufficiently low to allow temperatures up to 1100 C to be utilized for the drying air. The pneumatic-type dryer is particularly useful when most of the moisture to be removed from the product is removed during the constant-rate drying period. If the drying rate is low, the design can be changed sufficiently to allow additional passes of the material through the drying section.

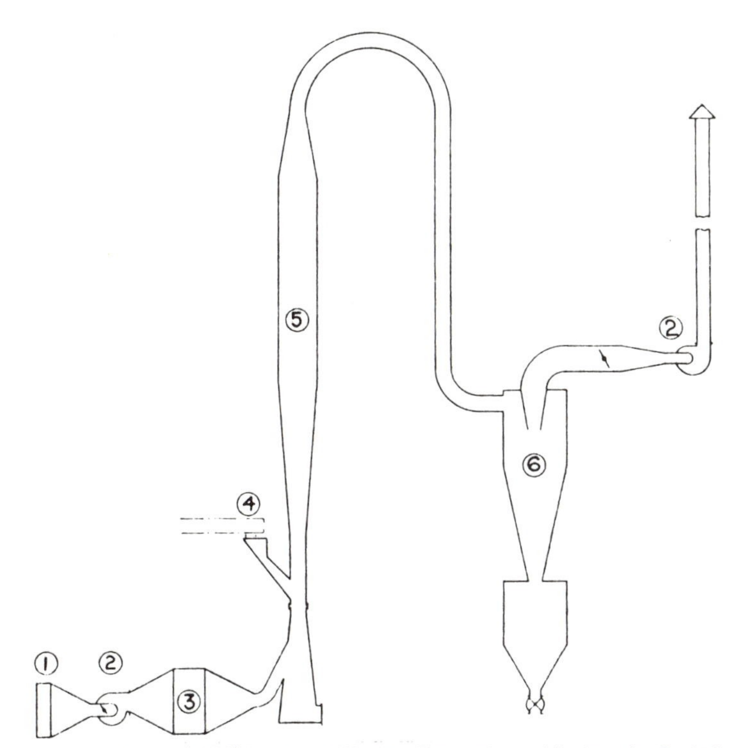


*From Forrest (1968) Courtesy of Kestner Evaporator and Engineering Co. Ltd.*

FIG. 6.26. SCHEMATIC ILLUSTRATION OF PNEUMATIC DRYER
1 Air filter
2 Fans
3 Heater
4 Product feed
5 Drying section
6 Separator

### 6.5.3 Fluidized-Bed Drying

The same intimate contact between product and drying air as illustrated in the pneumatic dryer can be achieved in what is usually called a fluidized-bed dryer. During the fluidized-bed process, air is forced through product particles at a sufficiently high velocity to overcome the gravitational forces on the product and maintain the particles in a suspended state. Even though the product is suspended throughout the drying process, it will move horizontally as drying proceeds and in most applications very uniform moisture contents are achieved. The air velocities required for maintaining the particles in a suspended state will vary with the product and more specifically with the particle size and density. Usually the air velocities are in the range of .05 m/s to 0.75 m/s. Variations of the normal design for products which are not easily fluidized include a vibratory conveyor which tends to assist in maintaining particle suspension in the air and may even allow lower air velocities. A spouted fluidized-bed dryer is a variation of the original and normal design. The product particles are blown into suspension by a jet of air directed vertically into a chamber. The particles in the center of the air jet are maintained in suspension to achieve the desired dehydration. Since the contact time is relatively short, the approach appears to be less effective than other fluidized-bed designs.

Forrest (1968) describes two additional types of suspended-product dryers—the ring dryer and the flowing-particle dryer. Although these types involve variations in design of the equipment, dehydration is accomplished in a manner very similar to the pneumatic dryer and the fluidized-bed dryer.

## 6.6. DRUM DEHYDRATION

A heated cylindrical drum which rotates around a horizontal axis can be used for continuous drying of product slurries or solutions. The product dries while adhering to the drum surface and is removed by the scraping action of a doctor knife. Drying occurs in the product film as a result of conduction heat transfer from the drum surface to the product. This results in relatively high heat-transfer coefficients but can introduce significant quality reduction when drying heat-sensitive products. This quality reduction can be minimized by assuring that the product film has uniform thickness and that the dry film is removed completely.

### 6.6.1 Types of Drum Dryers

In general, two types of drum dryers are utilzied in the food industry, namely, single- and double-drum dryers, as illustrated in Fig. 6.27. The single-drum type may establish a product film by having the heated

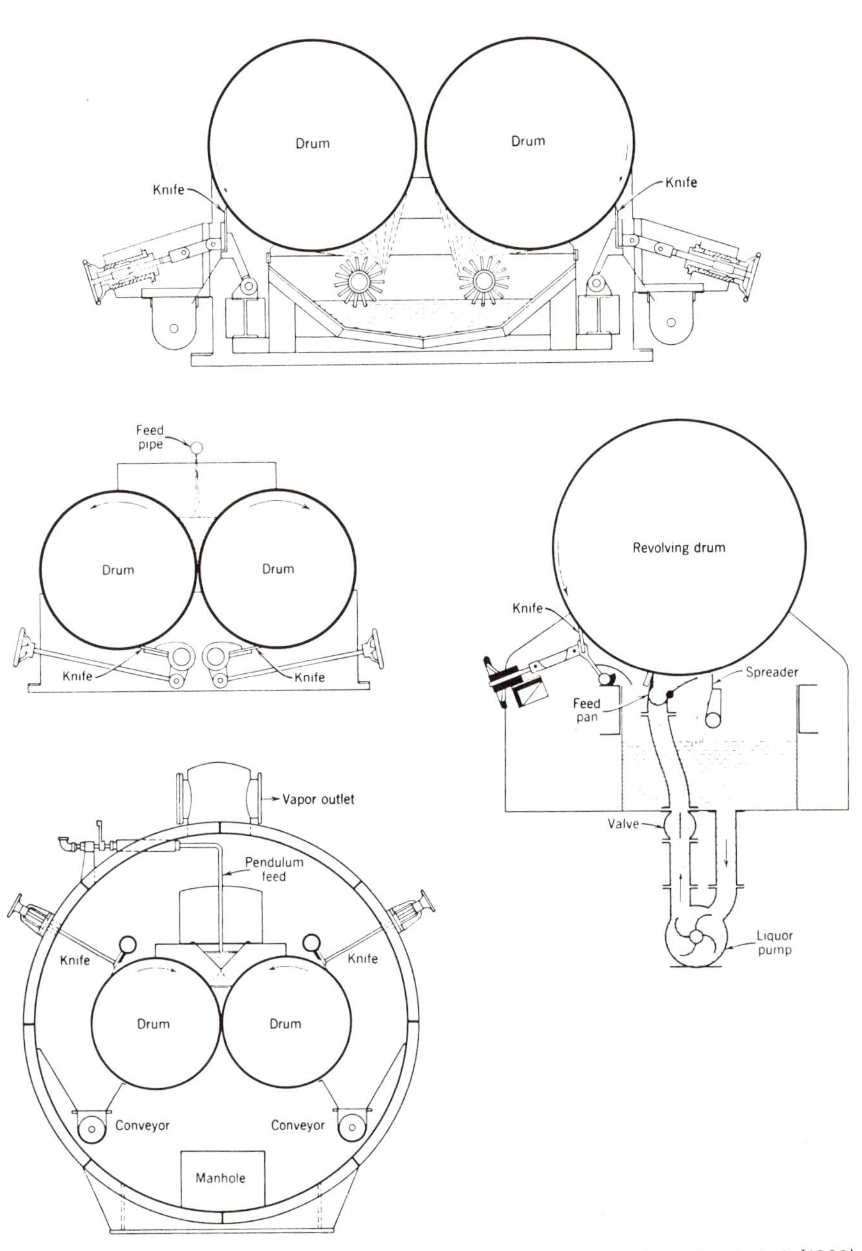


*From Foust et al. (1960)*

FIG. 6.27. SCHEMATIC ILLUSTRATION OF VARIOUS DRUM DRYER CONFIGURATIONS

drum dip directly into the product slurry or solution. Other types of single-drum dryers are designed with an auxiliary roller to be used in application of the product film to the heated drum.

Double-drum dryers are designed with two heated drums mounted parallel in a manner which allows the product to be fed into the region between and above the two drums. The drums rotate toward one another, product film thickness being controlled by the clearance between the two drums. In the twin-drum arrangement, the heated drums are rotating in opposite directions and the feed is normally accomplished by dipping directly into the product slurry.

### 6.6.2 Design Parameters

The basic equation which describes the rate of dehydration on a drum dryer is as follows:

$$\frac{dw}{dt} = \frac{UA\,\Delta T_m}{L} \tag{6.66}$$

where the temperature difference ($\Delta T_m$) is the mean temperature difference between the roller surface and the product. The overall heat-transfer coefficient ($U$) has been measured experimentally by Van Marle (1948); it usually ranges from 1000 $W/m^2$ C to 2000 $W/m^2$ C. It is obvious that the parameters in equation (6.66) will be influenced considerably by other operating parameters. Both the solids content of the slurry being fed to the dryer and drum speed will influence dehydration rate, as illustrated in Fig. 6.28. These data were obtained by Talburt and Smith (1967) while developing a procedure for drying potato slurry to obtain potato flakes. The results indicate that the production rate increased with drum speed and with increasing product solids. Talburt and Smith (1959) recommended an initial solids content of between 20 and 22% in the slurry in order to achieve good adhesion and a low moisture content in the final product without scorching. Similar results were obtained by Bakker-Arkema *et al.* (1967) while developing procedures for drum-drying pea beans. The same increase in drying rate with increasing drum speed was obtained, as shown in Fig. 6.29. These results were obtained, however, without addition of moisture to obtain a bean slurry. In addition, the number of product layers which were allowed to accumulate on the drum influenced the production rate. These layers were accumulated by lifting the doctor blade and allowing it to remove product only on selected cycles. It is obvious from Fig. 6.29 that increasing the number of layers increased the production rate. The influence of clearance between rollers cannot be overemphasized on this type of drum dryer. A clearance of $2.5 \times 10^{-4}$ to $3.0 \times 10^{-3}$m appears to be optimum.

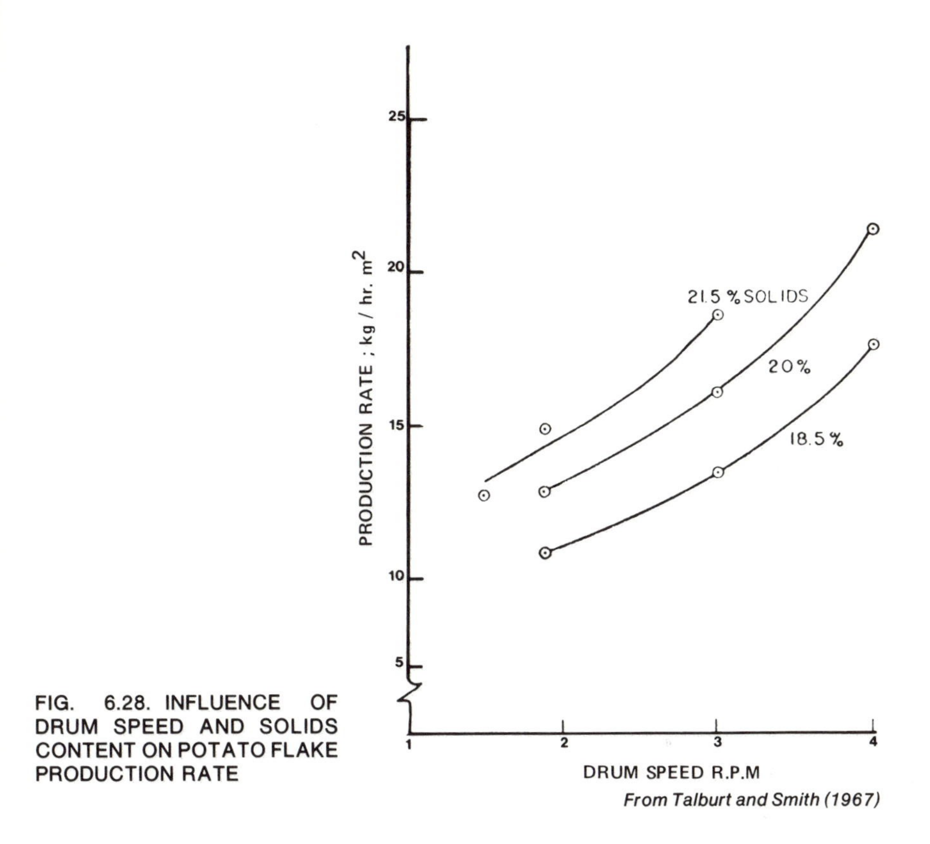


FIG. 6.28. INFLUENCE OF DRUM SPEED AND SOLIDS CONTENT ON POTATO FLAKE PRODUCTION RATE

EXAMPLE 6.10 A drum dryer is being designed for drying of a product from an initial total solids content of 12% to a moisture content of 4%. An overall heat-transfer coefficient ($U$) of 1700 W/m² C is being estimated for the product. An average temperature difference between the roller surface and the product of 85 C will be used for design purposes. Determine the surface area of the roller required to provide a production rate of 20 kg product/hr.

SOLUTION

(1) The production rate of 20 kg product/hr corresponds to:

$$20 \frac{\text{kg product}}{\text{hr}} \times 0.96 \frac{\text{kg solids}}{\text{kg product}} = 19.2 \frac{\text{kg solids}}{\text{hr}}$$

(2) The feed rate must be:

$$\frac{19.2 \text{ kg solids/hr}}{0.12 \text{ kg solids/kg product}} = 160 \text{ kg product feed/hr}$$

(3) The water must be removed from product at the following rate: 160 kg product feed/hr − 20 kg product/hr = 140 kg water/hr

(4) Using equation (6.66), using $L$ = 2420 kJ/kg:

$$140 \frac{\text{kg water}}{\text{hr}} = \frac{1700 A (85)}{2420}$$

$$A = 2.35 \text{ m}^2$$

(5) The effective area on the roller needed for drying is 2.35 $m^2$. Assuming that all the area on a 0.75 m diameter by 1 m long cylinder was used for drying, the size indicated would be adequate. Assuming that about one-third of the surface area would not be used, a 1 m diameter by 1 m length would be suggested.

## 6.7. MISCELLANEOUS DEHYDRATION PROCESSES

In the description of drying processes by Forrest (1968), several additional procedures for food dehydration were described. These processes include the rotary dryer, which revolves while the product moves through the unit and air is directed through the circular openings of the revolving cylinder. In addition, a classification of vacuum dryers was provided to describe dryers which specifically utilize vacuum in the dehydration of food products. Specific types of vacuum dryers include

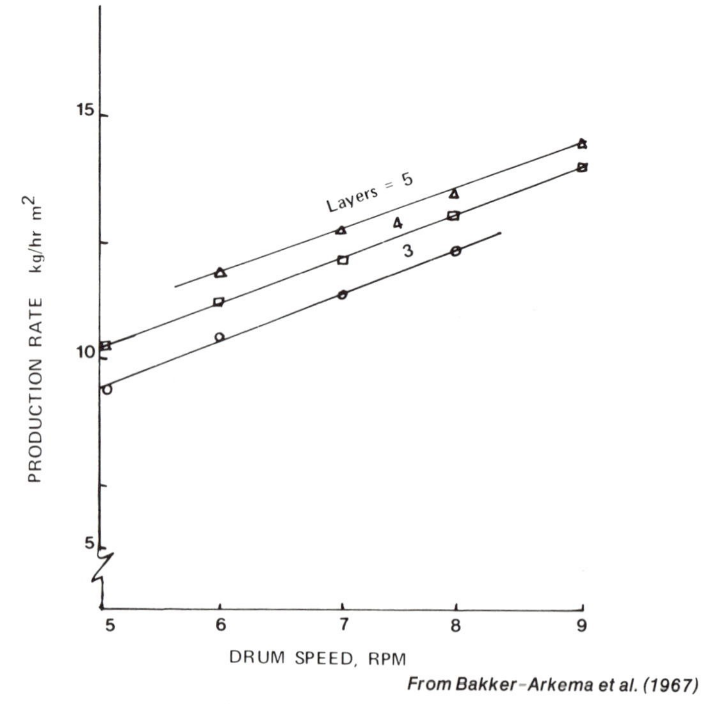


*From Bakker-Arkema et al. (1967)*

FIG. 6.29. INFLUENCE OF DRUM SPEED AND NUMBER OF PRODUCT LAYERS ON PRODUCTION RATE OF BEAN POWDER FROM NONPUREED BEANS

the shelf dryer, the agitated pan dryer, and the continuous dryer. Probably the most popular of these types is the continuous dryer in which the drying occurs on a moving belt inside a vacuum chamber.

## 6.8. FREEZE DEHYDRATION

Moisture removal from food products can be accomplished by freeze-drying, which involves the sublimation of water from the frozen state directly to the vapor state. The obvious advantage of the freeze-dehydration process is that moisture removal or dehydration can be accomplished without exposing the product to excessively high temperatures. In addition, it is normally possible to maintain the product structure in a more acceptable state, resulting in a higher-quality product. In general, the process involves initial freezing followed by application of heat to the product surfaces, resulting in sublimation of ice. The initial application of heat at the product surfaces results in sublimation of ice at that point and immediate vapor removal. This initial sublimation and water-vapor removal results in a receding of the ice front. As the ice front moves away from the product surface, two phenomena, heat transfer or vapor diffusion, establish the rate of freeze-drying for the product.

The basic components of a freeze-dryer include an evaporator, where heat is generated to be used as a source of energy for drying, and a condenser which collects the vapors produced by the product. Both the condenser and the evaporator are located in a vacuum chamber and the vacuum is maintained by means of a vacuum pump or steam ejector.

### 6.8.1 Heat and Mass Transfer

The basic transport phenomena involved in freeze-drying are heat transfer and mass transfer within the product as sublimation occurs. For the one-dimensional case illustrated in Fig. 6.30, heat- and mass-transfer

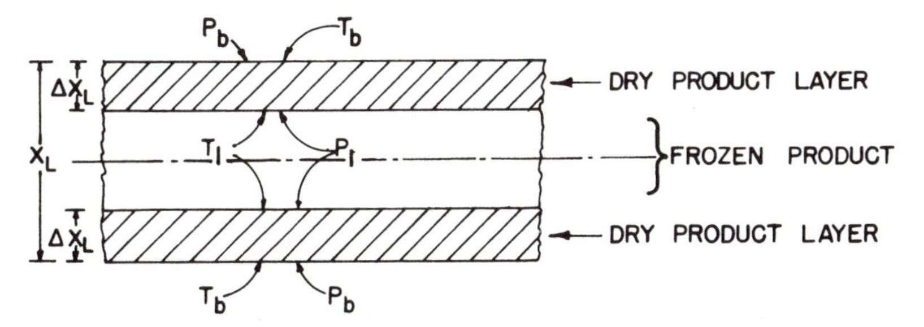


FIG. 6.30. SCHEMATIC ILLUSTRATION OF PRODUCT FREEZE-DRYING

equations can be written to describe the existing situation. Heat being transferred into the product can be described by the following expression:

$$q = hA(T_a - T_b) = \frac{kA}{\Delta X_L}(T_b - T_I) \quad (6.67)$$

where $\Delta X_L$ represents the thickness of dry material through which heat must transfer to reach the ice front. Equation (6.67) can be rewritten as a series expression as follows:

$$q = \frac{A(T_a - T_I)}{1/h + k/\Delta X_L} \quad (6.68)$$

where the temperature gradient represents the difference between the external temperature and the temperature at the ice front. The transfer of vapor from the ice front to the surrounding medium can be described in a similar way as follows:

$$N_A = \frac{k_m A}{R_g T_A}(p_b - p_a) = \frac{DA}{\Delta X_L R_g T_A}(p_I - p_b) \quad (6.69)$$

where $k_m$ represents the mass transfer coefficient which describes transfer of vapor from the surface to the surrounding medium. An expression similar to the series heat transfer equation can be written for mass transfer as follows:

$$N_A = \frac{A(p_I - p_a)}{(1/k_m + \Delta X_L/D)R_g T_A} \quad (6.70)$$

The thickness of the dried layer ($\Delta X_L$) will vary with drying time, but can be expressed as a function of the fraction of water remaining in the product, as illustrated by Sandall *et al.* (1967). The magnitude of the diffusion coefficient ($D$) in equation (6.69) and (6.70) depends on the mechanism of moisture diffusion in the dry product layer. In general, all the mechanisms discussed as mechanisms of dehydration earlier in the chapter (equations 6.9 to 6.17) can account for vapor diffusion in the freeze-dried product. According to Sandall *et al.* (1967), these mechanisms include bulk diffusion, Knudsen diffusion, slip flow, and Poiseuille flow. In most commercial freeze-dryers, heat transferred from a heated plate to the product surface is by radiation. Under these circumstances, the rate of drying is limited by the internal heat-transfer rate or the thermal conductivity of the dry product. In laboratory units, heat transfer may be by conduction from a heated plate to a product surface and conduction through the frozen product layer.

### 6.8.2 Freeze-drying Times

Several approaches have been proposed for the computation of drying time for freeze dehydration. Mink and Sachsel (1962) presented simplified expressions based on heat and mass transfer. For the case of heat transfer, the expression is:

$$t = \frac{L_s X_L^2}{4k(T_a - T_I)} \tag{6.71}$$

where $X_L$ is the thickness of the product being dried. If the expression is based on mass transfer, the following equation results:

$$t = \frac{B X_L^2}{16 k_m} \tag{6.72}$$

where $B$ represents the mass of water per unit volume of product. Although these expressions are rather easy to utilize in computations and to modify for various cases, the results are somewhat limited in accuracy. Charm (1978) utilized the solution to the Neumann problem (presented in Chapter 4) to compute drying times during freeze dehydration. This procedure is somewhat more complex to use and may not yield accurate answers due to assumptions made in the determination of average surface temperature. Forrest (1968) utilized the following expression:

$$\frac{dw}{dt} = pA\,(w_0 - w_e)\,\frac{dx}{dt} \tag{6.73}$$

to obtain the following equation for drying time:

$$t = \frac{p(w_0 - w_e)X_L^2}{4a'(p_a - b')dP} \tag{6.74}$$

where:

$$a' = \frac{K'M}{\mu T_A}$$

$$b' = \frac{0.0133\,\lambda^{0.8}}{\sqrt{K'}}$$

Equation (6.74) appears to be an acceptable approach for computing drying time during freeze dehydration, although its application to a specific situation has not been illustrated.

The uniformly retreating ice front (URIF) model developed for a slab geometry by King (1970) provides an alternate approach to predicting freeze-drying rates as well as detailed analysis of the process. Utilizing Fig. 6.30, the thickness of the dry product layer ($\Delta X_L$) can be related to the fraction remaining moisture ($Y$) by:

$$\frac{\Delta X_L}{X_L} = \frac{1 - Y}{2} \tag{6.75}$$

Use of equation (6.69) assumes that all moisture is removed from the product as the ice front passes. The relationships between mass transfer rate ($N_A$) and the rate of moisture-content change ($dY/dt$) is described by the following expression:

$$N_A = \frac{X_L A}{2M V_W}\left(-\frac{dY}{dt}\right) \tag{6.76}$$

where $M$ is the molecular weight of water and $V_w$ is the volume occupied by a unit mass of water in product initially. By equations (6.70) and (6.76):

$$\frac{X_L A}{2M V_W}\left(-\frac{dY}{dt}\right) = \frac{A(p_I - p_a)}{(1/k_m + \Delta X_L/D)R_g T_A} \tag{6.77}$$

and incorporating $\Delta X_L$ from equation (6.75):

$$(1 - Y) = \frac{4DMV_W(p_I - p_a)}{R_g T_A X_L^2(-dY/dt)} - \frac{2D}{k_m X_L} \tag{6.78}$$

an expression relating freeze-drying rate and factors which influence the rate is obtained. Equation (6.78) can be integrated, assuming $(p_I - p_a)$ is constant and $Y = 0$, to obtain:

$$t = \frac{R_g T_A X_L^2}{8DM V_W(p_I - p_a)}\left[1 + \frac{4D}{k_m X_L}\right] \tag{6.79}$$

Equation (6.79) can be used to predict freeze-drying rates for slab geometry when the rate is internal mass-transfer-limited. A similar equation to describe freeze-drying rate for internal heat-transfer-limited can be derived in a similar way [King (1970), (1973)]:

$$t = \frac{L_s X_L^2}{8kM V_W(T_b - T_I)} \tag{6.80}$$

Both equations (6.79) and (6.80) have been used in several experimental investigations as discussed in considerable detail by King (1973).

EXAMPLE 6.11 A concentrated liquid product with 50% total solids is being freeze-dried by placing the frozen product in a pan which is suspended over a heated platen in the freeze-dryer. In order to evaluate the optimum drying conditions for the product, the appropriate heat- and mass-transfer properties of the product must be known or evaluated. The product weight history during a freeze-drying trial was recorded after freezing to −75 C. The platen temperature was 30 C and the average thickness of product was 1.55 cm. The following results were obtained:

| Time, $t$ (hr) | Product Weight | Time, $t$ (hr) | Product Weight |
|---|---|---|---|
| 0 | 5.13 | 32 | 3.35 |
| 4 | 4.80 | 36 | 3.15 |
| 8 | 4.58 | 40 | 2.97 |
| 12 | 4.36 | 44 | 2.82 |
| 16 | 4.16 | 48 | 2.72 |
| 20 | 3.96 | 52 | 2.59 |
| 24 | 3.75 | 56 | 2.58 |
| 28 | 3.55 | 60 | 2.58 |

Determine the rate-controlling parameters for the product and develop expressions to be used in scale-up of the process.

SOLUTION

(1) Using the URIF model, the product properties can be determined as illustrated by King (1970). In the case being considered, the process will be mass transfer-controlled, since heat transfer will occur through the frozen product layer and water vapor must be transported through the dry product layer.

(2) In addition, equation (6.78) must be modified slightly since heat and mass transfer will be occurring in only one direction. In integrated form, the modified equation is:

$$1 - Y = \frac{2DM\ V_W(p_I - p_a)t}{R_g T_A X_L^2 (1 - Y)} - \frac{2D}{k_m X_L}$$

This equation can be used to develop a plot of $(1 - Y)$ versus $t/(1 - Y)$, as illustrated in Fig. 6.32.

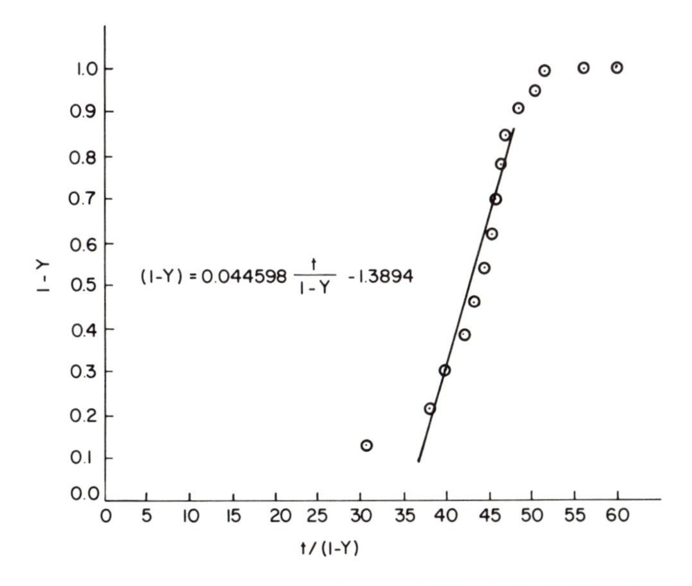


FIG. 6.31. PLOT USED TO EVALUATE MASS TRANSFER PARAMETERS IN EXAMPLE 6.11

(3) From Fig. 6.31:

$$\frac{2DM\ V_W(p_I - p_a)}{R_g T_A X_L^2} = 0.044598$$

where:

$$M = 18$$

$$V_w = \frac{1}{(w_0 - w_f)\rho_D} = \frac{1}{(1.0 - 0.035)(0.80)} = 1.3$$

where final moisture content ($w_f$) and density of dry solid ($\rho_D$) were measured.

$p_I$ = 0.592 mm Hg (for ice core temperature = −22.8 C)

$p_a$ = 0.00808 mm Hg (for condenser temperature −60 C)

$R$ = 62363.32 cm³ mm Hg/mole K (from Table A.13)

$T_A$ = 250.22 K

$X_L$ = 1.55 cm

so:

$$D = \frac{(0.044598)(62363.22)(250.22)(1.55)^2}{2(18)(1.3)(0.592 - 0.00808)}$$

$$= 61182.9\ \text{cm}^2/\text{hr}$$

$$D = 16.995\ \text{cm}^2/\text{sec}$$

In addition:

$$\frac{2D}{k_m X_L} = 1.38941$$

and:

$$k_m = \frac{2(16.995)}{(1.38941)(1.55)} = 15.78\ \text{g mole/sec cm}^2\ (\text{mm Hg})$$

(4) By rewriting the modified equation given in (2) with $Y = 0$, an expression relating freeze-drying time to product thickness can be developed:

$$t = \frac{R_g T_A}{2DM\ V_w(p_I - p_a)} X_L^2 + \frac{R_g T_A}{k_m M_t V_w (p_I - p_a)} X_L$$

$$t = 9.33 X_L^2 + 20.1 X_L$$

for freeze-drying time in hr and thickness in cm.

(5) The results illustrated the procedures for evaluating the mass-transfer properties of the product and the relationships between freeze-drying time and product thickness.

### 6.8.3 Influence of Parameters

Several product characteristics and other factors influence the rate of freeze-drying and other characteristics of the freeze-dehydration procedure. In some cases, attempts have been made to optimize these parameters so that the most rapid freeze-drying rate is obtained. In other situations, however, the parameters cannot be varied, as is the case with product characteristics. One example would be the percentage of water that remains unfrozen at the temperature to which the product is frozen before freeze-drying. As illustrated in Chapter 4, a relatively small percentage, although it may be significant, remains unfrozen at very low temperatures. Such would be the case when products are frozen to −25 C to −30 C before drying. This product characteristic results in a change in product temperature during the freeze-drying process. The initial temperature will be the sublimation temperature of the product; this temperature will normally be maintained until the frozen portion of the water is completely sublimed. The unfrozen water must be removed at a slightly higher temperature corresponding to the temperature at which vaporization will occur.

When the freeze-drying rate is determined by the rate of heat transfer from the surface of the ice front, the thermal conductivity of the dried product becomes very significant. The thermal conductivity value is normally relatively low, resulting in very low heat-transfer rates. Harper and Chichester (1960) have shown that pressure has a significant influence on the thermal conductivity of a freeze-dried product. As is illustrated in Fig. 6.32, the thermal conductivity decreases significantly as the pressure is decreased from 100 to 0.1 mm of Hg. This difference has been attributed to the contribution of the vapors at the higher pressure which do not affect the thermal conductivity at low pressures. Since heat transfer is normally accepted as the limiting factor in a rate of drying, several attemps have been made to improve or increase the heat-transfer rate during freeze-drying. Among these attempts have been the use of spiked plates in contact with the product by Mink and Sachsel (1962) and the use of microwave heating, as described by Leatherman and Stutz (1962). The latter approach has yet to become accepted on a commercial scale to any large extent.

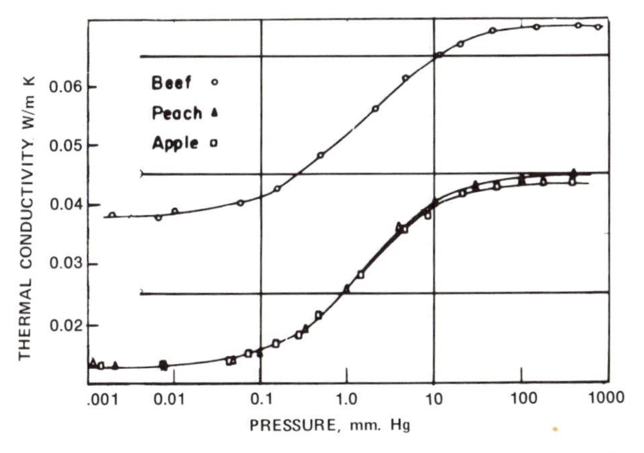


*From Harper and Chichester (1960)*

FIG. 6.32. INFLUENCE OF PRESSURE ON THERMAL CONDUCTIVITY OF FREEZE-DRIED PRODUCT

An additional factor which influences drying rate somewhat directly is the rate of freezing. Greaves (1954) suggests that the properties of the frozen material may influence drying rate considerably. This assumption results from the fact that more rapid freezing results in smaller ice crystals and a greater surface area for sublimation. If, however, the more popular assumption that heat transfer is the rate-controlling factor is accepted, then the availability of more surface area for sublimation cannot be a factor except at the very beginning of the drying. The initial rate of drying may be somewhat more rapid because of the exposure of greater ice surface near the product surface.

Probably one of the more significant factors in determining freeze-drying rate is the structure of the product. This factor cannot be controlled to any great extent and therefore may be much more significant than the other factors mentioned. The shape of the product tissue will influence the conduction heat transfer, and the structure will also determine the manner in which the vapor flows from the frozen ice front to the product surface. Accepting the assumption that heat transfer is the rate-determining factor, the structure and shape of product tissue will establish the freeze-drying rate.

### 6.8.4 Atmospheric Freeze-drying

By moving cold, desiccated air over frozen product, freeze-drying can be accomplished at atmospheric pressures. Several earlier investigations [Meryman (1966), Dunoyer and Larousse (1961), Woodward (1963), Lewin and Mateles (1962)] and more recently Hohner (1970) and Heldman and Hohner (1974) have illustrated the potential for atmospheric freeze-drying. In general, the obvious limiting factor is length of the drying process.

The most detailed investigation of atmospheric freeze-drying potential was based on numerical solution of the mathematical model which incorporates simultaneous heat and mass transfer, as presented by Hohner (1970). The analysis indicated that vapor diffusion parallel to fibers of precooked beef was significantly greater than perpendicular to the fibers. In addition, heat transfer in the dry portion of the product was influenced by counter-current flow of vapor in the product structure.

A computer analysis conducted by Hohner (1970) indicated that size of particle and the surface mass-transfer coefficient offer the best opportunity for reducing atmospheric freeze-drying time. The relationship among the various factors influencing freeze-drying time is illustrated in Fig. 6.33, where dimensionless time ($Dt/\delta^2$) is presented as a function of a

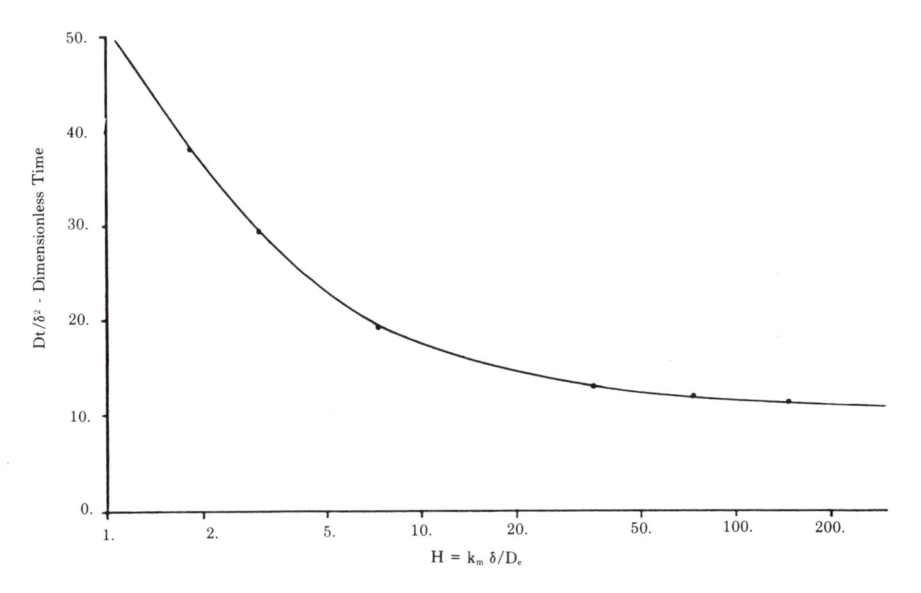


FIG. 6.33. DIMENSIONLESS TIME ELAPSED TO ATTAIN 97% MOISTURE REMOVAL FROM PRECOOKED CUBES OF BEEF AS A FUNCTION OF MASS TRANSFER DURING ATMOSPHERIC FREEZE-DRYING

dimensionless mass transfer number ($k_m$ $\delta/D_e$). This relationship indicates that the time to remove 90% of the product moisture ($t$) can be reduced by increasing the surface mass-transfer coefficient ($k_m$) and by reducing the half-thickness ($\delta$) of the product cube. There is an obvious limit to the freeze-drying time reduction to be achieved as illustrated for dimensionless mass-transfer numbers over 100. These results suggest that the fluidized-bed approach attempted by Malecki *et al.* (1969) offers the most potential for making atmospheric freeze-drying feasible.

## PROBLEMS

6.1 *Prediction of drying time.* A food product is being dried in heated air from an initial moisture content of 0.8 kg water/kg dry solids to 0.3 kg water/kg dry solids. The critical moisture content is 0.5 kg water/kg dry solids and the equilibrium moisture content for the product is 0.15 kg water/kg dry solids and the equilibrium moisture content for the product is 0.15 kg water/kg dry solids. Experimental data collected during the falling-rate drying period were as follows:

| Moisture Content, $w$ (kg water/kg dry solids) | Drying Rate, $N_F$ (kg water/hr m$^2$) |
|---|---|
| 0.5 | 0.57 |
| 0.46 | 0.53 |
| 0.42 | 0.46 |
| 0.38 | 0.40 |
| 0.35 | 0.35 |
| 0.32 | 0.29 |
| 0.29 | 0.22 |
| 0.27 | 0.18 |
| 0.25 | 0.12 |

If the rate of moisture removal is 6.5 kg water/hr m$^2$ during the constant-rate drying period, compute the time required to accomplish the drying for 200 kg product with 0.1 m$^2$/kg dry solids of moisture removal surface.

6.2 *Energy and mass balance of a dryer.* A dry product with a moisture content of 0.08 kg water/kg dry solids is being produced at a rate of 1000 kg dry solids/hr. The heat capacity of the dry product solids is 2 kJ/kgC and the product enters the dryer at 100°C with 0.007 kg water/kg dry air, and leaves at 55°C. Determine the air flow rate and moisture content of product at the exit to the dryer if the exit air is 15% RH.

6.3 *Construction of a sorption isotherm.* Construct an equilibrium moisture isotherm for a dry product between water activities of 0 and 0.45 when the product has the following BET constants: $w_m$ = 0.15 kg water/kg dry solids and $S$ = 15.

6.4 *Droplet size in a centrifugal pressure nozzle.* Compute the mean droplet size (SMD) produced by a swirl-plate centrifugal pressure nozzle for a liquid product with viscosity of $2 \times 10^{-3}$ kg/ms, density of 900 kg/m$^3$ and surface tension of 0.05 N/m. The volumetric flow rate for the product is 0.05 m$^3$/hr with a differential pressure of 750 kPa.

6.5 *Spray drying of foods.* Estimate the final moisture content of a dry food particle produced from an 80 micron liquid droplet in a spray dryer. The latent heat of vaporization is 2175 kJ/kg and the air enters the dryer at 110°C with 1.5% RH. The droplet diameter at the critical moisture content is 60 micron. The total residence time in the dryer is 4.2 s.

6.6 *Drum drying of foods.* A drum dryer is being used to dry 300 kg/hr of food product with an initial moisture content of 60% (wet basis). The drum has a surface area of 4 m$^2$ and an estimated overall heat-transfer coefficient of 1500 W/m$^2$C. If an average temperature difference between product and drum surface of 70 C is maintained and the latent heat of vaporization is 2500 kJ/kg, determine the moisture content of the dry product produced by the system.

## COMPREHENSIVE PROBLEM VI

Design Considerations for Spray Drying Systems

### Objectives

1. To become acquainted with factors involved in the design of a spray dryer for liquid foods.
2. To determine the influence of initial total solids for the product, ambient air conditions and product feed rate on spray dryer design.
3. To evaluate the importance of droplet or particle residence time in spray dryer design.

### Procedures

A. The spray drying system being designed is a co-current flow system using heated air at 120 C. The atomizer to be used produces droplet diameters ranging from 60 to 100 micron. The final moisture content of the dry product is 3.5% (wet basis) and the product should not be heated above 55 C. It can be assumed that the critical moisture content is 40% (wet basis) and the latent heat of vaporization for moisture removal is 2350 kJ/kg. The product feed temperature is 50 C.

B. The variables to be considered are as follows:
   1. Feed rate; evaluate 200, 275 and 350 kg/hr.
   2. Initiate total solids content for product feed; evaluate 30, 40 and 50%.
   3. Ambient air conditions; evaluate 2 C and 70% RH, 20 C and 50% RH and 35 C and 30% RH.

C. The design parameters to be evaluated include the following:
   1. Dry product production rate; kg/hr.
   2. Air flow requirements for drying; $m^3$/min.
   3. Drying time for particles, sec.
   4. Predicted dimensions of drying chamber.

D. Develop a computational program that will allow evaluation of all four design parameters. An effort should be made to select the optimum conditions for each situation.

### Discussion:

The following factors should be discussed based on the results of the computations:

A. The influence of ambient air conditions on air flow rate requirements for the process.
B. The relationships between initial total solids for the products and individual particle drying times.
C. The importance of product feed rate when estimating drying chamber dimensions.
D. The design considerations that may not be utilized when following the recommended procedures.

## NOMENCLATURE

$A$ = area, $m^2$
$a$ = $m^2$ of heat transfer area/$m^3$ of bed.
$a'$ = constant in equation (6.74).
$B$ = mass of water per unit product volume equation (6.72).
$B_1$ = constant in equation (6.40).
$b$ = adsorption coefficient in equation (6.32).
$b'$ = constant in equation (6.74).
$C$ = moisture concentration, kg/$m^3$.
$C_o$ = initial moisture concentration, kg/$m^3$.
$C_e$ = equilibrium moisture concentration, kg/$m^3$.
$C_d$ = drag coefficient defined by equation (6.50).
$C_Q$ = discharge coefficient for atomizer illustrated in Fig. 6.21.
$CMD$ = count or number mean diameter equation (6.37), microns.
$c_p$ = specific heat, kJ/kg C.
$c_s$ = humid heat, kJ/kg dry air, C.
$D$ = diffusivity, $m^2$/s.
$D_e$ = effective vapor transfer coefficient, kg mole/hr $m^2$ bar.
$d$ = diameter, m.
$d_s$ = harmonic mean diameter of spherical particles, m.
$d_c$ = diameter of cylinder, m.
$d_L$ = atomizer diameter, m.
$d_o$ = orifice diameter, m.
$E$ = energy, HP.
$e$ = void fraction.
$F$ = drag force on droplet or particle, $kg_f$.
$FN$ = Flow Number used in equation (6.40).
$G$ = mass flow rate, kg/s.
$g$ = acceleration of gravity, m/$s^2$.
$H_A$ = absolute humidity, kg water/kg dry air.
$H_G$ = enthalpy of air, kJ/kg.
$H_m$ = enthalpy of wet solids, kJ/kg.
$H_R$ = relative humidity.

$H_1, H_2$ = absolute humidity entering and leaving the dryer, kg water/kg dry air.
$h$ = convective heat transfer coefficient, $W/m^2C$.
$h'$ = height of vanes equation (6.46).
$j$ = coefficient in equation (6.15).
$K$ = constant in equation (6.46).
$K'$ = permeability.
$K_1, K_2$ = constants in equations (6.22), (6.48) and (6.49).
$k$ = thermal conductivity, W/m C.
$k_m$ = mass transfer coefficient, kg mole/s $m^2$.
$l$ = length of cylinder, m.
$L$ = latent heat of vaporization, kJ/kg.
$L_s$ = latent heat of sublimation, kJ/kg.
$N_{Le}$ = Lewis number.
$M$ = molecular weight.
$M_c$ = Mach number.
$M_m$ = average molecular weight of gas mixture for vapor transport in air equation (6.47).
$MMD$ = mass mean diameter equation (6.38), microns.
$m$ = mass, kg.
$m_t$ = mass flow rate of solids in a dryer, kg dry solids/s.
$N$ = number of particles or droplets.
$N'$ = rotation speed, rpm.
$N_A$ = vapor transport rate, kg/s.
$N_F$ = drying rate for constant-rate period, kg/s.
$N_c$ = drying rate for constant-rate period, kg/s.
$N_{Nu}$ = Nusselt number.
$n$ = number of molecules.
$n'$ = number of vanes equation (6.46).
$P$ = total pressure, kPa.
$N_{Pr}$ = Prandtl number.
$P$ = pressure gradient, kPa.
$p$ = partial pressure, kPa.
$Q$ = volumetric flow rate.
$q$ = heat flux, W.
$R_g$ = gas constant, kJ/kg K.
$R$ = radius of a sphere.
$R_o$ = radius of spray dryer chamber, m.
$N_{Re}$ = Reynolds number.
$r$ = radius, m.
$S$ = constant in equation (6.33).
$N_{Sc}$ = Schmidt number.
$N_{Sh}$ = Sherwood number.
$SMD$ = Sauter mean diameter equation (6.39), microns.

$T$ = temperature, C.
$T_G$ = temperature of air, C.
$T_A$ = absolute temperature, K.
$T_S$ = temperature of product solids, C.
$t$ = time, s.
$t_c$ = time for constant-rate drying, s.
$t_F$ = time for falling-rate drying, s.
$U$ = overall heat-transfer coefficient, W/m$^3$ C.
$u$ = velocity, m/s.
$u$ = mean velocity, m/s.
$u_R$ = radial velocity, m/s.
$u_r$ = relative velocity, m/s.
$u_t$ = tangential velocity, m/s.
$u_v$ = axial velocity, m/s.
$V$ = volume, m$^3$.
$VMD$ = volume mean diameter, micron.
$w$ = moisture content, dry basis, kg water/kg dry solids.
$w_o$ = initial moisture content, dry basis, kg water/kg dry solids.
$w_c$ = critical moisture content, dry basis, kg water/kg dry solids.
$w_e$ = equilibrium moisture content, dry basis, kg water/kg dry solids.
$w_m$ = moisture content for monomolecular layer of adsorbed moisture.
$w_1, w_2$ = free moisture content entering and leaving the dryer, kg water/kg dry solids.
$X_L$ = thickness of solid, m.
$x$ = orientation in regular coordinate system.
$Y$ = fraction of moisture in product.
$y$ = thickness of bed, m.
$\beta_o$ = constant defined in equation (6.59).
$\delta$ = half-thickness of an infinite plate, ft or cm.
$\gamma$ = surface tension, kg/m.
$\lambda$ = mean free path.
$\mu$ = viscosity, kg/ms.
$\rho$ = density.
$\rho_s$ = bulk density, kg dry solid/m$^3$.
$\theta$ = spray angle in equation (6.40).

## Subscripts

$a$ = air.
$b$ = boundary between air and product.
$c$ = critical moisture content.
$e$ = equilibrium condition.
$g$ = gas phase.
$L$ = liquid phase.

$I$ = frozen phase or ice.
$i$ = index.
$o$ = time equal, zero or temperature at 0 C.
$p$ = particle or droplet.
$s$ = saturation condition.
$v$ = vapor.
$W$ = water.
$w$ = wet-bulb condition or water.

## BIBLIOGRAPHY

BAKKER-ARKEMA, F.W., PATTERSON, R.J., and BEDFORD, C.L. 1967. Drying characteristics of pea beans on single and double-drum driers. Trans. Am. Soc. Agr. Engrs. *10,* No. 2, 154–156.

BRUNAUER, S., EMMETT, H.P., and TELLER, W. 1938. Adsorption of gases in multi-molecular layers. J. Am. Chem. Soc. *60,* 309.

CHARLESWORTH, D.H., and MARSHALL, W.R., JR. 1960. Evaporation from drops containing dissolved solids. Am. Inst. Chem. Engr. J. *6,* No. 1, 9–23.

CHARM, S.E. 1978. The Fundamentals of Food Engineering, 2nd Edition. AVI Publishing Co., Westport, Conn.

COULSON, J.M. and RICHARDSON, J.F. 1978. Chemical Engineering, Vol. II. Pergamon Press, New York.

DITTMAN, F.W. 1977. Classifying a drying process. Chemical Engineering, January 17, pp. 106–108.

DOMBROWSKI, N., and MUNDAY, G. 1968. Spray drying. In Biochemical and Biological Engineering Science, N. Blakebrough (Editor). Academic Press, New York.

DUNOYER, J.M. and J. LAROUSSE. 1961. Experiences Nouvelles sur la Lyophilisation. Trans. Eighth Vacuum Symposium and Second International Congress 2:1059–1063.

EDELING, C. 1950. Cited by W.R. Marshall, Jr. 1954. Atomization and spray drying. Chem. Eng. Progr. Monograph Ser. *50,* No. 2.

FISH, B.P. 1958. Diffusion and thermodynamics of water in potato starch gel. In Fundamental Aspects of Dehydration of Foodstuffs. Macmillan Co., New York.

FORD, R.E., and FURMIDGE, C.G.L. 1967. The formation of drops from viscous Newtonian liquids sprayed through fan-jet nozzles. Brit. J. Appl. Phys. *18,* 335–348.

FORREST, J.C. 1968. Drying processes. In Biochemical and Biological Engineering Science, N. Blakebrough (Editor). Academic Press, New York.

FOUST, A.S., L.A. WENZEL, C.W. CLUMP, L. MAUS and L.B. ANDERSON. 1960. Principles of Unit Operations. John Wiley & Sons. New York.

FROESSLING, N. 1938. Cited by W.R. Marshall, Jr. 1954. Atomization and spray-drying. Chem. Engr. Progr. Monograph Ser. *50,* No. 2.

FROESSLING, N. 1940. Cited by N. Dombrowski and G. Munday. 1968. In

Biochemical and Biological Engineering Science, N. Blakebrough (Editor). Academic Press, New York.

GORLING, P. 1958. Physical phenomena during the drying of foodstuffs. In Fundamental Aspects of the Dehydration of Foodstuffs, Macmillan Co., New York.

GREVES, R.I.W. 1954. The Biological Applications of Freezing and Drying, R.J.C. Harris (Editor). Academic Press, New York.

HARPER, J.C., and CHICHESTER, C.O. 1960. Freeze-dehydration of foods. A Military-Industry Meeting. Sept. 20–21, 1956. Chicago, Ill. Ed. Technical Services Office, QM Food and Container Institute for Armed Forces.

HELDMAN, D.R. and HOHNER, G.A. 1974. Atmospheric freeze-drying processes for food. J. Food Sci. *39,* 147.

HERRING, W.M. and W.R. MARSHALL. 1955. Cited by Masters, K. 1976. Spray Drying. 2nd Edition. John Wiley and Sons, Inc. New York.

HOHNER, G.A. 1970. An Analysis of Heat and Mass Transfer in Atmospheric Freeze-Drying. Ph.D. Thesis. Michigan State University, East Lansing, MI.

KING, C.J. 1970. Freeze-drying of foodstuffs. Criticial Reviews Food Technol. *1,* 379.

KING, C.J. 1973. Freeze-drying. In Food Dehydration, 2nd Edition, Vol. 1, W.B. Van Arsdel, M.J. Copley, and A.I. Morgan (Editors). AVI Publishing Co., Westport, Conn.

JASON, A.C. 1958. A study of evaporation and diffusion processes in the drying of fish muscle. In Fundamental Aspects of the Dehydration of Foodstuffs, Macmillan Co., New York.

LABUZA, T.P. 1968. Sorption phenomena in foods. Crit. Rev. Food Technology. 2.355.

LABUZA, T.P. 1980. The effect of water activity on reaction kinetics of food deterioration. Food Technology. 34(4):36–59.

LANGMUIR, I. 1918. The adsorption of gases on plane surface of glass, mica and platinum. J. Am. Chem. Soc. *40,* 1361.

LEATHERMAN, A.F., and STUTZ, D.E. 1962. The application of dielectric heating to freeze-drying. In Freeze-drying of Foods, F.R. Fisher (Editor). Natl. Res. Council-Nat. Acad. Sci. Washington, D.C.

LEWIN, L.M., and MATELES, R.I. 1962. Freeze-drying without vacuum: A preliminary investigation. Food Technol. *16,* no. 1, 94–96.

LOWE, E., RAMAGE, W.D., DURKEE, E.L., and HAMILTON, W.E. 1955. Belt-trough—a new continuous dehydrator. Food Engr. *27,* No. 7, 43–4.

MALECKI, G.J., SHINDLE, P., MORGAN, A.I., Jr., and FARKAS, D.F. 1969. Atmospheric fluidized-bed freeze-drying of apple juice and other liquid foods. Food Tech. *24* (5):93.

MARSHALL, W.R., JR. 1954. Atomization and spray drying. Chem. Eng. Progr. Monograph Ser. *50,* No. 2.

MARSHALL, W.R., JR., and FRIEDMAN, S.J. 1950. Drying. In Chemical Engineer's Handbook 3rd. Edition, J.H. Perry (Editor). McGraw-Hill Book Co., New York.

MASTERS, K. 1976. Spray Drying. 2nd Edition. John Wiley and Sons, Inc. New York.

MERYMAN, H.T. 1966. Freeze-drying. In Cryobiology, H.T. Meryman (Editor). Academic Press, New York.

MINK, W.H., and SACHSEL, G.F. 1962. Evaluation of freeze-drying mechanisms using mathematical models. In Freeze-drying of Foods, F.R. Fisher (Editor). Natl. Res. Council—Nat. Acad. of Sci., Washington, D.C.

MORGAN, A.I., GRAHAM, R.P., GINNETTE, L.F., and WILLIAMS, G.S. 1961. Recent developments in foam-mat drying. Food Technol. *15,* 37−39.

NUKIYAMA, S. and Y. TANASAWA. 1937. Cited by Masters, K. 1976. Spray Drying. 2nd Edition. John Wiley and Sons, Inc. New York.

PATSAVAS, A.C. 1963. The spray dryer. Chem. Eng. Progr. *59,* No. 4, 65−70.

PORTER, H.F., P.Y. McCORMICK, R.L. LUCAS and D.F. WELLS. 1973. Gas-solid systems, *In* Chemical Engineer's Handbook. Ed. Perry, R.H. and C.H. Chilton. McGraw-Hill Book Company, New York.

PROBERT, R.P. 1946. The influence of spray particle size and distribution in the combustion of oil droplets. Phil. Mag. *37,* 94.

RANZ, W.E., and MARSHALL, W.R., JR. 1952. Evaporation from drops. Chem. Engr. Progr. *48,* 141−180.

ROCKLAND, L.B. 1969. Water activity and storage stability. Food Technol. *23,* 1241.

ROCKLAND, L.B. and S.K. NISHI. 1980. Influence of water activity on food product quality and stability. Food Technology *34*(4):42−51.

SALWIN, H. 1959. Defining minimum moisture contents for dehydrated foods. Food Technol. *13,* No. 10, 594.

SANDALL, O.C., KING, C.J., and WILKE, C.R. 1967. The relationship between transport properties and rates of freeze-drying of poultry meat. Am. Inst. Chem. Engrs. J. *13,* No. 3, 428−438.

SHERWOOD, T.K. 1929. Drying of solids. I. Ind. Eng. Chem. *21,* 12.

SHERWOOD, T.K. 1931. Application of theoretical diffusion equations to the drying of solids. Trans. Am. Inst. Chem. Engr. *27,* 190.

TALBURT, W.F., and SMITH, O. 1967. Potato Processing, 2nd Edition. AVI Publishing Co., Westport, Conn.

TATE, R.W. 1965. Sprays and spraying for process use. I. Type and principles. II. Applications and selection. Chem. Eng. July, 1965. 157.

VAN ARSDEL, W.B. 1950. Convergence criteria in numerical solution of the diffusion equation. U.S. Dept. Agr., Western Reg. Res. Lab., Bull. AIC 287.

VAN ARSDEL, W.B., COPLEY, M.J. and MORGAN, A.I., JR. 1973. Food Dehydration, 2nd Edition, Vol. 1, AVI Publishing Co., Westport, Conn.

VAN MARLE, E.J. 1938. Drum Drying. Ind. Eng. Chem. *30,* 1006.

WOODWARD, H.T. 1963. Freeze-drying without vacuum. Food Engineering *35*(6)96−98.

7

# Contact Equilibrium Processes

Several operations involved in the processing of foods involve the separation of selected components from the food product. Often these operations must be conducted on a large scale to meet the demands of a large manufacturing plant. Mass transfer is the basic process involved in one group of these separation operations, which are normally referred to as contact equilibrium separations. These equilibrium processes normally include (a) gas absorption such as hydrogenation of oil or carbonation of beverages; (b) liquid-solid extraction as involved in the removal of edible oil from soy beans; (c) liquid extraction, commonly used in removing sucrose from cane or beets; and (d) distillation, used in removal of oil from solvents or in manufacturing of alcoholic beverages. All these processes are similar in that the separation of absorption process involves the transfer of mass from a component or phase of the product to a secondary phase or component. Due to the many factors which influence this transfer of mass, a fundamental understanding of the process is required.

## 7.1 BASIC PRINCIPLES

One of the basic principles involved in any contact equilibrium process is the equilibrium condition to be attained between the two phases or two insoluble components of the mixture. The equilibrium condition can be described in terms of variations in component concentrations in each phase as a function of time. At equilibrium, net transfer of a given component between phases will not exist and the component concentrations in each phase will be constant with time. The equilibrium condition can be attained in any type of separation system including liquid-liquid, liquid-gas, liquid-solid, or gas-solid types. The primary difference between these types of processes is the manner in which the equilibrium condition is described.

### 7.1.1 General Description of Transport Processes

The generally accepted approach for describing mass transfer between two phases is the two-film theory of molecular diffusion, as illustrated in Fig. 7.1. The basis for this approach is that mass transfer from one phase to another must overcome the resistance exerted by two films at the interface between the two phases. These films are regions within each phase which are described by the concentration gradient of a specific component. As illustrated in Fig. 7.1 by a liquid-gas interface, a film in each phase represents the change in component concentration from the interface to that at the bulk concentration in either the liquid or gas phase. The liquid-phase concentration of the component changes from $X$ (liquid) to $X_i$ at the liquid-gas interface, while the gas concentration of the same component changes from $Y_i$ at the interface to $Y$ in bulk stream of gas.

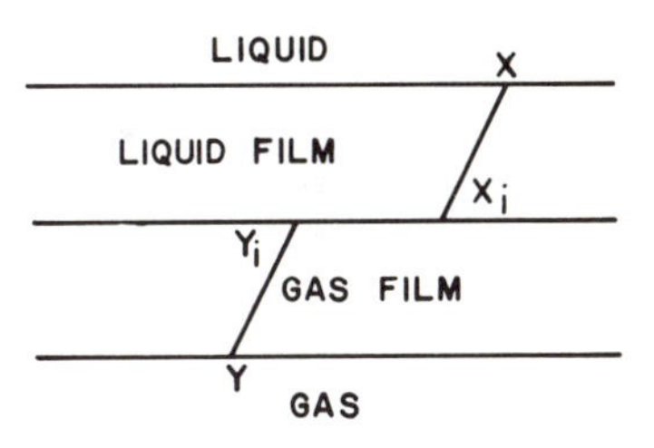


FIG. 7.1. MASS TRANSFER BETWEEN PHASES AS ILLUSTRATED BY THE TWO-FILM THEORY

A basic consideration in the description of mass transfer between two phases is the equilibrium condition. Any computation involved in the design of a contact equilibrium process will depend directly on the type of relationship illustrated in Fig. 7.2. The equilibrium curve presented for gas-liquid equilibria can be predicted by Raoult's law when an ideal solution is in equilibrium with an ideal gas. Situations which are nonideal require the use of experimentally determined constants in order to describe the relationship between the concentrations in liquid and gas phases. Similar expressions can be used to describe the equilibrium curve for liquid-liquid or liquid-solid equilibria. In most cases, however, these expressions are specific to the two phases being considered at the equilibrium condition.

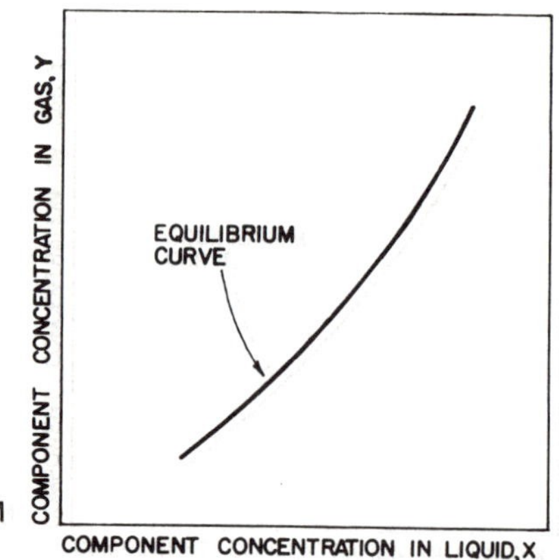


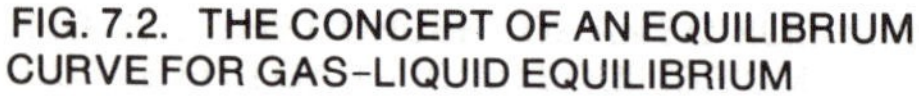
FIG. 7.2. THE CONCEPT OF AN EQUILIBRIUM CURVE FOR GAS–LIQUID EQUILIBRIUM

### 7.1.2 Material Balance

The general equilibrium separation process can be described by the contact equilibrium column presented schematically in Fig. 7.3. By conducting a material balance near the lower end of the column, equation (7.1) is obtained:

$$m_Y(Y - Y_1) = m_X(X - X_1) = \frac{\Delta N}{\Delta t} \tag{7.1}$$

where $Y$ and $X$ represent the concentration of the two phases at some location near the lower end of the column, and $m_Y$ and $m_X$ are the mass flow rates of the two component streams through the column. Equation (7.1) can be modified by conducting the mass balance on a differential element within the column to obtain:

$$m_Y\,dY = m_X\,dX = \frac{dN}{dt} \tag{7.2}$$

In both equations (7.1) and (7.2), the products of the mass flow rates and concentration gradients are equal to the rate of mass transfer ($dN/dt$). Both equations (7.1) and (7.2) describe an operating line relating the concentration in the $X$-phase stream to concentration in the $Y$-phase stream at the same location in the column.

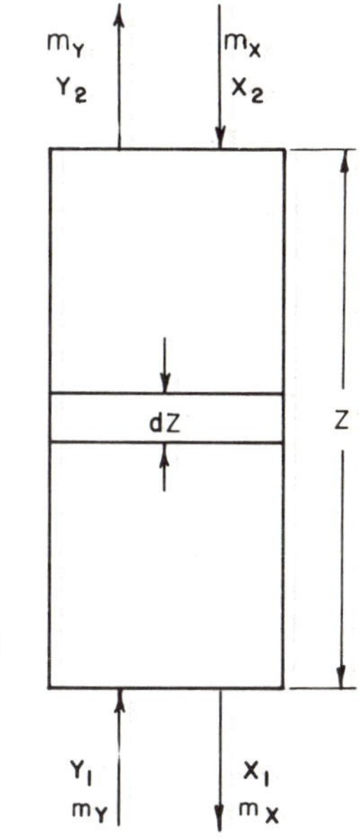


FIG. 7.3. SEPARATION COLUMN UTILIZED IN CONDUCTING A MATERIAL BALANCE

EXAMPLE 7.1 A volatile component is being removed from air by passing the mixture through an absorption column where it is brought into contact with a solvent. The solute and the air are flowing through the column at a rate of 50 kg/min, while the solvent is flowing at 10 kg/min. At the exit, the air contains 0.05 weight fraction of the volatile component and the solvent has a weight fraction of 0.3 for solute. Plot the operating line for the absorption process.

SOLUTION

(1) Using equation (7.1) as follows, with $m_Y$ = 10 kg solvent/min and $m_X$ = 50 kg air/min:

$$10(Y - 0.3) = 50(X - 0.05)$$

an equation for the operating line is obtained.

(2) By rearrangement of the above equation:

$$Y = 0.05 + 5X$$

which represents the equation of the operating line as illustrated in Fig. 7.4.

In addition, the mass-transfer rate will be a function of a mass-transfer coefficient ($k_X$), the area through which mass transfer occurs ($A$) and the concentration gradient, as illustrated by the following expression:

$$\frac{dN}{dt} = k_X\, dA(X - X_i) = m_X\, dX \tag{7.3}$$

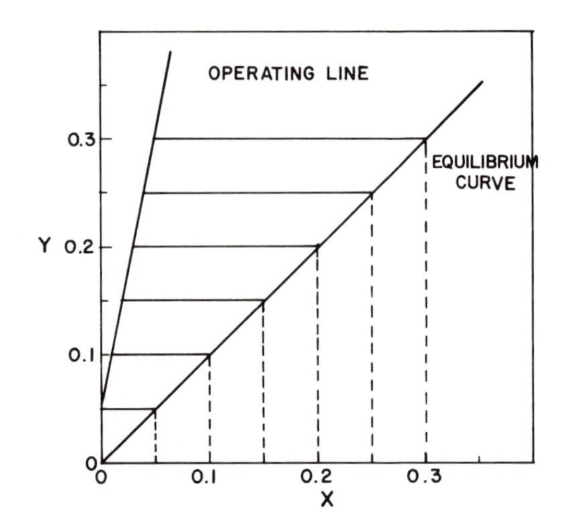


FIG. 7.4. OPERATING LINE DESCRIBING CONDITIONS IN EXAMPLE 7.1

In equation (7.3), the mass-transfer rate is described in terms of one film when utilizing Fig. 7.1 as a reference. The concentration gradient in this case is the difference between the bulk concentration ($X$) and the concentration at the interface ($X_i$). Equation (7.3) can be rewritten as follows:

$$m_X\, dX = k_X\, as(X - X_i)\, dZ \tag{7.4}$$

where the transfer area is written in terms of the differential column length ($dZ$), interfacial area per unit volume ($a$) and cross-sectional area of column ($s$). When equation (7.4) is written in a form ready for integration, the following expression results:

$$\int_{X_1}^{X_2} \frac{dX}{X - X_i} = \frac{k_X\, as}{m_X} \int_{Z_1}^{Z_2} dZ \tag{7.5}$$

where the concentrations $X_2$ and $X_1$ correspond to column locations of $Z_2$ and $Z_1$. A similar expression can be used to describe mass transfer in the gas film side of the interface and the following similar expression is obtained:

$$\int_{Y_1}^{Y_2} \frac{dY}{Y_i - Y} = \frac{k_Y\, as}{m_Y} \int_{Z_1}^{Z_2} dZ \tag{7.6}$$

When the operating line and the equilibrium curve are presented as illustrated in Fig. 7.5, one of the more basic diagrams necessary for a computation of equilibrium separation processes is obtained. An analysis of equation (7.3) and a similar equation for the film containing the $Y$-component would indicate that the ratio of the mass-transfer coefficients in the two films must be equal to the ratio of the concentration gradients:

$$\frac{k_X}{k_Y} = \frac{Y_i - Y}{X - X_i} \tag{7.7}$$

This ratio represents the slope of a line between the operating point on the operating line and the corresponding point on the equilibrium curve as shown in Fig. 7.5.

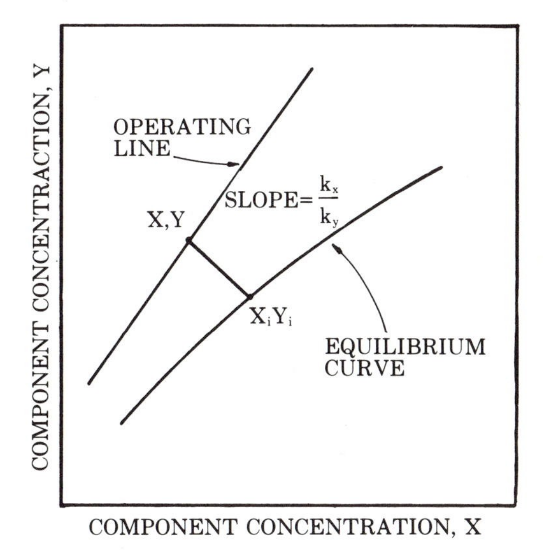


FIG. 7.5. THE OPERATING LINE-EQUILIBRIUM CURVE RELATIONSHIP WITH CONNECTING LINE BETWEEN OPERATING POINTS

For convenience, the mass-transfer coefficient is normally expressed as an overall coefficient based on either of the two phases. This approach to describing the overall mass-transfer coefficient can be presented using the following expression:

$$\frac{dN}{dt} = K_X\, dA(X - X_e) = K_Y\, dA(Y_e - Y) \tag{7.8}$$

If equation (7.8) describes mass transfer across a liquid-gas interface, the overall mass-transfer coefficient ($K_X$) is expressed in terms of component concentrations in liquid. The concentration gradient is between component concentration in the liquid bulk ($X$) and component concentration in equilibrium with the gas phase ($X_e$). The overall mass-transfer coefficient ($K_Y$) expressed in terms of the gas film is similar; the equilibrium component concentration is ($Y_e$). Since the subscript ($e$) represents the concentration along the equilibrium curve, each overall mass-transfer coefficient describes the rate of mass transfer independent of the existence of the opposite phase. Each overall mass-transfer coefficient is therefore very specific to the phase for which it describes the mass transfer, and the two coefficients will not normally be equal.

The design of most contact equilibrium processes involves computation of the height of the column needed to provide the required contact between the two streams carrying the component of interest. In addition, the design will normally determine the number of contact stages needed within the established column height. The accepted terms utilized to describe these design parameters are the height of a transfer unit (HTU) and the number of transfer units (NTU). The column height is determined from the product of HTU and NTU.

From expressions of the type given as equations (7.5) and (7.6), expressions for the height of a transfer unit (HTU) and the number of transfer units (NTU) can be developed. These can be expressed in terms of the overall mass-transfer coefficients as defined in equation (7.8). The number of transfer units is defined by the following equation:

$$\text{NTU} = \int_{Y_1}^{Y_2} \frac{dY}{Y_e - Y} = \int_{X_1}^{X_2} \frac{dX}{X - X_e} \tag{7.9}$$

where the driving forces or gradients are expressed in terms appropriate for use with the overall mass-transfer coefficient. The corresponding height of a transfer unit (HTU) is defined by the following equation:

$$\text{HTU} = \frac{m_X}{K_X as} = \frac{m_Y}{K_Y as} \tag{7.10}$$

where all parameters have been defined by previous equations.

EXAMPLE 7.2 Under ideal conditions, the equilibrium curve may be described by a linear relationship between solute concentrations in two streams considered. Asume the gas phase (air) resistance in Example 7.1 is negligible and the column height is 10 m. Compute HTU, NTU and the overall mass-transfer coefficient for the process.

SOLUTION

(1) A linear equilibrium curve is given in Fig. 7.4 for the situation where $X = Y$.

(2) Since the gas phase resistance is negligible, the concentration at the interface may be determined by drawing horizontal lines from the operating line to intersect with the equilibrium curve, as illustrated in Fig. 7.4.

(3) The number of transfer units (NTU) in equation (7.9) can be

determined by graphical integration as follows:

| $Y$ | $X - X_e$ | $\frac{1}{X - X_e}$ | $\Delta X$ | Ordinate Height (Avg) | Area |
|---|---|---|---|---|---|
| 0.05 | 0.05 | 20.0 | | | |
| 0.1 | 0.095 | 10.5 | 0.05 | 15.0 | 0.75 |
| 0.15 | 0.13 | 7.7 | 0.05 | 9.0 | 0.45 |
| 0.2 | 0.17 | 5.9 | 0.05 | 6.8 | 0.34 |
| 0.25 | 0.21 | 4.8 | 0.05 | 5.3 | 0.27 |
| 0.3 | 0.25 | 4.0 | 0.05 | 4.4 | 0.22 |
| | | | | | 2.03 |

(4) Based on the graphical integration:

$$\text{NTU} = 2.03$$

and since the column height is 10 m

$$\text{HTU} = 10/2 = 5 \text{ m}$$

(5) From equation (7.10)

$$\text{HTU} = \frac{m_Y}{K_Y as} = 5 = \frac{10\,\text{kg/min}}{K_Y as}$$

$$K_Y as = \frac{10}{5} = 2\,\text{kg/m min}$$

(6) For a column cross-section of 0.5 $m^2$

$$K_Y a = 2/0.5 = 4 \text{ kg/m}^3 \text{ min}$$

# 7.2 EXTRACTION

The term extraction might be applied very broadly to describe all separation processes. In most applications in the food industry, however, this term will refer to two rather distinct processes: liquid-solid leaching and liquid-liquid extraction. Leaching refers to the removal of some soluble component from a solid by using a liquid solvent. This process can be accomplished in a variety of ways but, as might be expected, its efficiency depends to a great extent on attaining intimate contact between the liquid solvent and the solid containing the solute. The leaching of sucrose from sugar beets utilizing hot water as a solvent and the removal of edible oils from soybeans are just two examples of leaching related to food processing. Liquid-liquid extraction involves the removal of a solute from one liquid phase and provides for removal or pickup of this solute by a second liquid phase. The two liquid phases are partially immiscible so that the efficiency of the separation process again depends on maintaining intimate contact between the two liquid phases. Probably the best example of liquid extraction related to food processing is in the purification of animal fat. In this particular application, the fat is one liquid phase and a solvent is utilized to remove undesirable components.

A variety of systems are used to accomplish the various extraction processes. In general, leaching will occur under unsteady-state or steady-state conditions. Probably the least complex form is the batch process for leaching, which falls into the unsteady-state classification. Processes involving countercurrent flow of liquid and solid increase multiple contact between the two phases. Steady-state leaching usually occurs in continuous systems which are countercurrent to obtain optimum efficiency. In addition, liquid extraction processes can be divided into single-stage and multi-stage equipment. The simplest form of single-stage process occurs in an agitated vessel. Single-stage efficiency is improved by obtaining more efficient mixing action between the two liquid phases in a flow-type mixer. Multi-stage liquid-liquid extraction is usually accomplished in columns of various types, which may be spray towers, packed towers or towers containing plates or agitators of various configurations.

### 7.2.1 Rate of Extraction

The rate of extraction is a measure of the rate at which the solute is transferred from one phase to another. This rate, for either leaching or liquid-liquid extraction, will depend on several factors including: (a) particle size, (b) the type of solvent, (c) temperature, and (d) the agitation of the fluid. Particle size will influence the rate of leaching in two ways: smaller particles create more interfacial area between the solid and liquid and a short distance for the solute to diffuse within the particle to reach the particle surface. Particle size can be reduced excessively, causing the circulation of liquid around the particles to be impeded. In addition, small particles are difficult to separate from the liquid after the extraction is completed.

The selection of a solvent is a significant factor in both leaching and liquid-liquid extraction. In general, a solvent should be chosen that is selective for the solute being removed, and the solvent should have low viscosity to promote circulation in both processes. Temperature is a factor because of the increased solubility resulting from increased temperature. In the case of leaching, diffusion coefficients for the solvent within the particles would be expected to increase with rising temperature and result in more rapid extraction rates. Agitation of the fluid or fluids is a factor in both types of extraction. Stronger agitation results in increased diffusion and a reduced resistance to mass transfer at the particle surface during the leaching process. For liquid-liquid extraction, an increase in agitation promotes more intimate contact between the two liquids involved in the process.

The rate of extraction can be described by an expression similar to equations (7.3) and (7.8) as follows:

$$\frac{dN}{dt} = K_L A(C_s - C) \tag{7.11}$$

where $K_LA$ represents the overall mass-transfer coefficient and $C_s$ represents the concentration of the solute in the solvent at the saturation condition. By assuming a batch process in which the total volume of solution is constant, the following equation would result:

$$dN = V\,dC \tag{7.12}$$

leading to the following integral form of equation (7.11):

$$\int \frac{dC}{C_s - C} = \int \frac{K_L A}{V}\,dt \tag{7.13}$$

The solution to equation (7.13) can be written in several ways. The following illustrates the form which best illustrates the rate at which extraction can occur:

$$C = C_s \left\{1 - \exp\left[-\frac{K_L A}{V}\,t\right]\right\} \tag{7.14}$$

Based on the solution presented, the mass-transfer coefficient, the area between solid and liquid and the total volume of the solution have a direct influence on the rate constant that determined the rate at which the solvent approaches the saturation concentration. Although equation (7.14) applies to the leaching process only, the rate at which liquid-liquid extraction will occur can be described in a similar way. In liquid-liquid extraction, two mass-transfer coefficients must be described, one for each of the two liquids involved in the process. The best available data on mass-transfer coefficients is for agitated vessels and appropriate expressions for predicting the mass-transfer coefficient are presented in Chapter 1.

EXAMPLE 7.3 The overall mass-transfer coefficient for leaching in an agitated vessel is to be estimated by measurement of the increase in solute concentration in the solvent with time. The following measurements were obtained:

| Time (min) | Solute Concentration (%) |
|---|---|
| 0 | 0 |
| 2 | 3 |
| 4 | 7 |
| 6 | 9 |

| Time (min) | Solute Concentration (%) |
|---|---|
| 8 | 11 |
| 10 | 13 |
| 13 | 15 |
| 16 | 16 |
| 19 | 17 |

If the saturation concentration for solute in the solvent is 20% and the solvent volume is constant at 60 $m^3$, estimate the mass-transfer coefficient.

SOLUTION

(1) Based on equation (7.14), the slope of $(1 - C/C_s)$ vs. time on semilog coordinates will be equal to $K_L A/V$.

(2) By plotting $(1 - C/C_s)$ vs. $t$ as illustrated in Fig. 7.6, a slope of 0.0447 is obtained.

(3) Using the above slope value:

$$K_L A/V = 0.0447$$

$$K_L A = 0.0447\ (60) = 2.68\ \text{m}^3\ \text{solution/min}$$

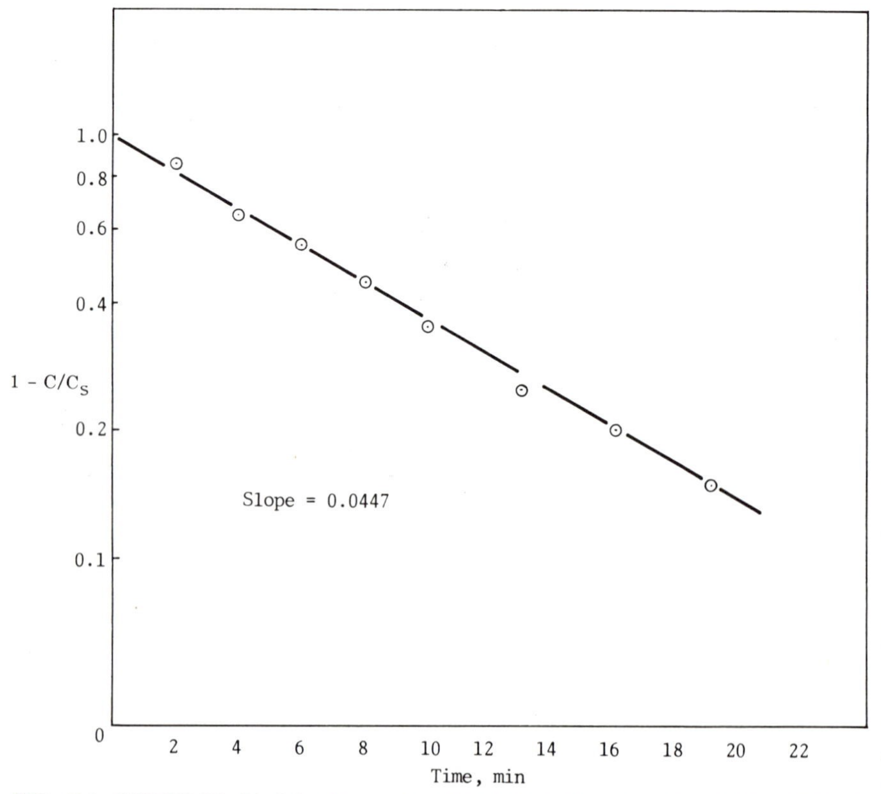


FIG. 7.6. INCREASE IN SOLUTE CONCENTRATION IN SOLVENT FOR DATA IN EXAMPLE 7.3

### 7.2.2 Leaching

The design of a system for leaching involves the determination of the amount of solvent required to produce a given amount of extract from a given quantity of solid material. Generally this is accomplished through the use of a material balance on the various components involved in the process. Three components must be considered in the leaching process: the solvent, the solute, and the insoluble solids. Initially the solute is contained within the insoluble solids, but during the leaching process it distributes or diffuses into the liquid solvent until the concentrations in the solvent and the insoluble solids are at equilibrium.

Since there are three components involved in the leaching process, a rather unique three-component system represented by a right triangle can be utilized for description. As illustrated in Fig. 7.7, each vertex of the triangle represents 100% concentration of one of the three components. In this illustration, the concentration of solvent increases along the vertical axis while the concentration of solute increases along the horizontal axis. Any location inside the triangle will represent some mixture of all three components, and therefore the triangle can be utilized to represent the process at any point during leaching.

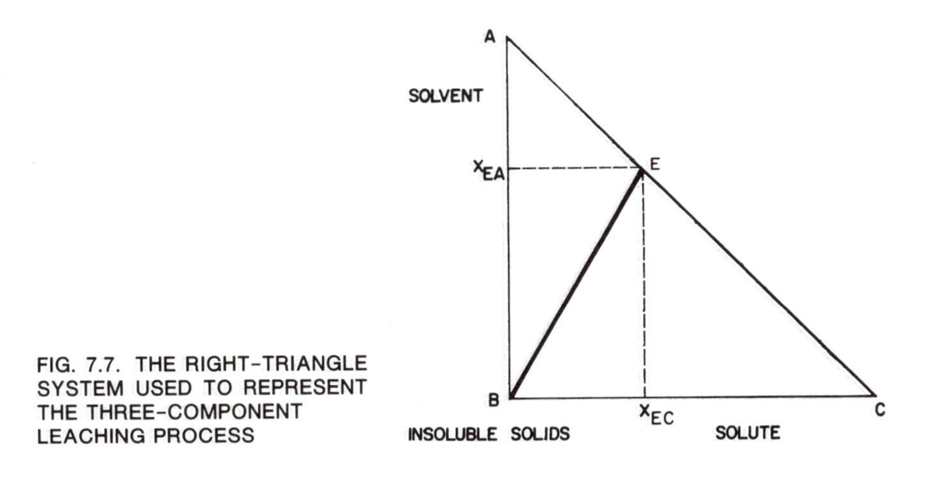


FIG. 7.7. THE RIGHT-TRIANGLE SYSTEM USED TO REPRESENT THE THREE-COMPONENT LEACHING PROCESS

The process parameters of significance in a leaching process become much more obvious after conducting a material balance. The illustration of a single-stage leaching process in Fig. 7.8 provides the necessary information with which to conduct a material balance. First of all, an overall material balance leads to the following expression:

$$F + S = E + W \tag{7.15}$$

where the incoming streams are product feed ($F$) and solvent ($S$). The streams leaving the process include the solute-containing solvent ($E$) and the stream-containing product solids plus solvent and solute ($W$). A material balance on the solute component of each stream leads to the following equation:

$$X_{FC}F + X_{SC}S = X_{EC}E + X_{WC}W \tag{7.16}$$

where the solute component fraction ($X_{SC}$) of the entering solvent ($S$) is non-existent. In most cases, the solute fraction of the feed ($X_{FC}$) will be known or specified and the solute fraction of the extract ($X_{EC}$) leaving the process will be known or specified. The solute fraction of the stream carrying the solids ($X_{WC}$) away from the process may be known or may be considered insignificant. Similar material balance equations can be written based on the fractions of the other two components (solvent and solids) involved in the leaching process.

FIG. 7.8. ILLUSTRATION OF SINGLE-STAGE LEACHING PROCESS

Several approaches are available when designing a leaching process. Under typical conditions, when the streams entering the leaching process include the feed (a mixture of solids and solute) and the pure solvent, an extract at the outlet of the leaching process will be predominantly solvent and solute, while the solids-carrying stream leaving the leaching process will contain all three components involved in the process. Under these conditions and when the more common parameters are known, the material-balance equations can be utilized to compute the fraction of solvent and solids in the solids-carrying stream at the outlet of the leaching process. These equations are given as equations (7.17) and (7.18):

$$X_{WA} = \frac{S - EX_{EA}}{W} \tag{7.17}$$

$$X_{WB} = \frac{X_{FB}F}{W} \tag{7.18}$$

It is evident that these equations provide the unknown in terms of other unknown factors and were obtained from material balance equations of the same form as equation (7.16).

An alternative approach utilizes the right-triangle system for determination of the unknown factors; by referring to Fig. 7.9, several factors can be illustrated. Point $E$ on Fig. 7.9 represents the extract leaving the leaching process ($E$) and assumes that the solids fraction ($X_{SC}$) is zero in this particular stream. The line passing through $W$ and parallel to the

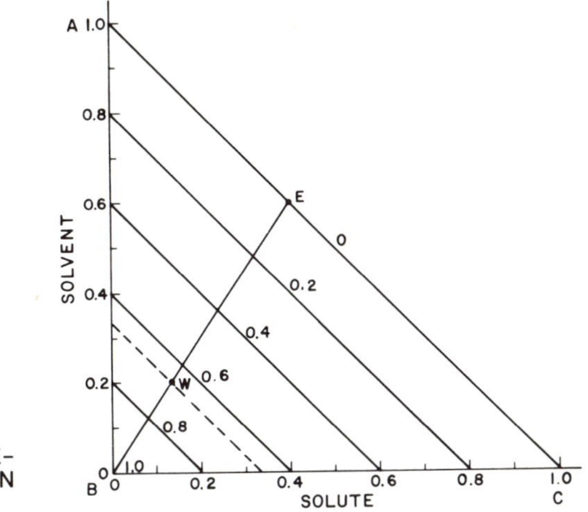


FIG. 7.9. RIGHT TRIANGLE DESCRIPTION OF EXTRACTION PROCESS IN EXAMPLE 7.4

hypotenuse of the right triangle represents the solute concentration ($X_{WC}$) of the solids-carrying stream leaving the leaching process ($W$). The line from the right angle vertex of the triangle to $E$ crosses the line describing the solids concentration at $W$ in the right triangle. This point represents the concentration of all three components in the solids-carrying stream leaving the leaching process. This approach leads to the same type of results as the material balance equations assuming all of the same factors are given or specified.

EXAMPLE 7.4 A countercurrent extraction system is being utilized to extract oil from 1000 kg soybeans per hour. The system is to be designed to extract oil from soybeans with 18% oil and provide 40% oil in the extract solution leaving at 800 kg per hour. If the weight of extract solution in the solids leaving the system is equal to 50% of the weight of solids, compute the composition of the stream containing solids leaving the first stage and the composition of solvent entering stage 1.

SOLUTION

(1) Utilizing equation (7.16), a mass balance can be conducted on the oil fraction in the first stage.

$$1000\ (0.18) + SX_{SC} = 800\ (0.4) + WX_{WC}$$

(2) A similar balance on solids fraction leads to:

$$1000\ (0.82) + S(0) = 800\ (0) + WX_{WB}$$

(3) In addition:

$$W = 1000\ (0.82) + 0.5\ (1000)(0.82) = 1230 \text{ kg/hr}$$

so:

$$X_{WB} = \frac{1000\,(0.82)}{1230} = 0.667$$

illustrating the use of equation (7.18).

(4) Utilizing Fig. 7.9, the composition of the stream carrying solids becomes (point $W$):

$$X_{WA} = 0.2$$
$$X_{WB} = 0.67$$
$$X_{WC} = 0.13$$

(5) A mass balance on the solvent fraction leads to:

$$F\,(0) + SX_{SA} = 800\,(0.6) + 1230\,(0.2)$$

(6) From a mass balance on oil fraction:

$$SX_{SC} = 800\,(0.4) + 1230\,(0.13) - 1000\,(0.18)$$

and overall mass balance:

$$1000 + S = 800 + 1230$$

then:

$$S = 1230 + 800 - 1000 = 1030 \text{ kg/hr}$$

(7) Using the balance on the oil fraction:

$$1030\,X_{SC} = 320 + 160 - 180$$

$$X_{SC} = \frac{300}{1030} = 0.29$$

(8) then: $X_{SA} = 0.71$

(9) The composition of solvent entering the first stage is 29% oil and 71% solvent.

EXAMPLE 7.5 A single-stage extraction system is being used to extract oil from codliver. The feed rate into the extractor is 1000 kg/hr and the solvent contains 0.99 mass fraction ether and 0.01 mass fraction oil. The codliver feed contains 0.326 mass fraction oil and 0.674 mass fraction solids. If a mass fraction of 0.183 oil is desired in the exit extract while a maximum of 0.579 mass fraction solids will be allowed in the exit stream carrying solids, determine the rate at which solvent will be required.

SOLUTION

(1) Using expressions of the form given in equations (7.15) and (7.16), the following equations can be developed:

Total mass: $1000 + S = E + W$
Solute fraction: $1000\,(0.326) + S(0.01) = E(0.183) + WX_{WC}$
Solvent fraction: $1000\,(0) + S(0.99) = E(0.817) + WX_{WA}$
Solid fraction: $1000\,(0.674) + S(0) = E(0) + W(0.579)$

(2) At this point, four equations have been presented and five unknowns remain. The right-triangle approach can be used to evaluate one of the unknowns, as illustrated in Fig. 7.10. From the illustration, the solute and solvent concentrations in the exit stream carrying solids are:

$$X_{WC} = 0.075$$
$$X_{WA} = 0.352$$

(3) Using the equation for solids fraction:

$$W = \frac{1000\,(0.674)}{0.579} = 1164\ \text{kg/hr}$$

(4) Using the equations for total mass and solvent fraction:
(a) $E = 1000 + S - 1164$
(b) $S(0.99) = E(0.817) + 1164\ (0.352)$

or

$$S(0.99) = (1000 + S - 1164)\ (0.817) + 1164\ (0.352)$$

$$0.99\,S = 0.817\,S + 817 - 951 + 409$$
$$0.173\,S = 275$$
$$S = 1590\ \text{kg/hr}$$

(5) For the extraction process presented, solvent is being used at rate of 1.590 kg for each kg of codliver feed.

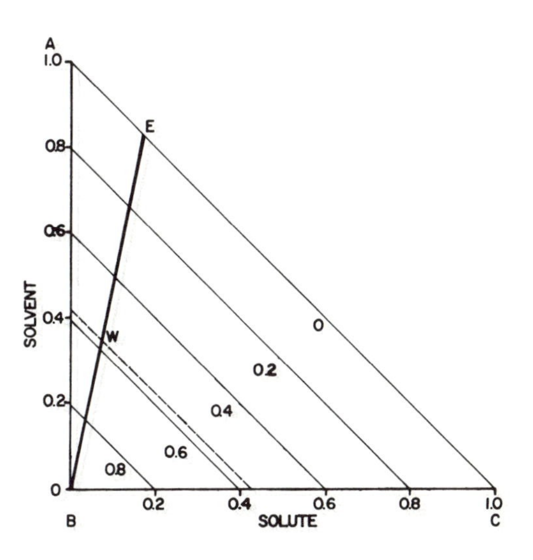


FIG. 7.10. RIGHT TRIANGLE DESCRIPTION OF EXTRACTION PROCESS IN EXAMPLE 7.5

### 7.2.3 Multiple-Stage Leaching

Since the desired levels of separation and refinement cannot be achieved in single-stage leaching processes, large-scale operations require multiple stages. The design of multiple-stage systems is accomplished in several ways and the following presentation is an attempt to illustrate the basic concepts associated with the process as would be applied to leaching of components from raw food products.

The multiple-stage leaching systems can be illustrated schematically as in Fig. 7.11. As indicated, the product feed ($F$) enters stage 1 and some fraction of the feed containing product solids moves progressively through the system. The solute component of the feed has been expressed as mass fraction ($X_{FC}$) and the quantity of solute becomes $FX_{FC} = C'_1$ in Fig. 7.11. The stream containing product solids leaving the final stage ($n$) of the system has a total quantity ($W$) and a solute quantity of $WX_{WC} = C'_{n+1}$.

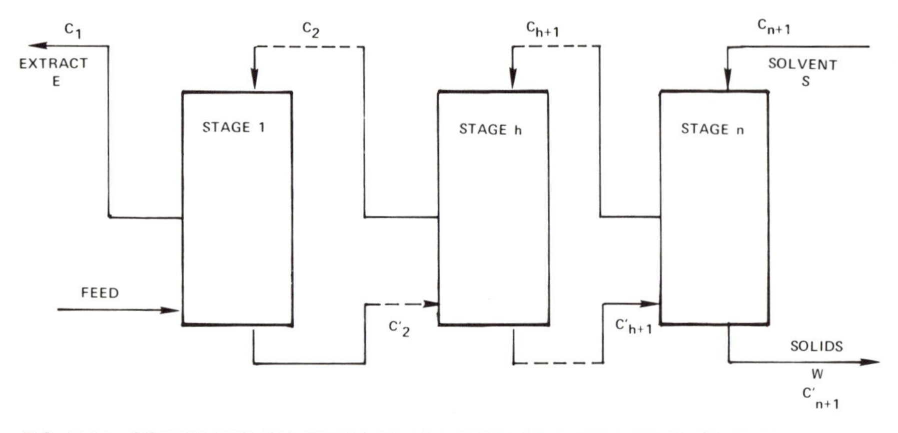


FIG. 7.11. SCHEMATIC DIAGRAM OF MULTISTAGE EXTRACTION SYSTEM

The total quantity of solvent ($S$) entering the system may contain a small quantity of solute; $SX_{SC} = C_{n+1}$, for a countercurrent flow situation. The solvent and solute stream leaves the system at stage 1 as extract ($E$) with the quantity of solute; $EX_{EC} = C_1$. The quantity of solvent-solute solution will be $(A+C)_{h+1}$ entering stage $h$ and the concentration of solute in solution will be:

$$x_h = \frac{C_h}{(A + C)_h} \tag{7.19}$$

In a similar manner, the concentration of solute in solution within the product solids stream may defined as:

$$x_h = \frac{C'_{h+1}}{(A' + C')_{h+1}} \tag{7.20}$$

The equations for determining the number of stages required in a multiple-stage leaching system are derived by conducting a materials balance on various components of the system with respect to various components of the streams. These equations will reflect changes in solute concentration in both primary streams in the system.

A material balance on solute for the entire system is:

$$C_{n+1} + C'_1 = C_1 + C'_{n+1} \tag{7.21}$$

while a similar material balance on the solute-solvent solution would be:

$$(A + C)_{n+1} + (A' + C')_1 = (A + C)_1 + (A' + C')_{n+1} \tag{7.22}$$

Based on these equations, the concentration of solute at the outlet from the system can be defined as:

$$x_1 = \frac{C_1}{(A + C)_1} = \frac{C_{n+1} + C'_1 - C'_{n+1}}{(A + C)_{n+1} + (A' + C')_1 - (A' + C')_{n+1}} \tag{7.23}$$

A more general set of equations is obtained by taking the material balance through the first intermediate stages ($h$) of the system. For solute:

$$C_{h+1} + C'_1 = C_1 + C'_{h+1} \tag{7.24}$$

and by incorporating equation (7.21):

$$C_{h+1} = C_{n+1} + C'_1 - C'_{n+1} + C'_{h+1} - C'_1$$

or

$$C_{h+1} = C_{n+1} - C'_{n+1} + C'_{h+1} \tag{7.25}$$

In a similar manner, a material balance on the solute-solvent solution gives:

$$(A + C)_{h+1} + (A' + C')_1 = (A + C)_1 + (A' + C')_{h+1} \tag{7.26}$$

and by incorporating equation (7.22):

$$(A + C)_{h+1} = (A + C)_{n+1} + (A' + C')_{h+1} - (A' + C')_{n+1} \tag{7.27}$$

Using equations (7.26) and (7.27), a general form of the equation for solute concentration in solution is obtained:

$$x_{h+1} = \frac{C_{n+1} - C'_{n+1} + C'_{h+1}}{(A + C)_{n+1} - (A' + C')_{n+1} + (A' + C')_{h+1}} \tag{7.28}$$

Using equations (7.23) and (7.28), successive calculations lead to determination of the number of stages required based on a desired level of solute in the product solids stream ($C'_{n+1}$).

EXAMPLE 7.6 A countercurrent multiple-stage leaching system is being used to extract the oil from 1000 kg soybeans per hour. The soybeans contain 18% oil and solvent is provided at a rate of 450 kg per hour. Determine the number of stages required if the solute-solvent solution represents 25% of the mass of solids leaving each stage in the system and solids leaving the system must have less than 1% oil.

SOLUTION

(1) An overall material balance on the system reveals:

| | *Solvent* | *Solute* | *Solids* |
|---|---|---|---|
| Feed ($F$) | — | 180 | 820 |
| Solvent ($S$) | 450 | — | — |
| Extract ($E$) | 255 | 170 | — |
| Solids ($W$) | 195 | 10 | 820 |

(2) Based on information given:

$$C_{n+1} = 0; \text{no oil in solvent entering system}$$

$$x_{n+1} = 0; \text{since } C_{n+1} = 0$$

$$(A + C)_{n+1} = \frac{450}{820} = 0.55 \text{ kg solvent/kg solids}$$

$$C'_{n+1} = \frac{10}{820} = 0.012 \text{ kg oil/kg solids}$$

$$(A' + C')_{n+1} = \frac{205}{820} = 0.25 \text{ kg solution/kg solids}$$

$$C'_1 = \frac{180}{820} = 0.22 \text{ kg oil/kg solids}$$

$$(\mathrm{A}' + C')_1 = 0.22 \text{ kg solution/kg solids}$$

(3) Using equation (7.23):

$$x_1 = \frac{0 + 0.22 - 0.012}{0.55 + 0.22 - 0.25} = 0.4$$

(4) Using equation (7.20):

$$x_1 = \frac{C'_2}{(A' + C')_2} = 0.4 = \frac{C'_2}{0.25}$$

$$C'_2 = 0.25\,(0.4) = 0.1 \text{ kg oil/kg solids}$$

(5) Using equation (7.28):

$$x_2 = \frac{C_{n+1} - C'_{n+1} + C'_2}{(A + C)_{n+1} - (A' + C')_{n+1} + (A' + C')_2}$$

$$x_2 = \frac{0 - 0.012 + 0.1}{0.55 - 0.25 + 0.25} = 0.16$$

(6) By repeating the step using equation (7.20):

$$C'_3 = 0.25\ (0.16) = 0.04 \text{ kg oil/kg solids}$$

(7) Then using equation (7.28):

$$x_3 = \frac{C_{n+1} - C'_{n+1} + C'_3}{(A + C)_{n+1} - (A' + C')_{n+1} + (A' + C')_3}$$

$$x_3 = \frac{0 - 0.012 + 0.04}{0.55 - 0.25 + 0.25} = 0.051$$

(8) Using equation (7.20):

$$C'_4 = 0.25\ (0.051) = 0.0127 \text{ kg oil/kg solids}$$

(9) Using equation (7.28)[

$$x_4 = \frac{0 - 0.012 + 0.0127}{0.55 - 0.25 + 0.25} = 0.0013$$

(10) By using equation (7.20):

$$C_5' = 0.25\ (0.0013) = 0.00032 \text{ kg oil/kg solids}$$

(11) Since $C_5' = 0.00032$ is well below the desired oil concentration in the solids at the system outlet, then *four* stages of the proposed extraction system would be required.

The number of stages required for countercurrent leaching can be determined using graphical procedures as introduced in Fig. 7.9. Point $E$ represents the composition of extract (solvent-solute solution) leaving the single-stage system and Point $W$ represents the composition of the stream containing product solids leaving the single-stage system. The line through $W$ and parallel to the hypotenuse represents the composition of all possible streams containing product solids at the exit of a designated stage for the system. The line $BE$ through $W$ represents the composition of potential solute-solvent solution mixture in equilibrium with the product solids.

The right triangle description in Fig. 7.12 illustrates the same points as would be used in analysis of a multiple-stage system. Point $F$ represents the composition of the product feed $[(X_{FC})_1]$ to the first stage of the system; the location along the base of the triangle indicates the proportion of solids and solute in the feed stream and the fact that it contains no solvent. The solvent entering a countercurrent system will have a composition designated at Point $A$ in Fig. 7.12 $[(X_{SC})_{n+1}]$ where the subscript refers to the number of stages required in the system. Point $E$ describes the composition of the solvent-solute solution $[(X_{EC})_1]$ produced by the system; the location along the hypotenuse of the triangle indicates the proportions of solvent and solute and the fact that this stream carries no product solids. The composition $[(X_{WC})_1]$ at Point $W$ is the proportion

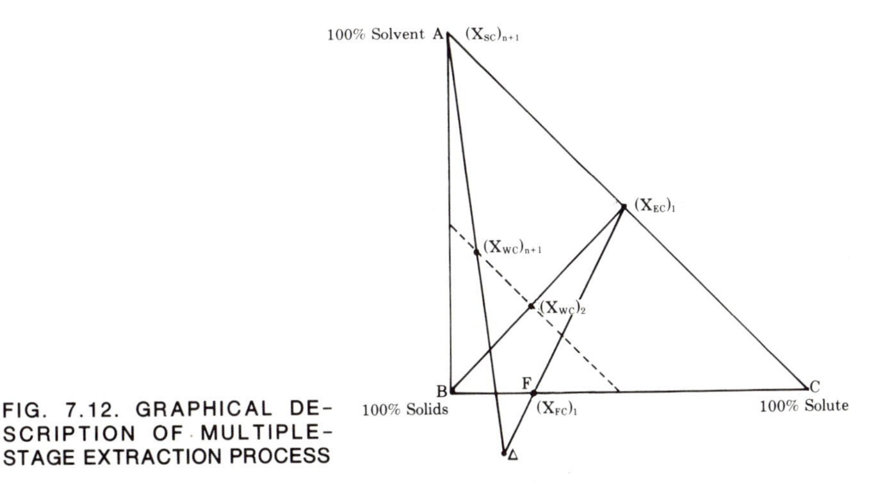


FIG. 7.12. GRAPHICAL DESCRIPTION OF MULTIPLE-STAGE EXTRACTION PROCESS

of product solids, solute and solvent in the product solids stream leaving the first stage of the system. The composition of solute in the product solids stream leaving the final stage of the system $[(X_{WC})_{n+1}]$ is designated on the line parallel to the hypotenuse through Point $W$.

The difference point ($\Delta$) represents the difference between the product solids stream and the solvent-solute solution stream at any point in the multiple-stage system. The location of the difference point ($\Delta$) is the intersection of two lines; one through Points $E$ and $F$ and a second line through $A$ and $W_{n+1}$. The line through $E$ and $F$ represents the differences between composition of product solids stream (feed) and solute-solvent stream at Stage 1. The line through $A$ and $W_{n+1}$ is the difference between product solids stream and the solvent stream at the final stage $(n+1)$ of the system.

The graphical procedure for determining number of stages required in a multiple-stage leaching system involves the following steps. Using the information presented in Fig. 7.12, the composition of the solvent-solute solution entering Stage 1 $[(X_{EC})_2]$ must be at a point on the hypotenuse of the right triangle intersecting with a line from $\Delta$ through $W_2$. This situation is established by recognizing that the difference point must represent the difference between the product solids stream and the solvent-solute solution stream between Stage 1 and Stage 2. The composition of the product solids stream $[(X_{WC})_3]$ leaving Stage 2 would be found by the intercept of lines between $[(X_{EC})_2]$ and $B$ and along the line parallel to the hypotenuse through $W_2$ and $W_{n+1}$. These procedures are continued until the composition of the solute in the product solids stream is less than $[(X_{WC})_{n+1}]$. The number of stages for the system is established by the number of extract streams required to achieve the desired compositions.

EXAMPLE 7.7 A multiple-stage leaching system is being designed to extract oil from codliver. The system is being designed to produce 150 kg per hour of extract containing 60% oil. The product solids stream should not contain more than 5% oil. Determine the number of stages required to achieve the desired extraction when the product feed contains 30% oil and the product solids stream leaving each stage does not contain more than 55% solvent-solute solution. Compute the quantities of pure solvent and product feed required for the process.

SOLUTION

(1) Based on the information given, inlet compositions would include:

$$(X_{FC})_1 = 0.3; \ (X_{SA})_{n+1} = 1.0; \ (X_{SC})_{n+1} = 0$$

(2) At the exit to the system, compositions would include:

$$(X_{EC})_1 = 0.6; \ (X_{WC})_{n+1} = 0.05$$

(3) In addition, since the outlet from each stage will contain 55% solvent-solute solution: $(X_{WA})_{n+1} = 0.5$; $(X_{WB})_2 = 0.45$

(4) By drawing lines using $(X_{FC})_1$ and $(E_{EC})_1$ as well as through $(X_{SA})_{n+1}$ and $(X_{WC})_{n+1}$, the difference Point ($\Delta$) is identified as illustrated in Fig. 7.13.

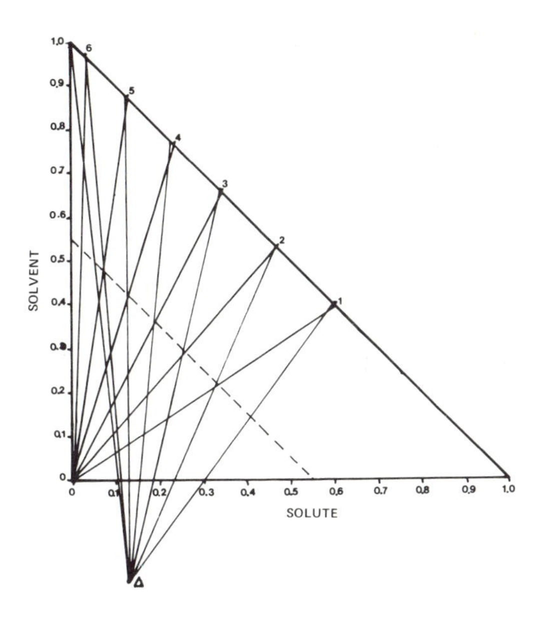


FIG. 7.13. GRAPHICAL SOLUTION OF MULTIPLE-STAGE EXTRACTION EXAMPLE 7.7

(5) By using the graphical procedure described, the number of extraction stages required can be determined. The first step involved use of the composition in the product solids stream at the exit of Stage 1:

$$(X_{WC})_2 = 0.325$$

to establish the composition of the solvent-solute solution entering Stage 1:

$$(X_{EC})_2 = 0.465$$

By continuing the procedure, it is found that the desired $(X_{WC})_{n+1} = 0.05$ is between $(X_{WC})_6 = 0.065$ and $(X_{WC})_7 = 0.015$ so *6 Stages* are required.

(6) From Fig. 7.13, the change in solvent-solute solution composition can be determined:

| | | |
|---|---|---|
| $(X_{SC})_{n+1}$ | $= 0$ | Enter |
| $(X_{EC})_6$ | $= 0.025$ | |
| $(X_{EC})_5$ | $= 0.12$ | |
| $(X_{EC})_4$ | $= 0.225$ | |
| $(X_{EC})_3$ | $= 0.34$ | |
| $(X_{EC})_2$ | $= 0.46$ | |
| $(X_{EC})_1$ | $= 0.6$ | Exit |

(7) In a similar manner, the change in oil concentration in the product solids stream can be determined from Fig. 7.13:

$$(X_{FC})_1 = 0.3 \quad \text{Feed}$$

$$(X_{WC})_2 = 0.325$$
$$(X_{WC})_3 = 0.25$$
$$(X_{WC})_4 = 0.185$$
$$(X_{WC})_5 = 0.12$$
$$(X_{WC})_6 = 0.065$$
$$(X_{WC})_7 = 0.015 \quad \text{Exit}$$

(8) A material balance on the entire system produces:

| | *Solids* | *Solvent* | *Oil* | *Total* |
|---|---|---|---|---|
| Feed | 2835 | — | 1215 | 4050 |
| Solvent | — | 3750 | — | 3750 |
| Extract | — | 600 | 900 | 1500 |
| Solids | 2835 | 3150 | 315 | 6300 |

Based on the material balance:

Feed = 4050 kg per hour
Solvent = 3750 kg per hour

As indicated earlier in this chapter, the design of an extraction system is a function of the rate of extraction as well as the number of stages required for the process. The rate of extraction will establish the required residence time within each stage and, in turn, the configuration of a given stage needed to achieve the extraction rate.

The rate of extraction in many food products will be controlled by diffusion of solute within the product solids as illustrated by Bomben *et al.* (1973). Since the diffusion rate may be small, large residence times will increase the number of stages required to achieve the desired levels of extraction. The diffusion of a solute within a product structure requires solution to unsteady-state partial differential equations. These equations have been presented and solved by Crank (1975) for standard geometrics (slab, cylinder, sphere). In addition, Treybal (1968) presents a chart (Fig. 7.14) to be used when dealing with problems requiring solutions to unsteady-state diffusion equations. As is evident, the chart presents relationships between an appropriate concentration ratio and a dimensionless group ($Dt/a^2$) to express time. The relationships can be used to determine time required to reduce concentrations by desired amounts or the concentration reduction achieved in some specified time.

EXAMPLE 7.8 A countercurrent extraction system is being used to remove salt from pickle brine stock and to concentrate the resulting brine stream. The brine stock enters the system at a rate of 300 kg per hour and pure water flows in the opposite direction through the system at the same rate. The brine stream leaving the system should have a concentration of 14.3% salt when the brine stock enters the system with 16.8% salt. Bomben *et al.* (1973) have shown that the diffusion coefficients for NaCl in pickle brine stock is $1.5 \times 10^{-5}$ $cm^2/s$ for 2.3 cm diameter stock at 50 C. Estimate the number of stages required to achieve the desired concentration in the outlet brine stream and reduce the salt concentration in the brine stock to below 2.5%.

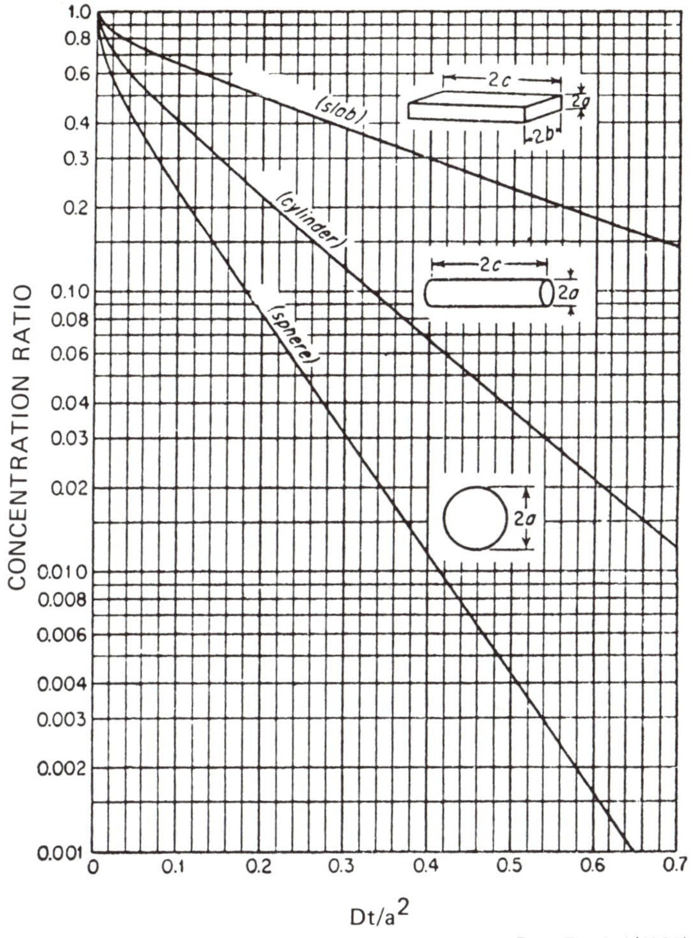


*From Treybal (1968)*

FIG. 7.14. CONCENTRATION HISTORIES FOR SOLUTE WITHIN SOLIDS FOR STANDARD GEOMETRICS

SOLUTION

(1) In order to estimate the number of stages, a residence time for brine stock in each stage of the system must be established. Using Fig. 7.14 and the dimensionless time:

$$\frac{Dt}{a^2} = \frac{(1.5 \times 10^{-5}\,\mathrm{cm^2/s})\,(900\,\mathrm{s})}{(1.15)^2} = 0.0102$$

where a residence time of 15 min has been assumed, the concentration ratio for a cylinder is 0.8. Since this ratio would indicate that approximately 20% reduction in stock concentration would occur in each stage, a long residence time will be required to keep the number of stages at a reasonable level. By increasing the residence time to 45 min, $Dt/a^2 = 0.0306$ and the concentration ratio becomes 0.65.

(2) The concentration ratio is a function of the solute concentrations within and outside the brine stock:

$$\text{Concentration Ratio} = 0.65 = \frac{x'_{h+1} - x_h}{x'_h - x_h}$$

where the concentrations can be identified by location on Fig. 7.11:

$x'_h$ = solute concentration in brine stock entering Stage $h$; kg salt/kg solution.

$x_h$ = solute concentration in solvent stream leaving Stage $h$; kg salt/kg solution.

$x'_{h+1}$ = solute concentration in brine stock leaving Stage $h$; kg salt/kg solution.

(3) Before using equations (7.20) and (7.23), an overall material balance can be conducted using information given and knowledge that the brine stock will be approximately 10% insoluble solids:

| | *Water* | *Salt* | *Insoluble Solids* |
|---|---|---|---|
| Feed | 219.6 kg | 50.4 kg | 30 kg |
| Water (solvent) | 300 kg | — | — |
| Salt solution | 257.1 kg | 42.9 kg | — |
| Brine stock | 262.5 kg | 7.5 kg | 30 kg |

(4) In order to use equation (7.20) and (7.23), equations will be expressed on a per unit insoluble solids basis:

$C_{n+1}$ = 0; pure water used as solvent

$x_{n+1}$ = 0; since $C_{n+1} = 0$

$$(A + C)_{n+1} = \frac{300}{30} = 10 \text{ kg solvent/kg solids}$$

$$C'_{n+1} = \frac{7.5}{30} = 0.25 \text{ kg salt/kg solids}$$

$$(A' + C')_{n+1} = \frac{262.5 + 7.5}{30} = 9 \text{ kg solution /kg solids}$$

$$C'_1 = \frac{50.4}{30} = 1.68 \text{ kg salt/kg solids}$$

$$(A' + C')_1 = \frac{219.6 + 50.4}{30} = 9 \text{ kg solution/kg solids}$$

(5) Using equation (7.23)

$$x_1 = \frac{0 + 1.68 - 0.25}{10 + 9 - 9} = 0.143 \text{ kg salt/kg solution}$$

and represents the desired brine concentration leaving the system.

(6) Using the concentration ratio:

$$\frac{x'_2 - x_1}{x'_2 - x_1} = \frac{x'_2 - 0.143}{0.168 - 0.143} = 0.65$$

$x'_2 = 0.159$

Then, using equation (6.20):

$C'_2 = 0.159\ (A' + C')_2 = 0.159\ (9) = 1.43$ kg salt/kg solids

(7) Using equation (7.28):

$$x_2 = \frac{0 - 0.25 + 1.43}{10 - 9 + 9} = 0.118 \text{ kg salt/kg solution}$$

(8) By using the concentration ratio for Stage 2:

$$\frac{x'_3 - 0.118}{0.159 - 0.118} = 0.65$$

$$x'_3 = 0.145$$

and: $C'_3 = 0.145\ (9) = 1.302$ kg salt/kg solids

(9) Repeating use of equation (7.28):

$$x_3 = \frac{0 - 0.25 + 1.302}{10 - 9 + 9} = 0.105 \text{ kg salt/kg solution}$$

(10) For Stage 3:

$$\frac{x'_4 - 0.105}{0.145 - 0.105} = 0.65$$

$$x'_4 = 0.131$$

and $C'_4 = 0.131\ (9) = 1.179$ kg salt/kg solids

(11) Using equation (7.28):

$$x_4 = \frac{0 - 0.25 + 1.179}{10 - 9 + 9} = 0.0929$$

(12) For Stage 4:

$$\frac{x'_5 - 0.0929}{0.131 - 0.0929} = 0.65$$

$$x'_5 = 0.1177$$

and $C'_5 = 0.1177\ (9) = 1.059$

(13) Using equation (7.28):

$$x_5 = \frac{0 - 0.25 + 1.059}{10 - 9 + 9} = 0.0809$$

(14) For Stage 5:

$$\frac{x'_6 - 0.0809}{0.1177 - 0.0809} = 0.65$$

$$x'_6 = 0.1048$$

and $C'_6 = 0.1048\ (9) = 0.9434$

(15) Using equation (7.28):

$$x_6 = \frac{0 - 0.25 + 0.9434}{10 - 9 + 9} = 0.06934$$

(16) For Stage 6:

$$\frac{x'_7 - 0.06934}{0.1048 - 0.06934} = 0.65$$

$$x'_7 = 0.0924$$

and $C'_7 = 0.0924\ (9) = 0.8315$

(17) Using equation (7.28):

$$x_7 = \frac{0 - 0.25 + 0.8315}{10 - 9 + 9} = 0.05815$$

(18) For Stage 7:

$$\frac{x'_7 - 0.05815}{0.0924 - 0.05815} = 0.65$$

$$x'_7 = 0.0804$$

and $C'_7 = 0.0804\ (9) = 0.7237$

(19) Using equation (7.28):

$$x_8 = \frac{0 - 0.25 + 0.7237}{10 - 9 + 9} = 0.04737$$

(20) For Stage 8:

$$\frac{x'_8 - 0.04737}{0.0804 - 0.04737} = 0.65$$

$$x'_8 = 0.06884$$

and $C'_8 = 0.06884\ (9) = 0.619$

(21) Using equation (7.28):

$$x_9 = \frac{0 - 0.25 + 0.619}{10 - 9 + 9} = 0.0369$$

(22) For Stage 9:

$$\frac{x'_9 - 0.0369}{0.06884 - 0.0369} = 0.65$$

$$x'_9 = 0.0576$$

and $C'_9 = 0.0576\ (9) = 0.5186$

(23) Using equation (7.28):

$$x_{10} = \frac{0 - 0.25 + 0.5186}{10 - 9 + 9} = 0.02686$$

(24) For Stage 10:

$$\frac{x'_{10} - 0.02686}{0.0576 - 0.02686} = 0.65$$

$$x'_{10} = 0.04684$$

and $C'_{10} = 0.04684\ (9) = 0.42155$

(25) Using equation (7.28):

$$x_{11} = \frac{0 - 0.25 - 0.42155}{10 - 9 + 9} = 0.0172$$

(26) For Stage 11:

$$\frac{x'_{12} - 0.0172}{0.04684 - 0.0172} = 0.65$$

$$x'_{12} = 0.0365$$

and $C'_{12} = 0.0365\ (9) = 0.3285$

(27) Using equation (7.28):

$$x_{12} = \frac{0 - 0.25 - 0.3285}{10 - 9 + 9} = 0.00785$$

(28) For Stage 12:

$$\frac{x'_{13} - 0.00785}{0.0365 - 0.00785} = 0.65$$

$$x'_{13} = 0.0265$$

and $C'_{13} = 0.0265\ (9) = 0.238$ kg salt/kg solids

(29) Since the concentration of salt in the brine stock leaving Stage 12 is below the desired level of 0.25 kg salt/kg solids, *12 Stages* would be the recommended design with 45 min residence time for each stage.

(30) The number of stages required by the situation in this example would be difficult to reduce. Requiring a lower concentration of salt in the brine stream would have a direct influence as would inceased residence time in each stage. Residence time is dependent on diffusion coefficient and brine stock radius and those parameters are difficult to change.

Although several other contact equilibrium processes have not been dealt with in this chapter, the concepts presented for single and multiple stage leaching should be useful when studying similar processes. Other contact equilibrium processes are covered in depth in many unit operations and mass transfer books (Coulson and Richardson, 1978; Foust *et al.* 1960; King, 1971).

## PROBLEMS

7.1. *Operating parameters for an absorption column.* A component of a liquid stream is being removed in an absorption column. The removal is described by the following operating line:

$$Y = 0.1 + 3.5X$$

If the mass transfer for the component is 5 kg/min and the exit conditions are 0.01 weight fraction of component in the liquid stream and 0.2 weight fraction of component in solvent, determine the flow rate of the liquid stream. The solvent flow rate is 20 kg/min.

7.2. *Design of an adsorption column.* An adsorption column is being designed for exchange of a component between two liquid streams. The number of transfer units (NTU) has been determined to be seven (7) with a mass transfer coefficient ($K_Y a$) of 10 kg/m$^3$ min. Determine the size of column required for a solvent flow rate of 5 kg/min. (Note that both height and cross-sectional area are variables.)

7.3. *Prediction of time for liquid-solid extraction.* A liquid-solid extraction process is being conducted in a batch container with volume of 20 m$^3$. If the overall mass transfer coefficient ($K_L A$) is 2 m$^3$ solution/min and the saturation concentration for this solvent is 18%, determine the time required for the solute concentration in the solvent to reach 10%.

7.4. *Operating conditions of a single-stage extraction system.* Soybean oil is being extracted from soybeans in a single-stage extraction system. The feed rate for the soybeans is 500 kg/hr and the feed contains 18% oil. The extract stream should contain a minimum of 15% oil and the solids stream should not contain more than 5% oil. Determine the extract production rate and the rate that solvent will be required.

7.5. *Designing a multiple-stage extraction system.* A multiple-stage extraction system is being used to extract codliver oil and produce extract with 40% oil at a rate of 350 kg/hr. The product feed contains 30% oil and stream containing solids leaving the system should contain less than 2% oil. Determine the number of stages required for the process and the rate that the solvent will be required to achieve the desired extraction when the feed rate is 500 kg/hr and the solvent-solute solution represents 20% of the solids stream leaving each stage.

## COMPREHENSIVE PROBLEM VII

Design of Multiple-Stage Extraction Process

### Objectives

1. To gain experience with procedures used in design of liquid-solid extraction systems.
2. To determine the influence of feed rate, initial solute concentration and allowable solute concentration in exit solids stream on extractor design.
3. To investigate relationships between inlet conditions for the extraction process and solvent requirements.

## Procedures

A. A multiple-stage extraction process is being designed for extraction of oil from soybeans.

B. Variables to be considered include:
   1. Feed rates of 300, 500 and 700 kg/hr.
   2. Oil content in feed of 14, 18 and 22%.
   3. Maximum oil content in exit solids stream of 5, 10 and 15%.

C. Additional information includes an effective diffusion coefficient for oil within the soybean of $5.7 \times 10^{-6}$ cm$^2$/s and a soybean diameter of 0.7 cm. Solvent content in the exit solids stream is equal to oil concentration and oil content in the extract leaving the system is 25%.

D. The design parameters to be determined include:
   1. The solvent requirement rates.
   2. The rates of extract production.
   3. The number of stages required for the process.

E. A computer program to allow determination of the design parameters based on the input variables should be developed.

## Discussion:

A. Influence of feed rate on solvent requirement rates.
B. Factors having significant influence on rate of extract production.
C. Influence of oil content in feed on solvent requirements.
D. Factors to be considered in design of multiple stage systems.

## NOMENCLATURE

$A$ = area, m$^2$.
$A+C$ = solvent-solute concentration in liquid stream of extraction process, kg/kg solids.
$A'+C'$ = solvent-solute concentration in product stream of extraction process, kg/kg solids.
$a$ = interfacial area per unit volume, m$^2$/m$^3$.
$C$ = component concentration in liquid stream of extraction process; kg component/kg solids.
$C'$ = component concentration in solids stream of extraction process; kg component/kg solids.
$D$ = diffusion coefficient, cm$^2$/s.
$E$ = mass in extract stream of leaching, kg/s.
$F$ = mass in feed stream to leaching process, kg/s.
HTU = height of transfer units.

$K$ = overall mass transfer coefficient, kg/m²s.
$k$ = mass transfer coefficient for individual phase, kg/m²s.
$m$ = mass or mass flow rate of individual phases in contact equilibrium process, kg/s.
$N$ = mass component being transferred in equilibrium contact process, kg.
NTU = number of transfer units.
$S$ = mass of solvent stream to leaching process, kg/s.
$s$ = cross-sectional area of contact equilibrium column, $m^2$.
$T$ = temperature, C.
$t$ = time, s.
$V$ = volume, $m^3$.
$W$ = mass in underflow stream of leaching process, kg/s.
$w$ = mass of individual component in contact equilibrium process, kg.
$X$ = concentration of component in liquid phase of general contact equilibrium process, kg component/kg product.
$x$ = concentration of component in solution for liquid stream of kg component/kg solution.
$X'$ = concentration of component in solution for solids stream of equilibrium process, kg component/kg solution.
$Y$ = concentration of component in gas phase of general contact equilibrium process, kg component/kg gas.
$Z$ = length of equilibrium contact process, m.
$\Delta$ = difference point in multiple stage extraction.

## Subscripts

$A,B,C,$ = component references in contact equilibrium process description.
$E$ = extract phase reference in extraction process.
$e$ = equilibrium condition reference in contact equilibrium process.
$F$ = feed phase reference in contact equilibrium process.
$h,h+1$ = reference for streams entering or leaving stage $h$ of contact equilibrium process.
$i$ = interface location.
$L$ = overall mass transfer coefficient in extraction process.
$n,n+1,n-1$ = reference for streams entering or leaving stage $n$ of contact equilibrium process.
$S$ = solvent phase reference in separation process.
$s$ = identifies saturation condition in extraction process.
$W$ = underflow stream of leaching.

$X, Y$ = component references in general equilibrium contact process.

$x,y$ = phase references to identify locations in equilibrium contact column.

## BIBLIOGRAPHY

BOMBEN, J.L., DURKEE, E.L., LOWE, E., and SECOR, G.E. 1974. Countercurrent desalting of pickle brine stock. J. Food Sci. Vol. *39* 260–263.

CHARM, S.E. 1978. The Fundamentals of Food Engineering, 3rd Edition. AVI Publishing Co., Westport, Conn.

COULSON, J.M., and RICHARDSON, J.F. 1978. Chemical Engineering. Volume II. 3rd Edition. Pergamon Press. Elmsford, New York.

CRANK, J. 1975. The Mathematics of Diffusion, 2nd Edition. Clarendon Press, Oxford.

EARLE, R.L. 1966. Unit Operations in Food Processing. Pergamon Press, Elmsford, N.Y.

FOUST, A.S., WENZEL, L.A., CLUMP, C.W., MANS, L., and ANDERSON, L.B. 1960. Principles of Unit Operations. John Wiley & Sons, New York.

KING, C.J. 1971. Separation Processes. McGraw-Hill Book Co., New York.

McCABE, W.L., and SMITH, J.C. 1967. Unit Operations of Chemical Engineering, 2nd Edition. McGraw-Hill Book Co., New York.

MOORE, W.J. 1962. Physical Chemistry, 3rd Edition. Prentice-Hall, Englewood Cliffs, N.J.

TREYBAL, R.E. 1968. Mass Transfer Operations. McGraw-Hill Book Co., New York.

# 8

# Mechanical Separation Processes

Processes which depend primarily on physical forces to accomplish the desired separation of components are used quite commonly in most phases of the food industry. These processes are normally referred to as mechanical separations and include filtration, sedimentation, and centrifugation. Each of these processes involves the application of a force to the fluid of concern, resulting in separation of the product components due to the different reaction of the components to the force applied. In the case of filtration, the force involved is that required to move the fluid through a filter medium. In the case of sedimentation, the forces are those of gravity and the influence of gravitational forces on the different components of the product. In the case of centrifugation, an induced force is applied resulting in separation of the product components due to centrifugal force.

The importance of these processes is not obvious because of the overshadowing importance of other processes utilized for food products. In many cases, however, the desired product composition could not be achieved without the use of some mechanical separation. The examples involving filtration range from simple ones, such as the separation of cheese curd from water, to rather sophisticated filtration of microscopic particles from air. The most prevalent use of sedimentation is in the separation of solids from waste water. Sophisticated processes, however, utilize sedimentation in the concentration of casein. The separation of cream from skim milk is the common use of centrifugal force in a separation process, while less obvious examples include use of cyclone separators to recover dry particles from air.

# 8.1 FILTRATION

The removal of solid particles from a fluid can be described by rather basic and accepted equations. The actual removal of solid particles from liquids is accomplished by somewhat different mechanisms than the removal of solid particles from air. No attempt will be made to discuss specific applications of filtration techniques until the basic equations have been developed and presented.

## 8.1.1 Operating Equations

In general, the filtration process can be described by the manner in which the fluid being filtered flows through the filter medium, the solids being deposited in the filter medium (Fig. 8.1). The removal of these

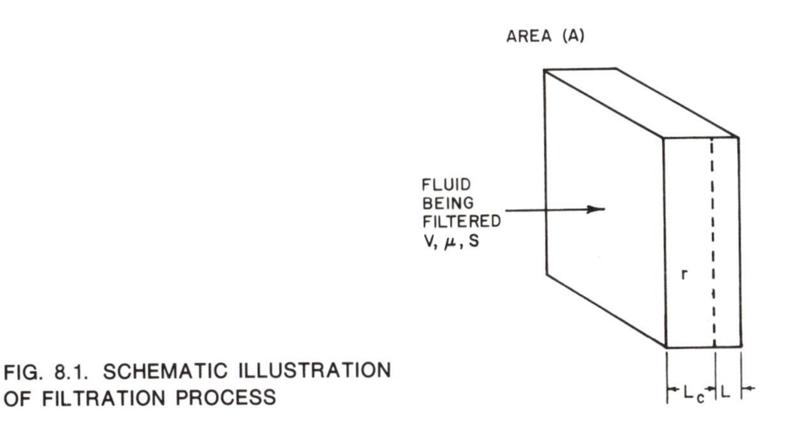


FIG. 8.1. SCHEMATIC ILLUSTRATION OF FILTRATION PROCESS

solids from the fluid results in build-up of the solids in the filter medium, resulting in an increased resistance to flow as the filtration process continues. All the preceding factors result in a description of filtration rate. In addition, this rate of filtration depends on several factors including (a) the pressure drop across the filter medium, (b) the area of the filtering surface, (c) the viscosity of the filtrate, (d) the resistance of the filter cake, which is made up by the solids removed from the fluid, and (e) the resistance of the filter media. The rate of filtration can be written as follows:

$$\text{Rate of Filtration} = \frac{\text{Driving Force}}{\text{Resistance}} \tag{8.1}$$

where the driving force is the pressure required to move the fluid through the filter media and the resistance is dependent on several factors. The overall resistance can be described by the following expression:

$$R = \mu r'(L_c + L) \tag{8.2}$$

where $L_c$ represents a thickness of the accumulated solids in the filter cake, $\mu$ is the fluid viscosity and $L$ is a fictitious thickness of the filter material or medium. The parameter $r'$ in equation (8.2) represents the specific resistance of the filter cake and will be a property of the particles forming the filter cake. Earle (1966) describes $L_c$ by the following expression:

$$L_c = \frac{SV}{A} \tag{8.3}$$

where $S$ is the solids content of the fluid being filtered and $V$ is the volume which has passed through the filter with cross-sectional area ($A$). The thickness of the filter cake represents a somewhat fictitious value which would describe the total thickness of all solids accumulated. In some filtration processes, this may approach the real situation. Utilizing equations (8.2) and (8.3) the total resistance can be written in the following manner:

$$R = \mu r'\left(\frac{SV}{A} + L\right) \tag{8.4}$$

Combining equations (8.1) and (8.4) an expression for the rate of filtration is obtained as follows:

$$\frac{dV}{dt} = \frac{A\Delta P}{\mu r'\left(\frac{SV}{A} + L\right)} \tag{8.5}$$

Equation (8.5) is an expression used to describe the filtration process and can be used for scale-up if converted to appropriate forms.

The filtration process may occur in two phases: (a) constant-rate filtration, normally occurring during the early stages of the process, and (b) constant-pressure filtration, occurring during the final stages of the process.

**8.1.1a Constant-rate filtration.**—Constant-rate filtration will be described by the following integrated form of equation (8.5):

$$\frac{V}{t} = \frac{A\Delta P}{\mu r'\left(\frac{SV}{A} + L\right)} \tag{8.6}$$

or:

$$\Delta P = \frac{\mu r'}{A^2 t}\left[SV^2 + LAV\right] \tag{8.7}$$

which can be used to determine pressure drop required as a function of filtration rate.

Equation (8.6) can be expressed in a different form if the thickness ($L$) of the filter medium can be considered negligible. The following equation for pressure drop as a function of time is obtained:

$$\Delta P = \frac{\mu r' S V^2}{A^2 t} \tag{8.8}$$

In many situations, equation (8.8) can be used to predict pressure drop requirements for a filter during the early stages of the process.

EXAMPLE 8.1 An air filter is being used to remove small particles from an air supply to a quality control laboratory. The air is being supplied at a rate of 0.5 $m^3/s$ through an air filter with a 0.5 $m^2$ cross-section. If the pressure drop across the filter is 0.25 cm water after 1 hour of use, determine the life of the filter if a change of the filter is required when the pressure drop is 2.5 cm.

SOLUTION

(1) Due to the nature of air filtration, constant-rate filtration can be assumed and the filter media thickness is negligible.

(2) Based on information given and equation (8.8), the product of the specific resistance ($r'$) and the particle content of air ($S$) can be established:

$$r'S = \frac{\Delta P A^2 t}{\mu V^2}$$

(3) The pressure drop of 0.25 cm water can be expressed in consistent units (Pascals) by:

$$0.25 \text{ cm of water} = 24.52 \text{ Pa}$$

(4) Using equation (8.8) (for 1 hr of filtration):

$$r'S = \frac{(24.52 \text{ Pa})(0.5 \text{ m}^2)^2 (3600 \text{ sec})}{(1.72 \times 10^{-5} \text{ kg/ms})(0.5 \times 3600)^2}$$

where $\mu = 1.72 \times 10^{-5}$ (viscosity of air at 0 C from Table A.3)

$$r'S = 396/\text{m}^2$$

(5) Using $r'S = 396$ and equation (8.8), the time required for $\Delta P$ to increase to 2.5 cm water can be computed.

(6) Converting $\Delta P$ to consistent units:

$$2.5 \text{ cm of water} = 245.16 \text{ Pa}$$

then:

$$\frac{t}{V^2} = \frac{(1.73 \times 10^{-5})(396)}{(0.5)^2 (245.16)} = 1.11 \times 10^{-4} \text{ s/m}^6$$

(7) Since:

$$\frac{V}{t} = .5 \text{ m}^3/\text{s}$$

Therefore:

$$V^2 = 0.25\ t^2$$

so:

$$t = 1.11 \times 10^{-4} \times .25\ t^2 = 2.778 \times 10^{-5}\ t^2$$

and:

$$t = \frac{1}{2.775 \times 10^{-5}} = 35{,}997 \text{ s}$$

$$t = 10 \text{ hr}$$

(8) Thus the filter should be changed after 10 hour of operation.

**8.1.1b. Constant-pressure filtration.**—An expression which describes constant-pressure filtration can be obtained from the following form of equation (8.5):

$$\frac{\mu r' S}{A} \int_0^V V dV + \mu r' L \int_0^V dV = A \Delta P \int_0^t dt \qquad (8.9)$$

Integration leads to the following design equation:

$$\frac{tA}{V} = \frac{\mu r' S}{2 \Delta P} \frac{V}{A} + \frac{\mu r' L}{\Delta P} \qquad (8.10)$$

or the following equation if filter media thickness ($L$) can be assumed negligible:

$$t = \frac{\mu r' S V^2}{2 A^2 \Delta P} \qquad (8.11)$$

Essentially, equation (8.11) indicates the time required to filter a given volume of fluid when a constant pressure is maintained. Various procedures are utilized in the use of this equation to obtain information in the equation which may not be readily available. Such parameters as the specific resistance of the filter cake ($r'$) may not be known for specific types of solids and must be determined experimentally. Earle (1966) presented procedures for determination of these parameters and Charm (1978) has discussed determination of filtration constants which are somewhat more difficult than those proposed in the above presentation.

EXAMPLE 8.2 A liquid is being filtered at a pressure of 200 kPa through 0.2 m² of filter. Initial results indicate that 5 min is required to filter 0.3 m³ of liquid. Determine the time which will elapse until the rate of filtration drops to $5 \times 10^{-5}$ m³/s.

SOLUTION

(1) Since the filtration is assumed to be in the constant-pressure regime, equation (8.11) will apply as follows:

$$\mu r' S = \frac{2A^2 \Delta P t}{V^2} = \frac{2(0.2)^2 (200{,}000)(5 \times 60)}{0.3^2}$$

$$\mu r' S = 53.33 \times 10^6 \text{ kg/m}^3\text{s}$$

which is based on data obtained at 5 min.

(2) Using equation (8.5) (where $L$ is assumed to be negligible):

$$5 \times 10^{-5} = \frac{(0.2)^2 (200{,}000)}{[53.33 \times 10^6]\, V}$$

$$V = 3 \text{ m}^3$$

(3) Using equation (8.11):

$$t = \frac{[53.33 \times 10^6]\,(3)^2}{2(0.2)^2 (200{,}000)} = 29{,}998 \text{ s}$$

$t$ = 499.9 min; indicates that 500 min of filtration at a constant pressure would occur before the rate dropped to $5 \times 10^{-3}$ m$^3$/s.

Reference to equation (8.5) reveals that the filtration process is directly dependent on two factors: the filter medium and the fluid being filtered. In equation (8.5) a filter medium is described in terms of the area ($A$) and the specific resistance ($r'$). The filter medium will depend considerably on the type of fluid being filtered. In the case of liquid filtration the filter medium will contain, to a large extent, the solids removed from the liquid. This filter cake must be supported by some type of structure which plays only a limited role in the filtration process. In some cases these supporting materials may be woven, such as wool, cotton and linen, or they may be granular materials for particular types of liquids. In any event the primary role of the material is to support the collected solids so that the solids can act as a filter medium for the liquid. The design of filters for the removal of particles from air is significantly different. In this application the entire filter medium is designed into the filter, and the collected solids play a very minor role in the filtration process. The filter medium is a very porous collection of filter fibers of the same magnitude in size as the particles to be removed from the air. This results in filtration processes that are nearly constant-flow rate in all situations.

The second factor of equation (8.5) which influences filtration rate is the fluid or liquid being filtered as described by the fluid viscosity ($\mu$). The rate of filtration and the viscosity of the fluid being filtered are inversely related; as the viscosity of the fluid increases, the rate of filtration must decrease. Fluid viscosity plays a very important role in the filtration process and must be accounted for in all design computations.

### 8.1.2 Mechanisms of Filtration

The mechanisms involved in the removal of small particles during air filtration are relatively well-defined. Whitby and Lundgren (1965) have listed four mechanisms as follows: (a) Brownian diffusion, (b) interception, (c) inertial impaction, and (d) electrical attraction. Decker *et al.* (1962) indicated that deposition according to Stoke's law could be considered as an additional mechanism. The mechanism of Brownian diffusion will have very little influence on removal of particles larger than 0.5 micron. This particular mechanism contributes to the particle collection by causing the particles to deviate from the streamline of air flow around the filter fibers and bring them into contact with the fiber. Somewhat larger particles may be removed by the mechanism of interception even when the particles do not deviate from the air streamline. When these particles follow the air streamline and the streamline brings them sufficiently close to the fiber, the particles will be removed by direct interception. The larger particles will not follow the air streamline and will be removed by inertial impaction. Particles larger than 1 micron will normally be removed by this mechanism. When the particles and fiber have different or opposite electrical charges, the particles will deviate from the air streamline and deposit on the fiber as a result of this mechanism. Very large particles may be influenced by gravity and deviate from the air streamline due to gravitational force, resulting in deposition on the filter fiber. The contribution of this mechanism to small-particle removal is probably very small.

In the case of removal of solids from liquids, the mechanisms of removal are not well-defined. In addition, the mechanism or mechanisms may be considerably different depending on the mode of filtration considered. It would appear that during the initial stages when the layer of solids is being collected from the filter medium, the mechanisms are most likely to be direct interception or inertial impaction. During this stage of filtration, the liquid is moving through the filter medium at a rapid rate. After the filter cake is established and constant-pressure filtration occurs, the liquid flows through the filter cake in a streamline fashion. It would appear that the main mechanism is direct interception, although the other mechanisms mentioned for removal of particles from air could contribute in a minor way.

### 8.1.3 Design of a Filtration System

Although the expressions presented for the description of filtration rate by a given system seem to be relatively straightforward, the use of these expressions requires knowledge of several system parameters. In most cases, the viscosity and specific resistance values may not be known. The approach normally followed is to make a small-scale filtration experiment to obtain some indication of the relationships involved, followed by a scale-up of the entire process. Expressions of the type presented as equations (8.7) and (8.11) are ideal for this purpose. A small or pilot-scale operation usually results in determination of the filtrate volume after given periods of time at a constant pressure on the small-scale filter. Earle (1966) showed that this information can be utilized to determine a plot of the type given as Fig. 8.2. The approach in obtaining the filtration graph is through the use of equation (8.10) in which $L$ is not assumed to be negligible. Use of pilot-scale data to obtain a relationship for Fig. 8.2 leads to the evaluation of appropriate constants to be utilized in the scale-up of the filtration process. In some situations, it is desirable to collect pilot-scale data at different pressures and/or different filter areas.

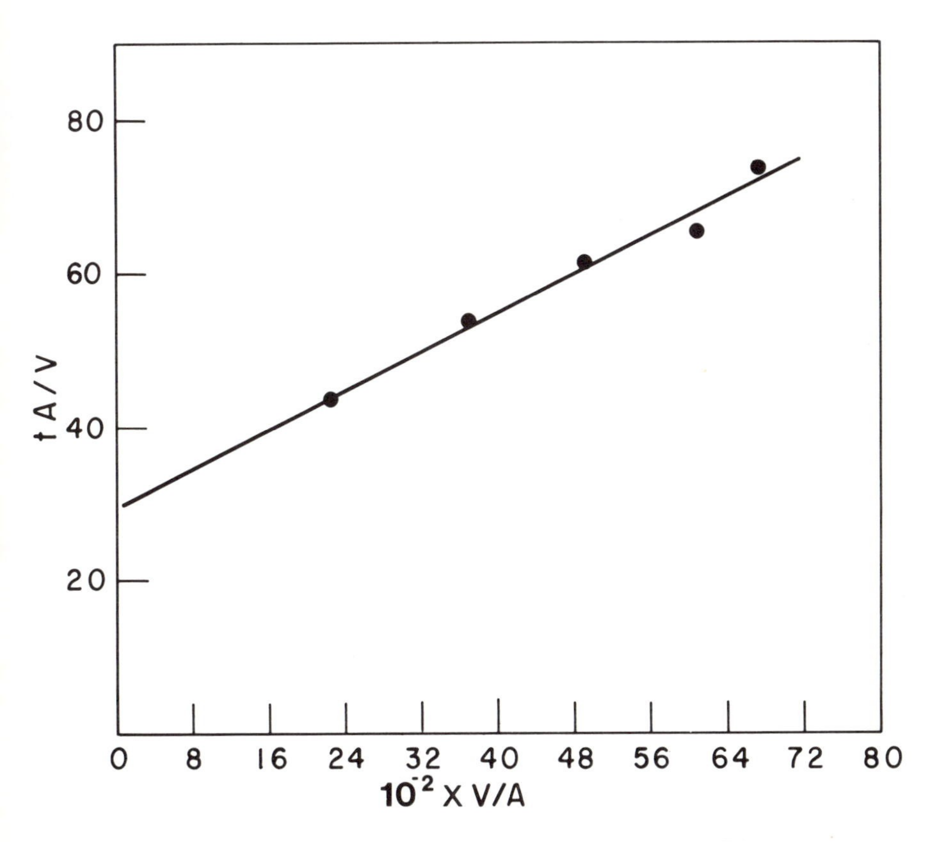


FIG. 8.2. FILTRATION GRAPH OBTAINED FROM PILOT-SCALE DATA

EXAMPLE 8.3 A filtration system is being designed to filter 4 $m^3$ of a slurry in 2 hr using a constant pressure of 400 kPa. The necessary design conditions were established on a laboratory scale using a filter with 0.1 $m^2$ surface area and 140 kPa constant pressure. The following results were obtained on a laboratory scale:

| Filtration Volume $10^{-2} \times m^3$ | Time (min) |
|---|---|
| 2.3 | 10 |
| 3.7 | 20 |
| 4.9 | 30 |
| 6.1 | 40 |
| 6.8 | 50 |

Determine the filter area required in the design situation which will provide the desired conditions.

SOLUTION

(1) Utilizing equation (8.10):

$$\frac{tA}{V} = \frac{\mu r'S}{2\Delta P}\left(\frac{V}{A}\right) + \frac{\mu r'L}{\Delta P}$$

A plot of $tA/V$ versus $V/A$ should provide a linear relationship with the slope equal to $\mu r'S/\Delta P$ and a vertical axis intercept of $\mu r'L/\Delta P$.

(2) From the experimental data provided:

| $10^{-2} \times V/A$ | $tA/V$ |
|---|---|
| 23 | 43.48 |
| 37 | 54.05 |
| 49 | 61.22 |
| 61 | 65.57 |
| 68 | 73.53 |

(3) From the results presented in Fig. 8.2, a slope of 62.5 and a vertical intercept of 29.83 provide a design equation as:

$$\frac{tA}{V} = 62.5\frac{V}{A} + 29.83$$

(4) Since pressure and area are the two variables existing between the laboratory and design situation, appropriate changes must be incorporated:

$$\text{Slope} = 62.5 = \frac{\mu r'S}{2(140{,}000)}\ ;\ \mu r'S = 1.75 \times 10^7$$

$$\text{Intercept} = 29.83 = \frac{\mu r'L}{140{,}000}\ ;\ \mu r'\text{L} = 4{,}176{,}200$$

so the design equation becomes:

$$\frac{tA}{V} = \frac{1.75 \times 10^7}{\Delta P}\left(\frac{V}{A}\right) + \frac{4{,}176{,}200}{\Delta P}$$

(5) For the filtration system being designed: $t$ = 2 hr = 120 min.; $V$ = 4 $m^3$; $\Delta P$ = 400 kPa

then:

$$\frac{(120)A}{4} = \frac{1.75 \times 10^7}{400{,}000}\left(\frac{4}{A}\right) + \frac{4{,}176{,}200}{400{,}000}$$

$$30A = \frac{175}{A} + 10.44$$

$$A = 3.09 \text{ m}^2$$

(6) An area of 3.09 $m^2$ is obtained when the positive solution of the quadratic equation is selected.

# 8.2 SEDIMENTATION

The separation of solids from fluid streams by using gravitational or centrifugal forces is referred to as sedimentation. In most processes in the food industry which involve sedimentation the objective is to remove particle solids from either liquid or gas. As will become obvious the use of gravity as a force in removal of solids from fluids can find considerable application in separation processes used in the food industry.

## 8.2.1 Sedimentation Velocities for Low-concentration Suspensions

Particles in a low-concentration suspension will settle at a rate representing the terminal velocity of the particle in the suspension fluid. The terminal velocity of each particle will be established as though each individual particle is the only particle in the suspension. This type of sedimentation is usually referred to as free settling, since there is no interaction between particles and the suspension. The terminal velocity of the particle can be predicted and established by examination of the forces acting on the particle. The force which resists the gravitational force is referred to as the drag force, and can be described by the following expression:

$$F_D = 3\pi\mu D u \tag{8.12}$$

where $(u)$ represents the relative velocity between the particle and the fluid, $D$ is particle diameter and $\mu$ is fluid viscosity. Equation (8.12) applies as long as the particle Reynolds number is less than 0.2, which will be the case of most applications in food processing. Expressions to use when particle Reynolds numbers are greater than 0.2 have been developed and are presented by Coulson and Richardson (1978). The gravitational force $(F_G)$ is a function of particle volume and density difference along with gravitational acceleration, as illustrated by the following equation:

$$F_G = \frac{1}{6} \mu D^3 (\rho_p - \rho_f) g \tag{8.13}$$

where ($D$) is the particle diameter, assuming a spherical shape. By setting equation (8.12) equal to equation (8.13) (the conditions which must exist at the terminal velocity of the particle), it is possible to solve and obtain an equation representing the terminal velocity as follows:

$$u_t = \frac{D^2 g}{18\mu} (\rho_p - \rho_f) \tag{8.14}$$

From equation (8.14), it is observed that the terminal velocity is directly related to the square of the particle diameter. In addition, the terminal velocity is dependent on the density of the particle and properties of the fluid. Equation (8.14) is the most common form of Stoke's law and should be applied only for streamline flow and spherical particles. When the particles are not spherical, the usual approach is to introduce a shape factor which will account for the irregular shape of the particle and the corresponding influence of this factor on terminal velocity.

EXAMPLE 8.4 The damage to blueberries and other fruits during handling immediately after harvest is closely related to the terminal velocity in air. Compute the terminal velocity of a blueberry with a diameter of 0.60 cm and density of 1120 kg/m$^3$ in air at 21/C and atmospheric pressure.

SOLUTION

(1) Equation (8.14) can be used with the following parameter values: $D$ = 0.60 cm, $g$ = 9.806 m/s$^2$, $\mu$ = 1.828 × 10$^{-5}$ kg/ms, $\rho_p$ = 1120 kg/m$^3$, $\rho_f$ = 1.2 kg/m$^3$.

(2) Using equation (8.14):

$$u_t = \frac{(0.006)^2 (9.806)}{(18)(1.828 \times 10^{-5})} [1120 - 1.2] = 1200 \text{ m/s}.$$

(3) A check of the particle Reynolds number:

$$N_{Re} = \frac{(1.2)(.006)(1200)}{1.828 \times 10^{-5}} = 4.73 \times 10^5$$

indicates that streamline flow does not exist and conditions for using equation (8.14) do not apply.

(4) By using drag force in the form of equation (6.33), the equation for terminal velocity becomes:

$$u_t = \left[\frac{4}{3} \frac{Dg}{C_d \rho_f} (\rho_p - \rho_f)\right]^{1/2}$$

(5) Using the drag coefficient from Fig. 6.23 for Re = 4.95 × 10$^5$; $C_d$ = 0.2, then

$$u_t = \left[\frac{4}{3} \frac{(0.006)(9.806)}{(0.2)(1.2)} (1120 - 1.2)\right]^{1/2} = 19.1 \text{ m/s}$$

(6) A check of the Reynolds number:

$$N_{Re} = \frac{(1.2)(0.006)(19.1)}{1.828 \times 10^{-5}} = 7523$$

indicates that $C_d$ = 0.4 and:

$$u_t = \left[\frac{4}{3}\,\frac{(0.006)(9.806)}{(0.4)(1.2)}(1120 - 1.2)\right]^{1/2} = 13.5\ \text{m/s}$$

(7) A final check of Reynolds number:

$$N_{Re} = \frac{(1.2)(.006)(13.5)}{1.828 \times 10^{-5}} = 5317$$

indicates that $C_d$ = 0.4 is appropriate and the terminal velocity is 13.5 m/s when the spherical geometry is assumed.

### 8.2.2 Sedimentation in High-concentration Suspensions

When the concentration of solids in the suspension becomes sufficiently high, Stoke's law no longer describes the velocity of settling. In these situations when the range of particle sizes is between 6 and 10 microns, the particles making up the solid suspension will fall at the same rate. This rate will correspond to the velocity predicted by Stoke's law, utilizing a particle size which is the mean of the smallest and the largest particle size in the suspension. Physically it is apparent that the rate at which the larger particles settle is reduced by the influence of the smaller particles on the properties of the fluid. The rate at which the smaller particles fall is accelerated by the movement of the larger particles.

The settling of particles results in very well-defined zones of solids concentrations. The movement of the solids at a constant rate results in a clear liquid at the top of the column suspension. Directly below the clear liquid is the constant concentration of solids which is moving at a constant rate. Below the solids which are settling is a zone of variable concentration representing the zone within which the solids are collecting. At the bottom of the suspension column is the collected sediment, which contains the largest particles in the suspension. The definition of these zones will be influenced somewhat by the size of particles involved and the range of these sizes.

The rates at which solids will settle in high-concentration suspensions have been investigated on a somewhat empirical basis and have been reviewed by Coulson and Richardson (1978). One approach is to modify Stoke's law by introduction of the density and viscosity of the suspension in place of the density and viscosity of the fluid. This may result in an equation for the settling rate of the suspension as follows:

$$u_s = \frac{KD^2\ (\rho_p - \rho_s)g}{\mu_s} \tag{8.15}$$

where $K$ is a constant to be evaluated experimentally. In most cases attempts are made to predict the density and viscosity of the suspension on the basis of the composition. A second approach attempts to account for the void between suspension particles which allows movement of fluid to the upper part of the suspension column and the expression for the particle settling velocity was obtained as follows:

$$u_p = \frac{D^2 (\rho_p - \rho_s) g}{18\mu} f(e) \tag{8.16}$$

where $f(e)$ is a function of the void in a suspension. Equation (8.16) is still a modification of Stoke's law where the density of the suspension and the viscosity of the fluid are utilized. The function of the void space in the suspension must be determined experimentally for each situation in which the expression is utilized. Although there are definite limitations to the use of expressions such as equation (8.15) and (8.16), they will provide practical results for high-concentration suspensions of large particles. Description of settling rates for small-particle suspensions is subject to considerable error and acceptable approaches need to be developed.

EXAMPLE 8.5 A sedimentation tank is used to remove larger-particle solids from waste water leaving a food-processing plant. The ratio of liquid mass to solids in the inlet to the tank is 9 kg liquid/kg solids and the inlet flow rate is 0.1 kg. The sediment leaving the tank bottom should have 1 kg liquid/kg solids. If the sedimentation rate for the solids in water is .0001 m/s, determine the area of sedimentation necessary.

SOLUTION

(1) Using the mass flow equation and the difference between inlet and outlet conditions on the waste water, the following equation for upward velocity of liquid in the tank is obtained:

$$u_f = \frac{(C_i - C_o)w}{A\rho}$$

where $C_i$ and $C_o$ are mass ratios for liquid to solids in waste water.

(2) Based on information given:

$$C_i = 9 \qquad w = 0.1$$
$$C_o = 1 \qquad \rho = 993$$

(3) Since the upward liquid velocity should equal sedimentation velocity for the solids, the area becomes:

$$A = \frac{(9-1)(0.1)}{(0.0001)(993)} = 8 \text{ m}^2$$

In addition to the properties of the solid and fluid in the suspension, other factors influence the sedimentation process. In general, the height of the suspension will not affect the rate of sedimentation or the consistency of the sediment produced. By experimentally determining the height of the liquid or fluid-sediment interface as a function of time for a given initial height, the results for any other initial height can be predicted. The diameter of the sedimentation column may influence the rate of sedimentation if the ratio of the diameter of the column to the diameter of the particle solids is less than 100. In this particular situation, the walls may have a retarding influence on the sedimentation rate. In general, the concentration of the suspension will influence the rate of sedimentation. Higher suspension concentrations tend to reduce the rate at which the sediment settles. The possibility exists of utilizing sedimentation as a means of removing solid particles from air, such as in the removal of solids after spray-drying processes. By introducing the particle suspension into a static air column the solid particles tend to settle to the floor surface. Relatively simple and straightforward computations utilizing expressions presented previously in this chapter will illustrate that such sedimentation processes are quite slow for particles of the type normally encountered. Due to this fact, the procedure of sedimentation is not used commonly, but procedures utilizing other forces on the particle to accelerate the removal are employed. Such processes will be discussed more thoroughly in the next section, which deals with centrifugal separation.

## 8.3 CENTRIFUGATION

In many processes the sedimentation separation of two liquids or of liquid and solid does not progress rapidly enough to accomplish separation efficiently. In these types of applications the separation can be accelerated through the use of centrifugal force.

### 8.3.1 Basic Equations

The first basic equation is one which describes the force acting on a particle moving in a circular path as follows:

$$F_c = \frac{mr\omega^2}{g_c} \tag{8.17}$$

where $\omega$ represents the angular velocity of the particle. Since $\omega$ can be expressed as the tangential velocity of the particle and at any radius ($r$), equation (8.17) can be presented as:

$$F_c = \left(\frac{m}{g_c}\right)\frac{u^2}{r} \tag{8.18}$$

Earle (1966) illustrates that if the rotational speed is expressed in revolutions per minute, equation (8.18) can be written as:

$$F_c = 0.011 \frac{mrN^2}{g_c} \tag{8.19}$$

where ($N$) represents the rotational speed in revolutions per minute. Equation (8.19) indicates that the centrifugal force acting on a particle is directly related to and dependent on the distance of the particle from the center of rotation ($r$), the centrifugal speed of rotation ($N$) and the mass of the particle considered ($m$). For example, if a fluid containing particles of different densities is placed in a rotating bowl the higher-density particles will move to the outside of the bowl as a result of the greater centrifugal force acting upon them. This will result in a movement of the lower-density particles toward the interior portion of the bowl, as illustrated in Fig. 8.3. This is the principle used in separation of liquid food products which contain components with different densities.

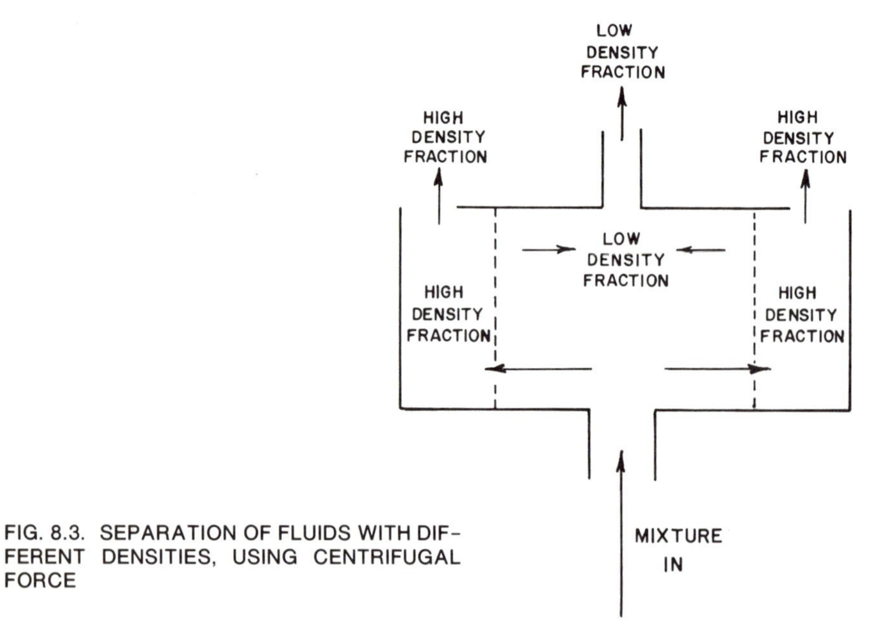


FIG. 8.3. SEPARATION OF FLUIDS WITH DIFFERENT DENSITIES, USING CENTRIFUGAL FORCE

### 8.3.2 Rate of Separation

The rate at which separation of materials of different densities can occur is usually expressed in terms of the relative velocity between the two phases. An expression for this velocity is the same as the equation for terminal velocity given as equation (8.14) in which the gravitational acceleration ($g$) is replaced by an acceleration parameter describing the influence of centrifugal force. This acceleration can be expressed in the following way:

$$a = r\left(\frac{2\pi N}{60}\right)^2 \tag{8.20}$$

where $(N)$ is the rotational speed of the centrifuge in revolutions per minute. Substitution of equation (8.20) into equation (8.14) results in the following expression:

$$u_c = \frac{D^2 N^2 r(\rho_p - \rho_s)}{1640\mu} \tag{8.21}$$

which describes the velocity of spherical particles in a centrifugal force field. Equation (8.21) can be applied to any situation in which there are two phases with different densities. The expression clearly illustrates that the rate of separation as expressed by velocity $(u_c)$ is directly related to the density between phases, the distance from the center of rotation, the speed of rotation, and the diameter of the particles in the higher-density phase.

EXAMPLE 8.6 The solid particles in a liquid-solid suspension are being separated by centrifugal force. The particles are 100 microns in diameter with a density of 800 kg/m$^3$. The liquid is water with a density of 993 kg/m$^3$ and the effective radius for separation is 7.5 cm. If the required velocity for separation is .03 m/s, determine the required rotation speed for the centrifuge.

SOLUTION

(1) Using equation (8.21) with the viscosity of water at $5.95 \times 10^{-4}$ kg/ms

$$N^2 = \frac{1640\,\mu u_c}{D^2\, r(\rho_p - \rho_s)} = \frac{1640(5.95 \times 10^{-4})(60\text{ s/min})}{(0.075)(100 \times 10^{-6})^2(993 - 800)}(.03)(60\text{ s/min})$$

$$N^2 = 7.28 \times 10^8$$
$$N = 26{,}983 \text{ rpm}$$

### 8.3.3 Liquid–liquid Separation

In the case of separation which involves two liquid phases it is usually easier to describe the process in terms of the surface which separates the two phases during separation. The differential centrifugal force acting on an annulus of the liquid in the separation cylinder can be written as:

$$dF_c = \frac{dm}{g_c} r\omega^2 \tag{8.22}$$

where $(dm)$ represents the mass in the annulus of liquid. Equation (8.22) can be rewritten as:

$$\frac{dF_c}{2\pi rb} = dP = \frac{\rho\omega^2\, rdr}{g_c} \tag{8.23}$$

where $(dP)$ represents the differential pressure across the annulus of liquid and $b$ represents the height of the separation bowl. By integration of equation (8.23) between two different radii in the separation cylinder, the different in pressure between these two locations can be computed from the following expression:

$$P_2 - P_1 = \frac{\rho \omega^2 (r_2^2 - r_1^2)}{2g_c} \tag{8.24}$$

At some point in the cylinder, the pressure of one phase must equal the other phase so that expressions of the type given by equation (8.24) may be written for each phase and can represent the radius of equal pressure in the following manner:

$$\frac{\rho_A \omega^2 (r_n^2 - r_1^2)}{2g_c} = \frac{\rho_B \omega^2 (r_n^2 - r_2^2)}{2g_c} \tag{8.25}$$

By solving for the radius of equal pressures for the two phases the following expression is obtained:

$$r_n^2 = \left[\frac{(\rho_A r_1^2 - \rho_B r_2^2)}{\rho_A - \rho_B}\right] \tag{8.26}$$

where $\rho_A$ equals the density of the heavy liquid phase and $\rho_B$ is the low-density liquid phase. Equation (8.26) is a basic expression to be utilized in the design of the separation cylinder. The radius of equal pressure which represents the radius at which the two phases may be or are separated is dependent of two radii ($r_1$ and $r_2$). These two values can be varied independently to provide optimum separation of the two phases involved and will account for the density of each phase as illustrated.

EXAMPLE 8.7 Design the inlet and discharge for a centrifugal separation of cream from whole milk. The density of the skim milk is 1025 kg/m$^3$. Illustrate the discharge conditions necessary if the skim milk outlet has a 2.5 cm radius and the cream outlet has a 5 cm radius. Suggest a desirable radius for the inlet. Cream density is 865 kg/m$^3$.

SOLUTION

(1) Using equation (8.26), the radius of the neutral zone can be computed:

$$r_n^2 = \frac{865\,(0.025)^2 - 1025\,(0.05)^2}{865 - 1025}$$

$$r_n^2 = .0126$$

$$r_n = 0.112 \text{ m} = 11.2 \text{ cm}$$

(2) Based on information given and the computation of neutral zone radius, the separator discharge should be designed as:

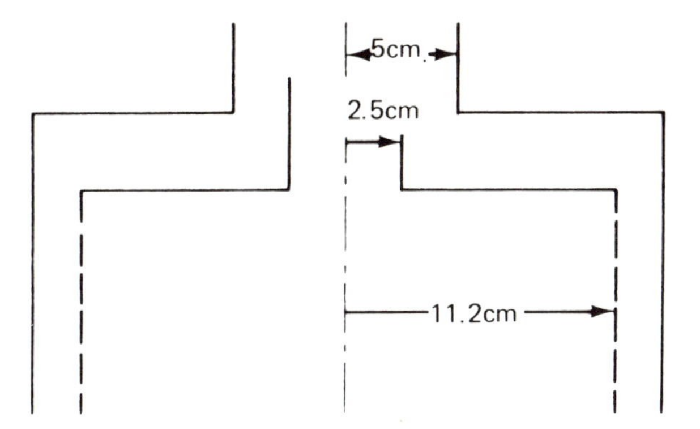


(3) The inlet would be designed to allow the product to enter with the least disturbance possible to the neutral zone:

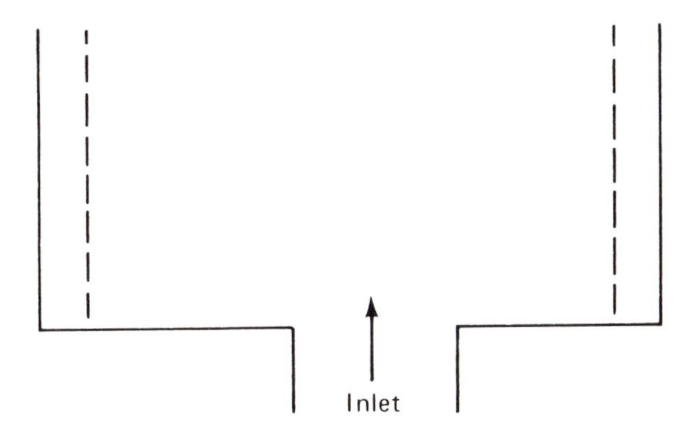


## 8.3.4 Particle–gas Separation

The separation of solid from a gas phase is a common operation in many food-processing operations. Probably the most common is the separation of a spray-dried product from the air stream after the drying operation is completed. This is usually accomplished in what is known as a cyclone separator. The basic equations presented earlier in this particular section will apply and equation (8.21) will provide some indication of the rate at which separation of solid particles from an air stream can be accomplished. It is obvious from this expression that the diameter of the particles must be known along with the density of the solid and the density of the air stream.

## PROBLEMS

8.1. *Constant-rate Air Filtration.* An air filter normally requires change after 100 hours of use when the pressure drop is 5 cm. The filter has dimensions of 1 m by 3 m and is designed for an air flow rate of 1.5 $m^3/s$. Compute the influence of increasing the air flow rate to 2.5 $m^3/s$ on the life of the filter.

8.2. *Terminal velocity of fruit.* The effective diameter of strawberry fruit is being estimated from terminal velocity measurements. Compute the effective diameter from a terminal velocity of 15 m/s and fruit density of 1150 $kg/m^3$.

8.3. *Separation by centrifugal force.* The 50-micron particles in a water suspension are being removed using a centrifugal separator operating at 70,000 rpm. Compute the effective radius for separation when the velocity of separation is 0.05 m/s and the particle density is 850 $kg/m^3$.

8.4. *Constant-pressure filtration.* A 0.5 $m^2$ filter is being used to filter a liquid at a pressure of 100 kPa. A 1 $cm^2$ laboratory-scale version of the filter media was used to filter 0.01 $m^3$ of the liquid in 1 min at the same pressure. Compute the rate of filtration after 2 hours through the commercial filter.

8.5. *Centrifugal separator design.* Cream is being separated from skim milk using a centrifugal separator. The separator inlet has a 25 cm. radius and the skim milk outlet is 10 cm. radius. Estimate the cream outlet radius when the cream density is 865 $kg/m^3$ and skim milk density is 1025 $kg/m^3$.

## COMPREHENSIVE PROBLEM VIII

Design of a Filtration System for a Liquid Suspension

### Objectives:

1. To become acquainted with procedures involved in the design of filtration systems for liquid foods.
2. To evaluate the influence of liquid volume on filtration time at constant filter surface area.
3. To investigate the influence of different constant filtration pressures on filtration time and volume.

### Procedures:

A. A constant-pressure filtration system is being designed for a liquid food product. The filter to be used has a cross-sectional area of 16 $m^2$ and will normally operate at a pressure of 350 kPa. A relationship between filtration time and flow volume is to be developed.

B. Laboratory-scale filtration data with the same liquid and filtration media were obtained using a constant pressure of 100 kPa and filter area of $5 \times 10^{-3}$ $m^2$. The following data were obtained:

| *Time* (min) | *Filtration Volume* ($cm^3$) |
|---|---|
| 15.0 | 800 |
| 22.0 | 1000 |
| 30.0 | 1200 |
| 42.0 | 1400 |
| 55.0 | 1600 |

C. Develop a relationship between filtration volume and time for the commercial filter with 16 $m^2$ area and pressure at 350 kPa.

D. Develop filtration volume-time relationships at pressures above 350 kPa with area of 16 $m^2$ and illustrate the use of larger area to achieve higher filtration volume without increasing pressure.

E. Discuss the following:

1. Potential errors associated with scale-up from laboratory to commercial-scale filtration systems.
2. Factors influencing the relationship between filtration volume and time.
3. Opportunities for achieving higher filtration volumes with increased pressure or larger filter area.

## NOMENCLATURE

$A$ = area, $m^2$.
$a$ = acceleration, $m/s^2$.
$b$ = height of centrifugal separation bowl, m.
$C$ = solids concentration in solution expressed as mass of liquid per unit mass solids, kg liquid/kg solid.
$D$ = particle diameter, micron, $10^{-6}$ m.
$F$ = force, N.
$f(e)$ = function used in equation (8.16).
$g$ = acceleration constant for gravity, $m/s^2$.
$g_c$ = constant in equations (8.17) and (8.18).
$K$ = experimental constant for gravity, $m/s^2$.
$L$ = thickness of filter material in liquid filtration system, m.
$L_c$ = thickness of filter cake in liquid filtration system, m.
$m$ = particle mass in centrifugal force field, kg.
$N$ = rotational speed of centrifugal separation system, revolutions/min.
$P$ = total pressure at some location in the separation, Pa.
$R$ = resistance in filtration separation system.
$N_{Re}$ = Reynolds number

$r$ = radius, m.
$r'$ = specific resistance of filter cake.
$S$ = solids content of liquid being filtered, kg solid/$m^3$ liquid.
$t$ = time, s.
$u$ = velocity, m/s.
$V$ = volume of liquid being filtered, $m^3$.
$w$ = mass flow rate, kg/s.
$\mu$ = viscosity, kg/ms.
$\omega$ = angular velocity.
$\rho$ = density, kg/$m^3$.

## Subscripts

$A$ = high density fraction in equation (8.25).
$B$ = low density fraction in equation (8.25).
$c$ = centrifugal separation system reference.
$D$ = drag in particle.
$f$ = fluid fraction reference.
$G$ = gravitation reference in separation system.
$i$ = input condition reference.
$n$ = location of equal pressure in centrifugal separation system.
$o$ = outlet condition reference.
$p$ = solid or particle fraction reference.
$s$ = suspension during separation.
$t$ = terminal condition.
*1,2,* = general location or time references.

## BIBLIOGRAPHY

CHARM, S.E. 1978. The Fundamentals of Food Engineering, 3rd Edition. AVI Publishing Co., Westport, Conn.

COULSON, J.M., and RICHARDSON, J.F. 1978. Chemical Engineering, 3rd Edition, Volume II. Pergamon Press, Elmsford, N.Y.

DECKER, H.M., BUCHANAN, L.M., HALL, L.B., and GODDARD, K.R. 1962. Air Filtration of Microbial Particles. Public Health Service. Publ. 593, U.S. Govt. Printing Office, Washington, D.C.

EARLE, R.L. 1966. Unit Operations in Food Processing. Pergamon Press, Elmsford, N.Y.

WHITBY, K.T., and LUNDGREN, D.A. 1965. The mechanics of air cleaning. Trans. Am. Soc. Chem. Engrs. *8*, No. 3, 342.

# APPENDIX: Useful Figures and Tables

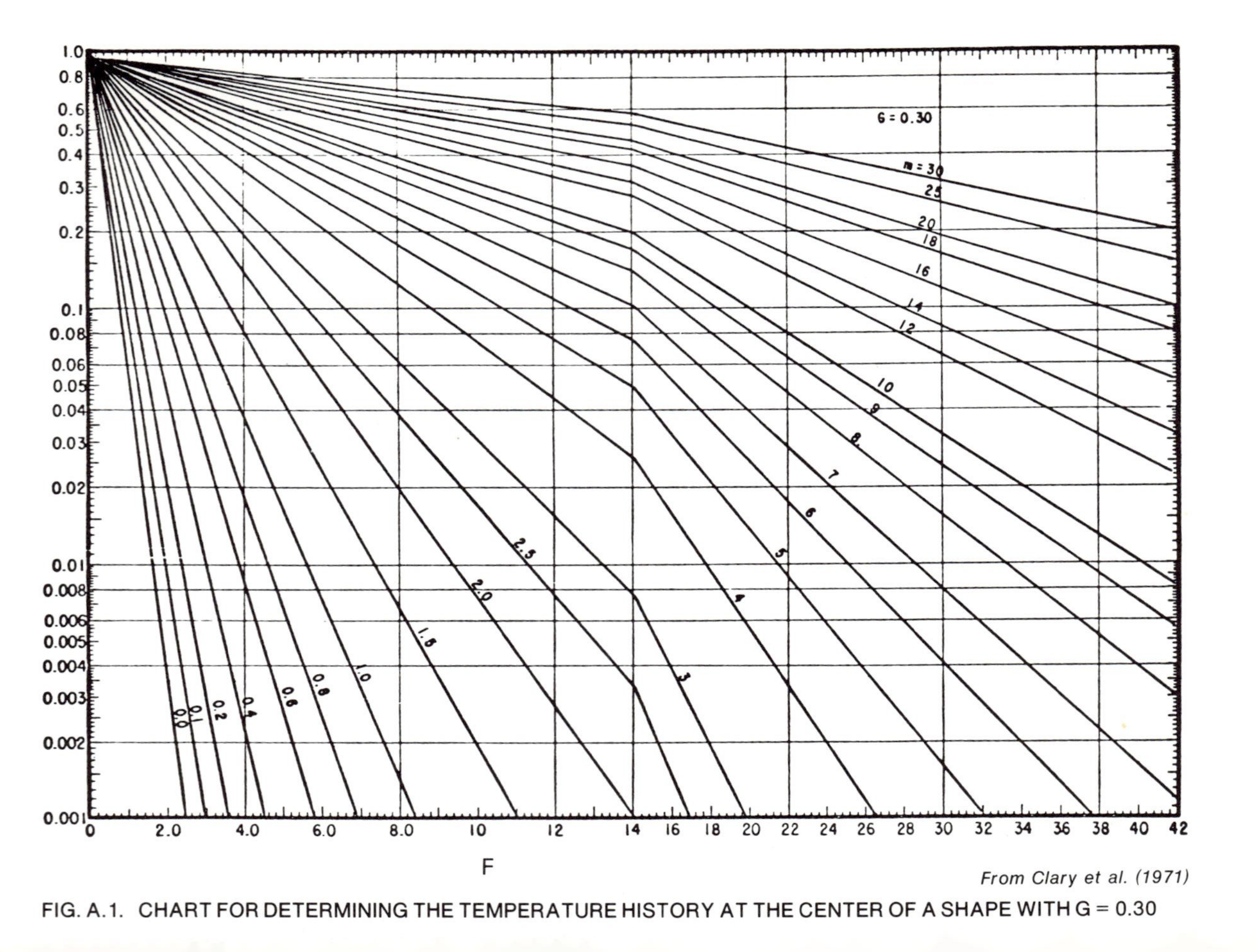


*From Clary et al. (1971)*

FIG. A.1. CHART FOR DETERMINING THE TEMPERATURE HISTORY AT THE CENTER OF A SHAPE WITH G = 0.30

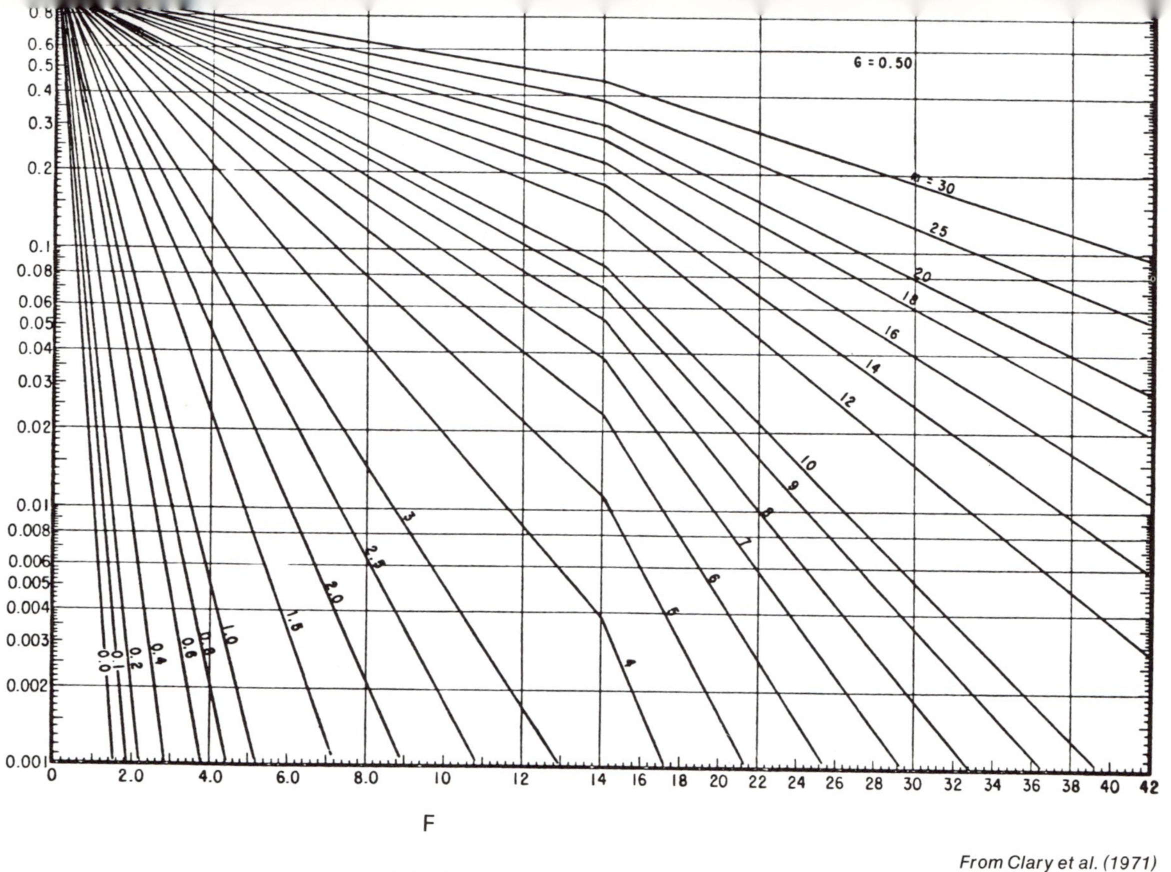


*From Clary et al. (1971)*

FIG. A.2. CHART FOR DETERMINING THE TEMPERATURE HISTORY AT THE CENTER OF A SHAPE WITH G = 0.50

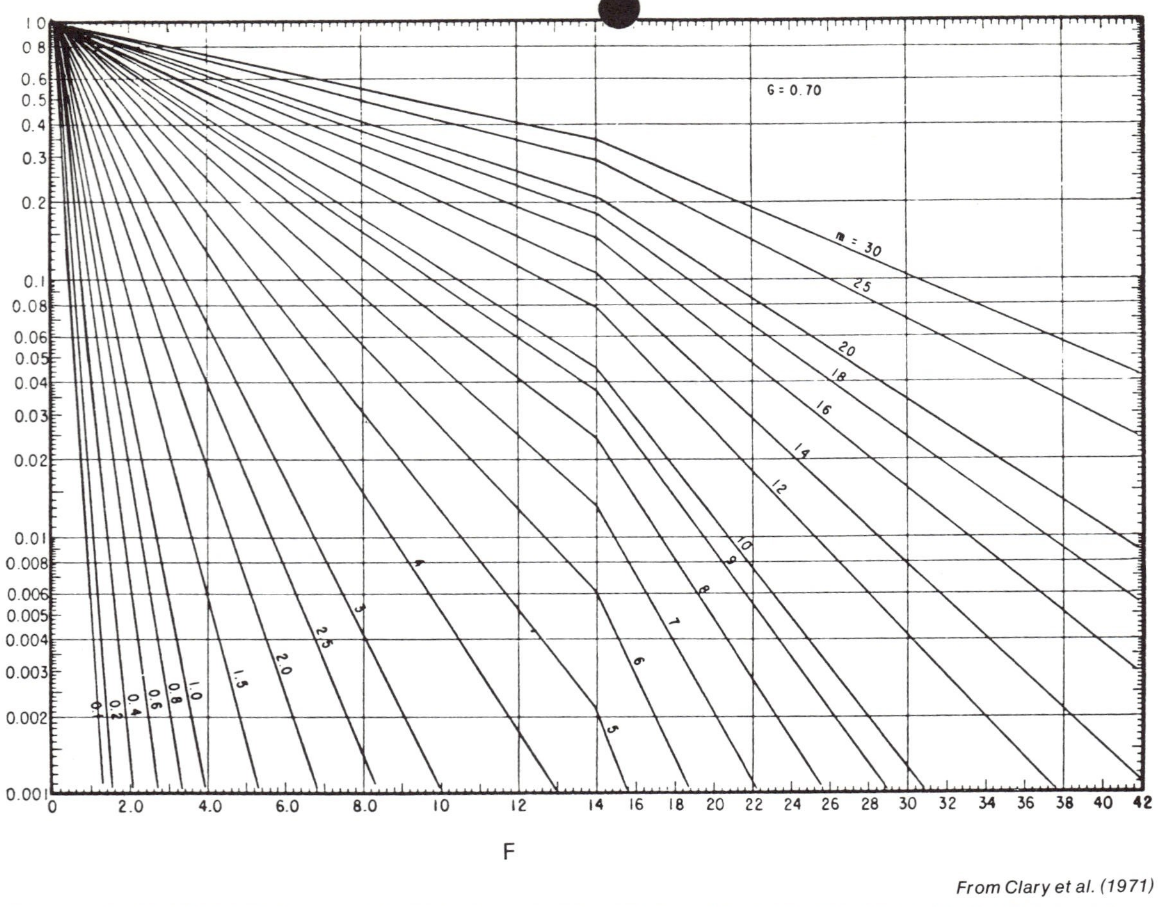


*From Clary et al. (1971)*

FIG. A.3. CHART FOR DETERMINING THE TEMPERATURE HISTORY AT THE CENTER OF A SHAPE WITH G = 0.70

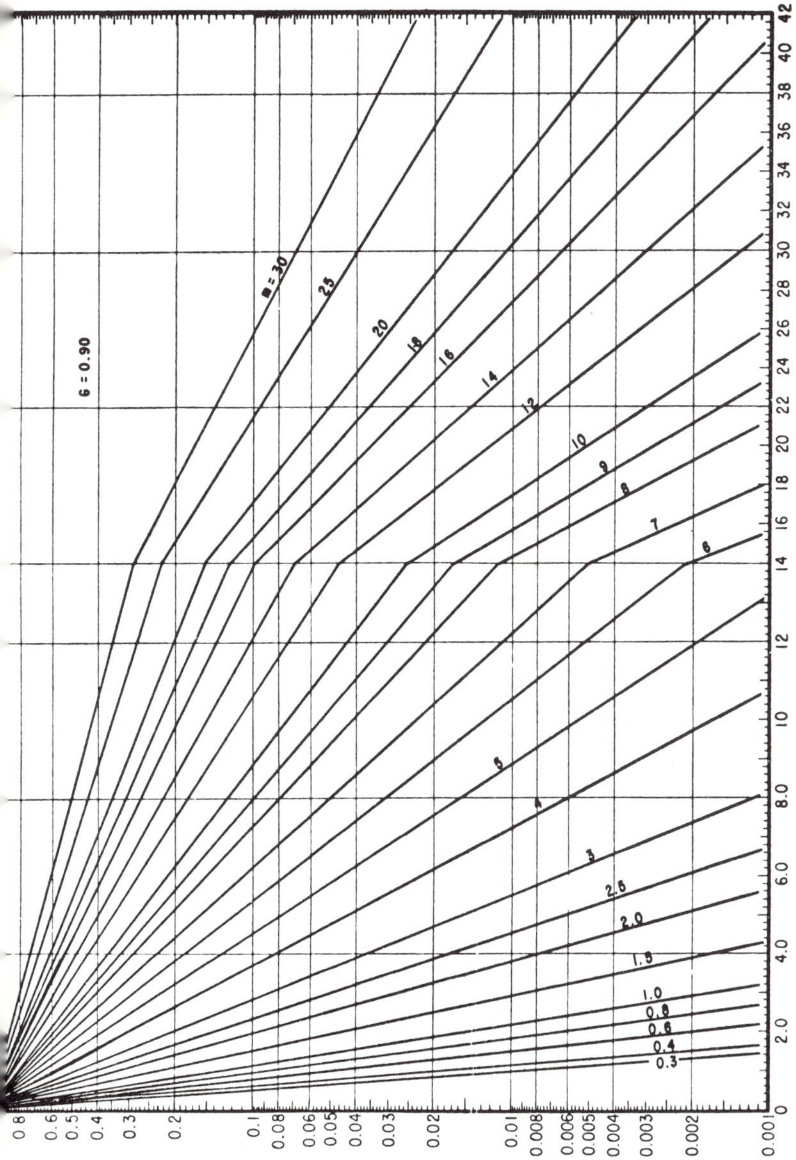

FIG. A.4. CHART FOR DETERMINING THE TEMPERATURE HISTORY AT THE CENTER OF A SHAPE WITH G = 0.90

*From Clary et al. (1971)*

FIG. A.5. PSYCHROMETRIC CHART
May be found at end of book.

TABLE A.1. DIFFUSIVITY DATA FOR GASES THROUGH PACKAGING MATERIALS

| Material | Gas | Diffusivity at 25°C | | |
|---|---|---|---|---|
| | | D ($cm^2/s$) | $D_o$ ($cm^2/s$) | $E_d$ (J/mole) |
| Polystyrene | He | $10.4 \times 10^{-6}$ | 0.0019 | 3.1 |
| | $H_2$ | 4.36 | 0.0036 | 4.0 |
| | $O_2$ | 0.11 | 0.125 | 8.3 |
| | $CO_2$ | 0.058 | 0.128 | 8.7 |
| Polyvinyl acetate (glassy) | He | 9.52 | 0.011 | 4.2 |
| | $H_2$ | 2.10 | 0.013 | 5.2 |
| | $O_2$ | 0.051 | 6.31 | 11.1 |
| | $CH_4$ | 0.0019 | $2.3 \times 10^5$ | 19.3 |
| Polyethylene (Density 0.964) | He | 3.07 | 0.037 | 5.6 |
| | $O_2$ | 0.17 | 0.43 | 8.8 |
| | $CO_2$ | 0.124 | 0.19 | 8.5 |
| | CO | 0.096 | 0.251 | 8.8 |
| | $N_2$ | 0.093 | 0.33 | 9.0 |
| | $CH_4$ | 0.057 | 2.19 | 10.4 |

Source: J. Crank and G. Park, *Diffusion in Polymers*, New York: Academic Press, 1968.
Note: $D = D_o \exp(-E_d/RT)$. D is in $cm^2/sec$. $E_d$ = activation energy.

TABLE A.2. PROPERTIES OF SOLID MATERIALS

| Material | Temperature (C) | Density ($kg/m^3$) | Specific Heat (kJ/kg K) | Thermal Conductivity (W/m K) |
|---|---|---|---|---|
| Metals | | | | |
| Aluminum | 20 | 2707 | 0.896 | 203 |
| Brass | 20 | 8522 | 0.385 | 98 |
| Cast iron | 20 | 7593 | 0.465 | 54 |
| Copper | 20 | 8954 | 0.383 | 386 |
| Lead | 20 | 1137 | 0.130 | 35 |
| Steel; 1%C | 20 | 7801 | 0.473 | 45 |
| Stainless steel (308) | 20 | 7849 | 0.461 | 15 |
| Stainless steel (304) | 0 | 7817 | 0.461 | 14 |
| Tin | 20 | 7304 | 0.227 | 61 |
| Insulating Materials | | | | |
| Asbestos | 37.8 | 577 | — | 0.168 |
| Corkboard | 30 | 160 | — | 0.0433 |
| Fiber insulation board | 21 | 237 | — | 0.048 |
| Glass wool | 37.8 | 64.1 | — | 0.0414 |
| Polystyrene foam | 0 | 24 | — | 0.0364 |
| Polyurethane foam | 0 | 32 | — | 0.026 |
| Building materials | | | | |
| Brick, building | 20 | — | — | 0.69 |
| Concrete, stone | 21 | 2307 | 0.837 | 0.935 |
| Glass, window | 21 | 2723 | 0.837 | 0.779 |
| Wood, oak ⊥ grain | 21 | 817 | 2.386 | 0.208 |
| Wood, oak ‖ grain | 21 | 817 | 2.386 | 0.346 |
| Wood, pine ⊥ grain | 21 | 497 | 2.805 | 0.104 |
| Wood, pine ‖ grain | 21 | 497 | 2.805 | 0.242 |

TABLE A.3. PROPERTIES OF DRY AIR AT ATMOSPHERIC PRESSURE

| Temperature | | Density | Volumetric coefficient of expansion | Specific heat | Thermal conductivity | Thermal diffusivity | Viscosity | Kinematic viscosity | Prandtl number |
|---|---|---|---|---|---|---|---|---|---|
| t | T | $\rho$ | $\beta \times 10^3$ | $c_p$ | $k$ | $\alpha \times 10^6$ | $\mu \times 10^6$ | $\upsilon \times 10^6$ | Pr |
| C | K | $kg/m^3$ | 1/K | kJ/kg K | W/m K | $m^2s$ | $Ns/m^2$ | $m^2/s$ | |
| −20 | 253.15 | 1.365 | 3.97 | 1.005 | 0.0226 | 16.8 | 16.279 | 12.0 | 0.71 |
| 0 | 273.15 | 1.252 | 3.65 | 1.011 | 0.0237 | 19.2 | 17.456 | 13.9 | 0.71 |
| 10 | 283.15 | 1.206 | 3.53 | 1.010 | 0.0244 | 20.7 | 17.848 | 14.66 | 0.71 |
| 20 | 293.15 | 1.164 | 3.41 | 1.012 | 0.0251 | 22.0 | 18.240 | 15.7 | 0.71 |
| 30 | 303.15 | 1.127 | 3.30 | 1.013 | 0.0258 | 23.4 | 18.682 | 16.58 | 0.71 |
| 40 | 313.15 | 1.092 | 3.20 | 1.014 | 0.0265 | 24.8 | 19.123 | 17.6 | 0.71 |
| 50 | 323.15 | 1.057 | 3.10 | 1.016 | 0.0272 | 26.2 | 19.515 | 18.58 | 0.71 |
| 60 | 333.15 | 1.025 | 3.00 | 1.017 | 0.0279 | 27.6 | 19.907 | 19.4 | 0.71 |
| 70 | 343.15 | 0.996 | 2.91 | 1.018 | 0.0286 | 29.2 | 20.398 | 20.65 | 0.71 |
| 80 | 353.15 | 0.968 | 2.83 | 1.019 | 0.0293 | 30.6 | 20.790 | 21.5 | 0.71 |
| 90 | 363.15 | 0.942 | 2.76 | 1.021 | 0.0300 | 32.2 | 21.231 | 22.82 | 0.71 |
| 100 | 373.15 | 0.916 | 2.69 | 1.022 | 0.0307 | 33.6 | 21.673 | 23.6 | 0.71 |
| 120 | 393.15 | 0.870 | 2.55 | 1.025 | 0.0320 | 37.0 | 22.555 | 25.9 | 0.71 |
| 140 | 413.15 | 0.827 | 2.43 | 1.027 | 0.0333 | 40.0 | 23.340 | 28.2 | 0.71 |
| 150 | 423.15 | 0.810 | 2.37 | 1.028 | 0.0336 | 41.2 | 23.732 | 29.4 | 0.71 |
| 160 | 433.15 | 0.789 | 2.31 | 1.030 | 0.0344 | 43.3 | 24.124 | 30.6 | 0.71 |
| 180 | 453.15 | 0.755 | 2.20 | 1.032 | 0.0357 | 47.0 | 24.909 | 33.00 | 0.71 |
| 200 | 473.15 | 0.723 | 2.11 | 1.035 | 0.0370 | 49.7 | 25.693 | 35.5 | 0.71 |
| 250 | 523.15 | 0.653 | 1.89 | 1.043 | 0.0400 | 60.0 | 27.557 | 42.2 | 0.71 |

Adapted from: Raznjevic, K. 1978. Handbook of Thermodynamic Tables and Charts. Hemisphere Publishing Corporation, Washington.

TABLE A.4. THERMODYNAMIC PROPERTIES OF WATER AT THE SATURATION PRESSURE

| Temperature | | Density | Coefficient of volumetric thermal expansion | Specific heat | Thermal conductivity | Thermal diffusivity | Absolute viscosity | Kinematic viscosity | Prandtl number |
|---|---|---|---|---|---|---|---|---|---|
| t | T | $\rho$ | $\beta \times 10^4$ | $c_p$ | $k$ | $\alpha \times 10^6$ | $\mu \times 10^6$ | $\upsilon \times 10^6$ | Pr |
| C | K | kg/m³ | 1/K | kJ/kg K | W/m K | m²/s | Pa·s | m²/s | |
| 0 | 273.15 | 999.9 | −0.7 | 4.226 | 0.558 | 0.131 | 1793.636 | 1.789 | 13.7 |
| 5 | 278.15 | 1000.0 | — | 4.206 | 0.568 | 0.135 | 1534.741 | 1.535 | 11.4 |
| 10 | 283.15 | 999.7 | 0.95 | 4.195 | 0.577 | 0.137 | 1296.439 | 1.300 | 9.5 |
| 15 | 288.15 | 999.1 | — | 4.187 | 0.587 | 0.141 | 1135.610 | 1.146 | 8.1 |
| 20 | 293.15 | 998.2 | 2.1 | 4.182 | 0.597 | 0.143 | 993.414 | 1.006 | 7.0 |
| 25 | 298.15 | 997.1 | — | 4.178 | 0.606 | 0.146 | 880.637 | 0.884 | 6.1 |
| 30 | 303.15 | 995.7 | 3.0 | 4.176 | 0.615 | 0.149 | 792.377 | 0.805 | 5.4 |
| 35 | 308.15 | 994.1 | — | 4.175 | 0.624 | 0.150 | 719.808 | 0.725 | 4.8 |
| 40 | 313.15 | 992.2 | 3.9 | 4.175 | 0.633 | 0.151 | 658.026 | 0.658 | 4.3 |
| 45 | 318.15 | 990.2 | — | 4.176 | 0.640 | 0.155 | 605.070 | 0.611 | 3.9 |
| 50 | 323.15 | 988.1 | 4.6 | 4.178 | 0.647 | 0.157 | 555.056 | 0.556 | 3.55 |
| 55 | 328.15 | 985.7 | — | 4.179 | 0.652 | 0.158 | 509.946 | 0.517 | 3.27 |
| 60 | 333.15 | 983.2 | 5.3 | 4.181 | 0.658 | 0.159 | 471.650 | 0.478 | 3.00 |
| 65 | 338.15 | 980.6 | — | 4.184 | 0.663 | 0.161 | 435.415 | 0.444 | 2.76 |
| 70 | 343.15 | 977.8 | 5.8 | 4.187 | 0.668 | 0.163 | 404.034 | 0.415 | 2.55 |
| 75 | 348.15 | 974.9 | — | 4.190 | 0.671 | 0.164 | 376.575 | 0.366 | 2.23 |
| 80 | 353.15 | 971.8 | 6.3 | 4.194 | 0.673 | 0.165 | 352.059 | 0.364 | 2.25 |
| 85 | 358.15 | 968.7 | — | 4.198 | 0.676 | 0.166 | 328.523 | 0.339 | 2.04 |
| 90 | 363.15 | 965.3 | 7.0 | 4.202 | 0.678 | 0.167 | 308.909 | 0.326 | 1.95 |
| 95 | 368.15 | 961.9 | — | 4.206 | 0.680 | 0.168 | 292.238 | 0.310 | 1.84 |

TABLE A.4. *(Continued)*

| Temperature | | Density | Coefficient of volumetric thermal expansion | Specific heat | Thermal conductivity | Thermal diffusivity | Absolute viscosity | Kinematic viscosity | Prandtl number |
|---|---|---|---|---|---|---|---|---|---|
| t | T | $\rho$ | $\beta \times 10^4$ | $c_p$ | $k$ | $\alpha \times 10^6$ | $\mu \times 10^6$ | $\upsilon \times 10^6$ | Pr |
| C | K | kg/m$^3$ | 1/K | kJ/kg K | W/m K | m$^2$/s | Pa·s | m$^2$/s | |
| 100 | 373.15 | 958.4 | 7.5 | 4.211 | 0.682 | 0.169 | 277.528 | 0.294 | 1.75 |
| 110 | 383.15 | 951.0 | 8.0 | 4.224 | 0.684 | 0.170 | 254.973 | 0.268 | 1.57 |
| 120 | 393.15 | 943.5 | 8.5 | 4.232 | 0.685 | 0.171 | 235.360 | 0.244 | 1.43 |
| 130 | 403.15 | 934.8 | 9.1 | 4.250 | 0.686 | 0.172 | 211.824 | 0.226 | 1.32 |
| 140 | 413.15 | 926.3 | 9.7 | 4.257 | 0.684 | 0.172 | 201.036 | 0.212 | 1.23 |
| 150 | 423.15 | 916.9 | 10.3 | 4.270 | 0.684 | 0.173 | 185.346 | 0.201 | 1.17 |
| 160 | 433.15 | 907.6 | 10.8 | 4.285 | 0.680 | 0.173 | 171.616 | 0.191 | 1.10 |
| 170 | 443.15 | 897.3 | 11.5 | 4.396 | 0.679 | 0.172 | 162.290 | 0.181 | 1.05 |
| 180 | 453.15 | 886.6 | 12.1 | 4.396 | 0.673 | 0.172 | 152.003 | 0.173 | 1.01 |
| 190 | 463.15 | 876.0 | 12.8 | 4.480 | 0.670 | 0.171 | 145.138 | 0.166 | 0.97 |
| 200 | 473.15 | 862.8 | 13.5 | 4.501 | 0.665 | 0.170 | 139.254 | 0.160 | 0.95 |
| 210 | 483.15 | 852.8 | 14.3 | 4.560 | 0.655 | 0.168 | 131.409 | 0.154 | 0.92 |
| 220 | 493.15 | 837.0 | 15.2 | 4.605 | 0.652 | 0.167 | 124.544 | 0.149 | 0.90 |
| 230 | 503.15 | 827.3 | 16.2 | 4.690 | 0.637 | 0.164 | 119.641 | 0.145 | 0.88 |
| 240 | 513.15 | 809.0 | 17.2 | 4.731 | 0.634 | 0.162 | 113.757 | 0.141 | 0.86 |
| 250 | 523.15 | 799.2 | 18.6 | 4.857 | 0.618 | 0.160 | 109.834 | 0.137 | 0.86 |

Adapted from: Raznjevic, K. 1978. Handbook of Thermodynamic Tables and Charts. Hemisphere Publishing Corporation, Washington.

TABLE A.5. PROPERTIES OF ICE AS A FUNCTION OF TEMPERATURE

| *Temperature* (C) | *Thermal Conductivity* (W/m K) | *Specific Heat* (kJ/kg K) | *Density* (kg/m$^3$) |
|---|---|---|---|
| −101 | 3.50 | 1.382 | 925.8 |
| − 73 | 3.08 | 1.587 | 924.2 |
| − 45.5 | 2.72 | 1.783 | 922.6 |
| − 23 | 2.41 | 1.922 | 919.4 |
| − 18 | 2.37 | 1.955 | 919.4 |
| − 12 | 2.32 | 1.989 | 919.4 |
| − 7 | 2.27 | 2.022 | 917.8 |
| 0 | 2.22 | 2.050 | 916.2 |

*Adapted from Dickerson (1969)*

TABLE A.6. PROPERTIES OF SATURATED STEAM

| Temperature (C) | Vapor Pressure (kPa) | Specific Volume ($m^3/kg$) | | Enthalpy (kJ/kg) | | Entropy (kJ/kg·K) | |
|---|---|---|---|---|---|---|---|
| | | Liquid | Sat'd Vapor | Liquid | Sat'd Vapor | Liquid | Sat'd Vapor |
| 0.01 | 0.6113 | 0.0010002 | 206.136 | 0.00 | 2501.4 | 0.0000 | 9.1562 |
| 3 | 0.7577 | 0.0010001 | 168.132 | 12.57 | 2506.9 | 0.0457 | 9.0773 |
| 6 | 0.9349 | 0.0010001 | 137.734 | 25.20 | 2512.4 | 0.0912 | 9.0003 |
| 9 | 1.1477 | 0.0010003 | 113.386 | 37.80 | 2517.9 | 0.1362 | 8.9253 |
| 12 | 1.4022 | 0.0010005 | 93.784 | 50.41 | 2523.4 | 0.1806 | 8.8524 |
| 15 | 1.7051 | 0.0010009 | 77.926 | 62.99 | 2528.9 | 0.2245 | 8.7814 |
| 18 | 2.0640 | 0.0010014 | 65.038 | 75.58 | 2534.4 | 0.2679 | 8.7123 |
| 21 | 2.487 | 0.0010020 | 54.514 | 88.14 | 2539.9 | 0.3109 | 8.6450 |
| 24 | 2.985 | 0.0010027 | 45.883 | 100.70 | 2545.4 | 0.3534 | 8.5794 |
| 27 | 3.567 | 0.0010035 | 38.774 | 113.25 | 2550.8 | 0.3954 | 8.5156 |
| 30 | 4.246 | 0.0010043 | 32.894 | 125.79 | 2556.3 | 0.4369 | 8.4533 |
| 33 | 5.034 | 0.0010053 | 28.011 | 138.33 | 2561.7 | 0.4781 | 8.3927 |
| 36 | 5.947 | 0.0010063 | 23.940 | 150.86 | 2567.1 | 0.5188 | 8.3336 |
| 40 | 7.384 | 0.0010078 | 19.523 | 167.57 | 2574.3 | 0.5725 | 8.2570 |
| 45 | 9.593 | 0.0010099 | 15.258 | 188.45 | 2583.2 | 0.6387 | 8.1648 |
| 50 | 12.349 | 0.0010121 | 12.032 | 209.33 | 2592.1 | 0.7038 | 8.0763 |
| 55 | 15.758 | 0.0010146 | 9.568 | 230.23 | 2600.9 | 0.7679 | 7.9913 |
| 60 | 19.940 | 0.0010172 | 7.671 | 251.13 | 2609.6 | 0.8312 | 7.9096 |
| 65 | 25.03 | 0.0010199 | 6.197 | 272.06 | 2618.3 | 0.8935 | 7.8310 |
| 70 | 31.19 | 0.0010228 | 5.042 | 292.98 | 2626.8 | 0.9549 | 7.7553 |
| 75 | 38.58 | 0.0010259 | 4.131 | 313.93 | 2635.3 | 1.0155 | 7.6824 |
| 80 | 47.39 | 0.0010291 | 3.407 | 334.91 | 2643.7 | 1.0753 | 7.6122 |
| 85 | 57.83 | 0.0010325 | 2.828 | 355.90 | 2651.9 | 1.1343 | 7.5445 |
| 90 | 70.14 | 0.0010360 | 2.361 | 376.92 | 2660.1 | 1.1925 | 7.4791 |
| 95 | 84.55 | 0.0010397 | 1.9819 | 397.96 | 2668.1 | 1.2500 | 7.4159 |
| 100 | 101.35 | 0.0010435 | 1.6729 | 419.04 | 2676.1 | 1.3069 | 7.3549 |
| 105 | 120.82 | 0.0010475 | 1.4194 | 440.15 | 2683.8 | 1.3630 | 7.2958 |
| 110 | 143.27 | 0.0010516 | 1.2102 | 461.30 | 2691.5 | 1.4185 | 7.2387 |
| 115 | 169.06 | 0.0010559 | 1.0366 | 482.48 | 2699.0 | 1.4734 | 7.1833 |
| 120 | 198.53 | 0.0010603 | 0.8919 | 503.71 | 2706.3 | 1.5276 | 7.1296 |
| 125 | 232.1 | 0.0010649 | 0.7706 | 524.99 | 2713.5 | 1.5813 | 7.0775 |
| 130 | 270.1 | 0.0010697 | 0.6685 | 546.31 | 2720.5 | 1.6344 | 7.0269 |
| 135 | 313.0 | 0.0010746 | 0.5822 | 567.69 | 2727.3 | 1.6870 | 6.9777 |
| 140 | 316.3 | 0.0010797 | 0.5089 | 589.13 | 2733.9 | 1.7391 | 6.9299 |
| 145 | 415.4 | 0.0010850 | 0.4463 | 610.63 | 2740.3 | 1.7907 | 6.8833 |

TABLE A.6. *(Continued)*

| Temperature (C) | Vapor Pressure (kPa) | Specific Volume ($m^3/kg$) | | Enthalpy (kJ/kg) | | Entropy (kJ/kg·K) | |
|---|---|---|---|---|---|---|---|
| | | Liquid | Sat'd Vapor | Liquid | Sat'd Vapor | Liquid | Sat'd Vapor |
| 150 | 475.8 | 0.0010905 | 0.3928 | 632.20 | 2746.5 | 1.8418 | 6.8379 |
| 155 | 543.1 | 0.0010961 | 0.3468 | 653.84 | 2752.4 | 1.8925 | 6.7935 |
| 160 | 617.8 | 0.0011020 | 0.3071 | 675.55 | 2758.1 | 1.9427 | 6.7502 |
| 165 | 700.5 | 0.0011080 | 0.2727 | 697.34 | 2763.5 | 1.9925 | 6.7078 |
| 170 | 791.7 | 0.0011143 | 0.2428 | 719.21 | 2768.7 | 2.0419 | 6.6663 |
| 175 | 892.0 | 0.0011207 | 0.2168 | 741.17 | 2773.6 | 2.0909 | 6.6256 |
| 180 | 1002.1 | 0.0011274 | 0.19405 | 763.22 | 2778.2 | 2.1396 | 6.5857 |
| 190 | 1254.4 | 0.0011414 | 0.15654 | 807.62 | 2786.4 | 2.2359 | 6.5079 |
| 200 | 1553.8 | 0.0011565 | 0.12736 | 852.45 | 2793.2 | 2.3309 | 6.4323 |
| 225 | 2548 | 0.0011992 | 0.07849 | 966.78 | 2803.3 | 2.5639 | 6.2503 |
| 250 | 3973 | 0.0012512 | 0.05013 | 1085.36 | 2801.5 | 2.7927 | 6.0730 |
| 275 | 5942 | 0.0013168 | 0.03279 | 1210.07 | 2785.0 | 3.0208 | 5.8938 |
| 300 | 8581 | 0.0010436 | 0.02167 | 1344.0 | 2749.0 | 3.2534 | 5.7045 |

Source: Abridged from J.H. Keenan, F.G. Keyes, P.G. Hill, and J.G. Moore, *Steam Tables—Metric Units*. New York: John Wiley & Sons, Inc., 1969. With permission of the authors and publishers.

**TABLE A.7. PROPERTIES OF SUPERHEATED STEAM (STEAM TABLE), SI UNITS ($v$, specific volume, $m^3/kg$; $H$, enthalpy, kJ/kg; $s$, entropy, kJ/kg·K)**

| Absolute Pressure, kPa (Sat. Temp., C) | | Temperature (C) 100 | 150 | 200 | 250 | 300 | 360 | 420 | 500 |
|---|---|---|---|---|---|---|---|---|---|
| 10 (45.81) | $v$ | 17.196 | 19.512 | 21.825 | 24.136 | 26.445 | 29.216 | 31.986 | 35.679 |
| | $H$ | 2687.5 | 2783.0 | 2879.5 | 2977.3 | 3076.5 | 3197.6 | 3320.9 | 3489.1 |
| | $s$ | 8.4479 | 8.6882 | 8.9038 | 9.1002 | 9.2813 | 9.4821 | 9.6682 | 9.8978 |
| 50 (81.33) | $v$ | 3.418 | 3.889 | 4.356 | 4.820 | 5.284 | 5.839 | 6.394 | 7.134 |
| | $H$ | 2682.5 | 2780.1 | 2877.7 | 2976.0 | 3075.5 | 3196.8 | 3320.4 | 3488.7 |
| | $s$ | 7.6947 | 7.9401 | 8.1580 | 8.3556 | 8.5373 | 8.7385 | 8.9249 | 9.1546 |
| 75 (91.78) | $v$ | 2.270 | 2.587 | 2.900 | 3.211 | 3.520 | 3.891 | 4.262 | 4.755 |
| | $H$ | 2679.4 | 2778.2 | 2876.5 | 2975.2 | 3074.9 | 3196.4 | 3320.0 | 3488.4 |
| | $s$ | 7.5009 | 7.7496 | 7.9690 | 8.1673 | 8.3493 | 8.5508 | 8.7374 | 8.9672 |
| 100 (99.63) | $v$ | 1.6958 | 1.9364 | 2.172 | 2.406 | 2.639 | 2.917 | 3.195 | 3.565 |
| | $H$ | 2676.2 | 2776.4 | 2875.3 | 2974.3 | 3074.3 | 3195.9 | 3319.6 | 3488.1 |
| | $s$ | 7.3614 | 7.6134 | 7.8343 | 8.0333 | 8.2158 | 8.4175 | 8.6042 | 8.8342 |
| 150 (111.37) | $v$ | | 1.2853 | 1.4443 | 1.6012 | 1.7570 | 1.9432 | 2.129 | 2.376 |
| | $H$ | | 2772.6 | 2872.9 | 2972.7 | 3073.1 | 3195.0 | 3318.9 | 3487.6 |
| | $s$ | | 7.4193 | 7.6433 | 7.8438 | 8.0720 | 8.2293 | 8.4163 | 8.6466 |
| 400 (143.63) | $v$ | | 0.4708 | 0.5342 | 0.5951 | 0.6458 | 0.7257 | 0.7960 | 0.8893 |
| | $H$ | | 2752.8 | 2860.5 | 2964.2 | 3066.8 | 3190.3 | 3315.3 | 3484.9 |
| | $s$ | | 6.9299 | 7.1706 | 7.3789 | 7.5662 | 7.7712 | 7.9598 | 8.1913 |
| 700 (164.97) | $v$ | | | 0.2999 | 0.3363 | 0.3714 | 0.4126 | 0.4533 | 0.5070 |
| | $H$ | | | 2844.8 | 2953.6 | 3059.1 | 3184.7 | 3310.9 | 3481.7 |
| | $s$ | | | 6.8865 | 7.1053 | 7.2979 | 7.5063 | 7.6968 | 7.9299 |
| 1000 (179.91) | $v$ | | | 0.2060 | 0.2327 | 0.2579 | 0.2873 | 0.3162 | 0.3541 |
| | $H$ | | | 2827.9 | 2942.6 | 3051.2 | 3178.9 | 3306.5 | 3478.5 |
| | $s$ | | | 6.6940 | 6.9247 | 7.1229 | 7.3349 | 7.5275 | 7.7622 |
| 1500 (198.32) | $v$ | | | 0.13248 | 0.15195 | 0.16966 | 0.18988 | 0.2095 | 0.2352 |
| | $H$ | | | 2796.8 | 2923.3 | 3037.6 | 3.1692 | 3299.1 | 3473.1 |
| | $s$ | | | 6.4546 | 6.7090 | 6.9179 | 7.1363 | 7.3323 | 7.5698 |
| 2000 (212.42) | $v$ | | | | 0.11144 | 0.12547 | 0.14113 | 0.15616 | 0.17568 |
| | $H$ | | | | 2902.5 | 3023.5 | 3159.3 | 3291.6 | 3467.6 |
| | $s$ | | | | 6.5453 | 6.7664 | 6.9917 | 7.1915 | 7.4317 |
| 2500 (223.99) | $v$ | | | | 0.08700 | 0.09890 | 0.11186 | 0.12414 | 0.13998 |
| | $H$ | | | | 2880.1 | 3008.8 | 3149.1 | 3284.0 | 3462.1 |
| | $s$ | | | | 6.4085 | 6.6438 | 6.8767 | 7.0803 | 7.3234 |
| 3000 (233.90) | $v$ | | | | 0.07058 | 0.08114 | 0.09233 | 0.10279 | 0.11619 |
| | $H$ | | | | 2855.8 | 2993.5 | 3138.7 | 3276.3 | 3456.5 |
| | $s$ | | | | 6.2872 | 6.5390 | 6.7801 | 6.9878 | 7.2338 |

Source: Abridged from J.H. Keenan, F.G. Keyes, P.G. Hill, and J.G. Moore, *Steam Tables—Metric Units*. New York: John Wiley & Sons, Inc., 1969. With permission of the authors and publishers.

TABLE A.8. RHEOLOGICAL PROPERTIES OF FLUID FOODS

| Product | Temperature (C) | Composition | Consistency Coefficient (m) ($Pa\ s^n$) | Flow Behavior Index (n) | Measurement Method | Reference |
|---|---|---|---|---|---|---|
| Olive oil | 20 | normal | 0.084 | 1.0 | unknown | Mohsenin (1970) |
| Honey | 24 | normal | 5.6 | 1.0 | Capillary tube | Charm (1978) |
| Soy bean oil | 30 | normal | 0.04 | 1.0 | unknown | Mohsenin (1970) |
| Honey | 24 | normal | 6.18 | 1.0 | single cylinder | Charm (1978) |
| Whole milk | 20 | normal | 0.0212 | 1.0 | unknown | Mohsenin (1970) |
| Skim milk | 25 | normal | 0.0014 | 1.0 | unknown | Mohsenin (1970) |
| Cream | 3 | 20% fat | 0.0062 | 1.0 | unknown | Mohsenin (1970) |
| Cream | 3 | 30% fat | 0.0138 | 1.0 | unknown | Mohsenin (1970) |
| Apple juice | 27 | 20° Brix | 0.0021 | 1.0 | Capillary tube | Saravacos (1968) |
| Apple juice | 27 | 60° Brix | 0.03 | 1.0 | Capillary tube | Saravacos (1968) |
| Grape juice | 27 | 20° Brix | 0.0025 | 1.0 | Capillary tube | Saravacos (1968) |
| Grape juice | 27 | 60° Brix | 0.11 | 1.0 | Capillary tube | Saravacos (1968) |
| Tomato concentrate | 32 | 5.8% T.S. | 0.223 | 0.59 | Coaxial cylinder | Harper and El Sahrigi (1965) |
| Tomato concentrate | 32 | 30% T.S. | 18.7 | 0.4 | Coaxial cylinder | Harper and El Sahrigi (1965) |
| Tomato puree | unknown | unknown | 0.92 | 0.554 | Coaxial cylinder | Charm (1978) |
| Corn syrup | 27 | 48.4% T.S. | 0.053 | 1.0 | Coaxial cylinder | Harper (1960) |
| Apricot puree | 21 | 17.7% T.S. | 5.4 | 0.29 | Coaxial cylinder | Harper (1960) |
| Apricot puree | 25 | 19% T.S. | 20.0 | 0.3 | Coaxial cylinder narrow gap | Watson (1968) |
| Apricot puree | 27 | 13.8% T.S. | 7.2 | 0.41 | Capillary tube | Saravacos (1968) |
| Apricot conc. | 25 | 26% T.S. | 67.0 | 0.3 | Coaxial cylinder narrow gap | Watson (1968) |

TABLE A.8. *(Continued)*

| Product | Temperature (C) | Composition | Consistency Coefficient (m) (Pa $s^n$) | Flow Behavior Index (n) | Measurement Method | Reference |
|---|---|---|---|---|---|---|
| Apple sauce | 24 | unknown | 0.66 | 0.408 | Capillary tube | Charm (1978) |
| Apple sauce | 25 | 31.7% T.S. | 22.0 | 0.4 | Coaxial cylinder narrow gap | Watson (1968) |
| Apple sauce | 27 | 11.6% T.S. | 12.7 | 0.28 | Capillary tube | Saravacos (1968) |
| Apple sauce | 24 | unknown | 0.5 | 0.645 | Coaxial cylinder | Charm (1978) |
| Apple sauce | unknown | unknown | 5.63 | 0.47 | Coaxial cylinder | Charm (1978) |
| Pear puree | 27 | 14.6% T.S. | 5.3 | 0.38 | Capillary tube | Saravacos (1968) |
| Pear puree | 27 | 15.2% T.S. | 4.25 | 0.35 | Coaxial cylinder | Harper (1960) |
| Pear puree | 32 | 18.31% T.S. | 2.25 | 0.486 | Coaxial cylinder | Harper and Lebermann (1964) |
| Pear puree | 32 | 45.75% T.S. | 35.5 | 0.479 | Coaxial cylinder | Harper and Lebermann (1964) |
| Peach puree | 27 | 10.0% T.S. | 4.5 | 0.34 | Capillary tube | Saravacos (1968) |
| Peach puree | 27 | 10.0% T.S. | 0.94 | 0.44 | Coaxial cylinder | Harper (1960) |
| Banana puree | 24 | unknown | 6.5 | 0.458 | Coaxial cylinder | Charm (1978) |
| Banana puree | 24 | unknown | 10.7 | 0.333 | Capillary tube | Charm (1978) |
| Banana puree | 20 | unknown | 6.89 | 0.46 | Capillary tube | Charm (1978) |
| Banana puree | 42 | unknown | 5.26 | 0.486 | Capillary tube | Charm (1978) |
| Banana puree | 49 | unknown | 4.15 | 0.478 | Capillary tube | Charm (1978) |

Note: This table is not intended to provide a complete list of rheological properties available for liquid food products. Variabilities with temperature and concentration may be illustrated more accurately in references provided.

TABLE A.9. SPECIFIC HEATS OF FOODS

| | Composition | | | | | Specific Heat | | | |
|---|---|---|---|---|---|---|---|---|---|
| Produce | Water (%) | Protein (%) | Carbohydrate (%) | Fat (%) | Ash (%) | Eq. (3.31) (kJ/kg K) | Eq. (3.32) (kJ/kg K) | Eq. (3.33) (kJ/kg K) | Experimental[1] (kJ/kg K) |
| Beef (Hamburger) | 68.3 | 20.7 | 0.0 | 10.0 | 1.0 | 3.39 | 3.35 | 3.35 | 3.52 |
| Fish, canned | 70.0 | 27.1 | 0.0 | 0.3 | 2.6 | 3.43 | 3.31 | 3.35 | |
| Starch | 12.0 | 0.5 | 87.0 | 0.2 | 0.3 | 1.976 | 1.612 | 1.754 | |
| Orange juice | 87.5 | 0.8 | 11.1 | 0.2 | 0.4 | 3.873 | 3.818 | 3.822 | |
| Liver, raw beef | 74.9 | 15.0 | 0.9 | 9.1 | 1.1 | 3.554 | 3.521 | 3.525 | |
| Dry milk, nonfat | 3.5 | 35.6 | 52.0 | 1.0 | 7.9 | 1.763 | 1.365 | 1.520 | |
| Butter | 15.5 | 0.6 | 0.4 | 81.0 | 2.5 | 2.064 | 2.390 | 2.043 | 2.051–2.135 |
| Milk, whole pasteurized | 87.0 | 3.5 | 4.9 | 3.9 | 0.7 | 3.860 | 3.768 | 3.831 | 3.852 |
| Blueberries, syrup pack | 73.0 | 0.4 | 23.6 | 0.4 | 2.6 | 3.508 | 3.073 | 3.445 | |
| Cod, raw | 82.6 | 15.0 | 0.0 | 0.4 | 2.0 | 3.751 | 3.630 | 3.697 | |
| Skim milk | 90.5 | 3.5 | 5.1 | 0.1 | 0.8 | 3.948 | 3.935 | 3.935 | 3.977–4.019 |
| Tomato soup, concentrate | 81.4 | 1.8 | 14.6 | 1.8 | 0.4 | 3.718 | 3.471 | 3.676 | |
| Beef, lean | 77.0 | 22.0 | — | — | 1.0 | 3.559 | 3.512 | 3.579 | |
| Egg yolk | 49.0 | 13.0 | — | 11.0 | 1.0 | 2.905 | 2.457 | 2.449 | 2.810 |
| Fish, fresh | 76.0 | 19.0 | — | — | 1.4 | 3.617 | 3.437 | 3.500 | 3.600 |
| Beef, lean | 71.7 | 21.6 | 0.0 | 5.7 | 1.0 | 3.458 | 3.404 | 3.437 | 3.433 |
| Potato | 79.8 | 2.1 | 17.1 | 0.1 | 0.9 | 3.680 | 3.596 | 3.634 | 3.517 |
| Apple, raw | 84.4 | 0.2 | 14.5 | 0.6 | 0.3 | 3.793 | 3.734 | 3.759 | 3.726–4.019 |
| Bacon | 49.9 | 27.6 | 0.3 | 17.5 | 4.7 | 2.926 | 2.864 | 2.851 | 2.01 |
| Cucumber | 96.1 | 0.5 | 1.9 | 0.1 | 1.4 | 4.090 | 4.073 | 4.061 | 4.103 |
| Blackberry, syrup pack | 76.0 | 0.7 | 22.9 | 0.2 | 0.2 | 3.588 | 3.487 | 3.521 | |
| Potato | 75.0 | 0.0 | 23.0 | 0.0 | 2.0 | 3.559 | 3.429 | 3.483 | 3.517 |
| Veal | 68.0 | 21.0 | 0.0 | 10.0 | 1.0 | 3.383 | 3.056 | 3.349 | 3.223 |
| Fish | 80.0 | 15.0 | 4.0 | 0.3 | 0.7 | 3.684 | 3.408 | 3.651 | 3.60 |
| Cheese, cottage | 65.0 | 25.0 | 1.0 | 2.0 | 7.0 | 3.307 | 2.776 | 3.215 | 3.265 |
| Shrimp | 66.2 | 26.8 | 0.0 | 1.4 | 0.0 | 3.337 | 3.111 | 3.404 | 3.014 |
| Sardines | 57.4 | 25.7 | 1.2 | 11.0 | 0.0 | 3.115 | 2.972 | 3.002 | 3.014 |
| Beef, roast | 60.0 | 25.0 | 0.0 | 13.0 | 0.0 | 3.081 | 3.098 | 3.115 | 3.056 |
| Carrot, fresh | 88.2 | 1.2 | 9.3 | 0.3 | 1.1 | 3.889 | 3.831 | 3.864 | 3.81–3.935 |

[1]Reidy, G.A. 1968. Thermal Properties of Foods and Methods of Their Determination. M.S. Thesis Food Science Department, Michigan State University, East Lansing, Michigan.

TABLE A.10. THERMAL CONDUCTIVITY OF SELECTED FOOD PRODUCTS

| Product | Moisture Content (%) | Temperature (C) | Thermal Conductivity (W/m K) |
|---|---|---|---|
| Apple | 85.6 | 2 to 36 | 0.393 |
| Applesauce | 78.8 | 2 to 36 | 0.516 |
| Beef, freeze dried | | | |
| – 1000 mm Hg pressure | — | 0 | 0.065 |
| – 0.001 mm Hg pressure | — | 0 | 0.037 |
| Beef, lean | | | |
| – perpendicular to fibers | 78.9 | 7 | 0.476 |
| – perpendicular to fibers | 78.9 | 62 | 0.485 |
| – parallel to fibers | 78.7 | 8 | 0.431 |
| – parallel to fibers | 78.7 | 61 | 0.447 |
| Beef fat | — | 24 to 38 | 0.19 |
| Butter | 15 | 46 | 0.197 |
| Cod | 83 | 2.8 | 0.544 |
| Corn, yellow dent | 0.91 | 8 to 52 | 0.141 |
| | 30.2 | 8 to 52 | 0.172 |
| Egg, frozen whole | — | –10 to –6 | 0.97 |
| Egg, white | — | 36 | 0.577 |
| Egg, yolk | — | 33 | 0.338 |
| Fish muscle | — | 0 to 10 | 0.557 |
| Grapefruit, whole | — | 30 | 0.45 |
| Honey | 12.6 | 2 | 0.502 |
| | 80 | 2 | 0.344 |
| | 14.8 | 69 | 0.623 |
| | 80 | 69 | 0.415 |
| Juice, apple | 87.4 | 20 | 0.559 |
| | 87.4 | 80 | 0.632 |
| | 36.0 | 20 | 0.389 |
| | 36.0 | 80 | 0.436 |
| Lamb | | | |
| – perpendicular to fiber | 71.8 | 5 | 0.45 |
| | | 61 | 0.478 |
| – parallel to fiber | 71.0 | 5 | 0.415 |
| | | 61 | 0.422 |
| Milk | — | 37 | 0.530 |
| Milk, condensed | 90 | 24 | 0.571 |
| | — | 78 | 0.641 |
| | 50 | 26 | 0.329 |
| | — | 78 | 0.364 |
| Milk, skimmed | — | 1.5 | 0.538 |
| | — | 80 | 0.635 |
| Milk, nonfat dry | 4.2 | 39 | 0.419 |
| Olive oil | — | 15 | 0.189 |
| | — | 100 | 0.163 |
| Oranges, combined | — | 30 | 0.431 |
| Peas, black-eyed | — | 3 to 17 | 0.312 |
| Pork | | | |
| – perpendicular to fibers | 75.1 | 6 | 0.488 |
| | | 60 | 0.54 |
| – parallel to fibers | 75.9 | 4 | 0.443 |
| | | 61 | 0.489 |
| Pork fat | — | 25 | 0.152 |
| Potato, raw flesh | 81.5 | 1 to 32 | 0.554 |
| Potato, starch gel | — | 1 to 67 | 0.04 |
| Poultry, broiler muscle | 69.1 to 74.9 | 4 to 27 | 0.412 |
| Salmon | | | |
| – perpendicular to fibers | 73 | 4 | 0.502 |
| Salt | — | 87 | 0.247 |
| Sausage mixture | 64.72 | 24 | 0.407 |
| Soybean oil meal | 13.2 | 7 to 10 | 0.069 |
| Strawberries | — | –14 to 25 | 0.675 |

TABLE A.10. *(Continued)*

| Product | Moisture Content (%) | Temperature (C) | Thermal Conductivity (W/m K) |
|---|---|---|---|
| Sugars | — | 29 to 62 | 0.087 to 0.22 |
| Turkey, breast | | | |
| – perpendicular to fibers | 74 | 3 | 0.502 |
| – parallel to fibers | 74 | 3 | 0.523 |
| Veal | | | |
| – perpendicular to fibers | 75 | 6 | 0.476 |
| | | 62 | 0.489 |
| – parallel to fibers | 75 | 5 | 0.441 |
| | | 60 | 0.452 |
| Vegetable & Animal oils | — | 4 to 187 | 0.169 |
| Wheat flour | 8.8 | 43 | 0.45 |
| | | 65.5 | 0.689 |
| | | 1.7 | 0.542 |
| Whey | | 80 | 0.641 |

Reidy, G.A. 1968. Thermal Properties of Foods and Methods of Their Determination. M.S. Thesis Food Science Dept. Michigan State University, East Lansing, Michigan.

TABLE A.11. ENTHALPY OF FROZEN FOODS[a]

| Product | Water Content %(wt) | Mean Specific Heat[b] 4 to 32C kJ/kgC | Temp. C | −40 | −30 | −20 | −18 |
|---|---|---|---|---|---|---|---|
| Fruits and Vegetables | | | | | | | |
| Applesauce | 82.8 | 3.73 | Enthalpy kJ/kg | 0 | 23 | 51 | 58 |
| | | | % water unfrozen[c] | — | 6 | 9 | 10 |
| Asparagus, peeled | 92.6 | 3.98 | Enthalpy kJ/kg | 0 | 19 | 40 | 45 |
| | | | % water unfrozen | — | — | — | — |
| Bilberries | 85.1 | 3.77 | Enthalpy kJ/kg | 0 | 21 | 45 | 50 |
| | | | % water unfrozen | — | — | — | 7 |
| Carrots | 87.5 | 3.90 | Enthalpy kJ/kg | 0 | 21 | 46 | 51 |
| | | | % water unfrozen | — | — | — | 7 |
| Cucumbers | 95.4 | 4.02 | Enthalpy kJ/kg | 0 | 18 | 39 | 43 |
| | | | % water unfrozen | — | — | — | — |
| Onions | 85.5 | 3.81 | Enthalpy kJ/kg | 0 | 23 | 50 | 55 |
| | | | % water unfrozen | — | 5 | 8 | 10 |
| Peaches without stones | 85.1 | 3.77 | Enthalpy kJ/kg | 0 | 23 | 50 | 57 |
| | | | % water unfrozen | — | 5 | 8 | 9 |
| Pears, Barlett | 83.8 | 3.73 | Enthalpy kJ/kg | 0 | 23 | 51 | 57 |
| | | | % water unfrozen | — | 6 | 9 | 10 |
| Plums without stones | 80.3 | 3.65 | Enthalpy kJ/kg | 0 | 25 | 57 | 65 |
| | | | % water unfrozen | — | 8 | 14 | 16 |
| Raspberries | 82.7 | 3.73 | Enthalpy kJ/kg | 0 | 20 | 47 | 53 |
| | | | % water unfrozen | — | — | 7 | 8 |
| Spinach | 90.2 | 3.90 | Enthalpy kJ/kg | 0 | 19 | 40 | 44 |
| | | | % water unfrozen | — | — | — | — |
| Strawberries | 89.3 | 3.94 | Enthalpy kJ/kg | 0 | 20 | 44 | 49 |
| | | | % water unfrozen | — | — | 5 | — |
| Sweet cherries without stones | 77.0 | 3.60 | Enthalpy kJ/kg | 0 | 26 | 58 | 66 |
| | | | % water unfrozen | — | 9 | 15 | 17 |
| Tall peas | 75.8 | 3.56 | Enthalpy kJ/kg | 0 | 23 | 51 | 56 |
| | | | % water unfrozen | — | 6 | 10 | 12 |
| Tomato pulp | 92.9 | 4.02 | Enthalpy kJ/kg | 0 | 20 | 42 | 47 |
| | | | % water unfrozen | — | — | — | — |
| Eggs | | | | | | | |
| Egg white | 86.5 | 3.81 | Enthalpy kJ/kg | 0 | 18 | 39 | 43 |
| | | | % water unfrozen | — | — | 10 | — |
| Egg yolk | 40.0 | 2.85 | Enthalpy kJ/kg | 0 | 19 | 40 | 45 |
| | | | % water unfrozen | 20 | — | — | 22 |
| Whole egg with shell[d] | 66.4 | 3.31 | Enthalpy kJ/kg | 0 | 17 | 36 | 40 |
| Fish and Meat | | | | | | | |
| Cod | 80.3 | 3.69 | Enthalpy kJ/kg | 0 | 19 | 42 | 47 |
| | | | % water unfrozen | 10 | 10 | 11 | 12 |
| Haddock | 83.6 | 3.73 | Enthalpy kJ/kg | 0 | 19 | 42 | 47 |
| | | | % water unfrozen | 8 | 8 | 9 | 10 |
| Perch | 79.1 | 3.60 | Enthalpy kJ/kg | 0 | 19 | 41 | 46 |
| | | | % water unfrozen | 10 | 10 | 11 | 12 |
| Beef, lean Fresh[e] | 74.5 | 3.52 | Enthalpy kJ/kg | 0 | 19 | 42 | 47 |
| | | | % water unfrozen | 10 | 10 | 11 | 12 |
| Beef, lean dried | 26.1 | 2.47 | Enthalpy kJ/kg | 0 | 19 | 42 | 47 |
| | | | % water unfrozen | 96 | 96 | 97 | 98 |
| Bread | | | | | | | |
| White bread | 37.3 | 2.60 | Enthalpy kJ/kg | 0 | 17 | 35 | 39 |
| Whole wheat bread | 42.4 | 2.68 | Enthalpy kJ/kg | 0 | 17 | 36 | 41 |

[a] Above −40C.
[b] Temperature range limited to 0 to 20C for meats and 20 to 40C for egg yolk.
[c] Total weight of unfrozen water = (total weight of food)(% water content/100)(water unfrozen/100)

| -14 | -12 | -10 | -9 | -8 | -7 | -6 | -5 | -4 | -3 | -2 | -1 | 0 |
|---|---|---|---|---|---|---|---|---|---|---|---|---|
| 73 | 84 | 95 | 102 | 110 | 120 | 132 | 152 | 175 | 210 | 286 | 339 | 343 |
| 14 | 17 | 19 | 21 | 23 | 27 | 30 | 37 | 44 | 57 | 82 | 100 | — |
| 55 | 61 | 69 | 73 | 77 | 83 | 90 | 99 | 108 | 123 | 155 | 243 | 381 |
| 5 | 6 | — | 7 | 8 | 10 | 12 | 15 | 17 | 20 | 29 | 58 | 100 |
| 64 | 73 | 82 | 87 | 94 | 101 | 110 | 125 | 140 | 167 | 218 | 348 | 352 |
| 9 | 11 | 14 | 15 | 17 | 18 | 21 | 25 | 30 | 38 | 57 | 100 | — |
| 64 | 72 | 81 | 87 | 94 | 102 | 111 | 124 | 139 | 166 | 218 | 357 | 361 |
| 9 | 11 | 14 | 15 | 17 | 18 | 20 | 24 | 29 | 37 | 53 | 100 | — |
| 51 | 57 | 64 | 67 | 70 | 74 | 79 | 85 | 93 | 104 | 125 | 184 | 390 |
| — | — | — | 5 | — | — | — | — | 11 | 14 | 20 | 37 | 100 |
| 71 | 81 | 91 | 97 | 105 | 115 | 125 | 141 | 163 | 196 | 263 | 349 | 353 |
| 14 | 16 | 18 | 19 | 20 | 23 | 26 | 31 | 38 | 49 | 71 | 100 | — |
| 72 | 82 | 93 | 100 | 108 | 118 | 129 | 146 | 170 | 202 | 274 | 348 | 352 |
| 13 | 16 | 18 | 20 | 22 | 25 | 28 | 33 | 40 | 51 | 75 | 100 | — |
| 73 | 83 | 95 | 101 | 109 | 120 | 132 | 150 | 173 | 207 | 282 | 343 | 347 |
| 14 | 17 | 19 | 21 | 23 | 26 | 29 | 35 | 43 | 54 | 80 | 100 | — |
| 84 | 97 | 111 | 119 | 129 | 142 | 159 | 182 | 214 | 262 | 326 | 329 | 333 |
| 20 | 23 | 27 | 29 | 33 | 37 | 42 | 50 | 61 | 78 | 100 | — | — |
| 65 | 75 | 85 | 90 | 97 | 105 | 115 | 129 | 148 | 174 | 231 | 340 | 344 |
| 10 | 13 | 16 | 17 | 18 | 20 | 23 | 27 | 33 | 42 | 61 | 100 | — |
| 54 | 60 | 66 | 70 | 74 | 79 | 86 | 94 | 103 | 117 | 145 | 224 | 371 |
| — | 6 | 7 | — | — | 9 | 11 | 13 | 16 | 19 | 28 | 53 | 100 |
| 60 | 67 | 76 | 81 | 88 | 95 | 102 | 114 | 127 | 150 | 191 | 318 | 367 |
| 7 | 9 | 11 | 12 | 14 | 16 | 18 | 20 | 24 | 30 | 43 | 86 | 100 |
| 87 | 100 | 114 | 123 | 133 | 149 | 166 | 190 | 225 | 276 | 317 | 320 | 324 |
| 21 | 26 | 29 | 32 | 36 | 40 | 47 | 55 | 67 | 86 | 100 | — | — |
| 73 | 84 | 95 | 102 | 111 | 121 | 133 | 152 | 176 | 212 | 289 | 319 | 323 |
| 16 | 18 | 21 | 23 | 26 | 28 | 33 | 39 | 48 | 61 | 90 | 100 | — |
| 57 | 63 | 71 | 75 | 81 | 87 | 93 | 103 | 114 | 131 | 166 | 266 | 382 |
| — | 6 | 7 | 8 | 10 | 12 | 14 | 16 | 18 | 24 | 33 | 65 | 100 |
| | | | | | | | | | | | | |
| 53 | 58 | 65 | 68 | 72 | 75 | 81 | 87 | 96 | 109 | 134 | 210 | 352 |
| — | — | 13 | — | — | — | 18 | 20 | 23 | 28 | 40 | 82 | 100 |
| 56 | 62 | 68 | 72 | 76 | 80 | 85 | 92 | 99 | 109 | 128 | 182 | 191 |
| 24 | — | 27 | 28 | 29 | 31 | 33 | 35 | 38 | 45 | 58 | 94 | 100 |
| 50 | 55 | 61 | 64 | 67 | 71 | 75 | 81 | 88 | 98 | 117 | 175 | 281 |
| | | | | | | | | | | | | |
| 59 | 66 | 74 | 79 | 84 | 89 | 96 | 105 | 118 | 137 | 177 | 298 | 323 |
| 13 | 14 | 16 | 17 | 18 | 19 | 21 | 23 | 27 | 34 | 48 | 92 | 100 |
| 59 | 66 | 73 | 77 | 82 | 88 | 95 | 104 | 116 | 136 | 177 | 307 | 337 |
| 11 | 12 | 13 | 14 | 15 | 16 | 18 | 20 | 24 | 31 | 44 | 90 | 100 |
| 58 | 65 | 72 | 76 | 81 | 86 | 93 | 101 | 112 | 129 | 165 | 284 | 318 |
| 13 | 14 | 15 | 16 | 17 | 18 | 20 | 22 | 26 | 32 | 44 | 87 | 100 |
| 58 | 65 | 72 | 76 | 81 | 88 | 95 | 105 | 113 | 138 | 180 | 285 | 304 |
| 14 | 15 | 16 | 17 | 18 | 20 | 22 | 24 | 31 | 40 | 55 | 95 | 100 |
| 62 | 66 | 70 | — | 74 | — | 79 | — | 84 | — | 89 | — | 93 |
| 100 | — | — | — | — | — | — | — | — | — | — | — | — |
| | | | | | | | | | | | | |
| 49 | 56 | 67 | 75 | 83 | 93 | 104 | 117 | 124 | 128 | 131 | 134 | 137 |
| 56 | 66 | 78 | 86 | 95 | 106 | 119 | 135 | 150 | 154 | 157 | 160 | 163 |

[d] Calculated for a weight composition of 58% white (86.5% water) and 32% yolk (50% water).
[e] Data for chicken, veal, and venison very nearly matched the data for beef of the same water content.[3]

*From Dickerson (1981)*

TABLE A.12. INITIAL FREEZING TEMPERATURE OF FRUITS, VEGETABLES AND JUICES

| Product | Water content (% by weight) | Initial freezing temperature (C) |
|---|---|---|
| Apple juice | 87.2 | −1.44 |
| Apple juice concentrate | 49.8 | −11.33 |
| Applesauce | 82.8 | −1.67 |
| Asparagus | 92.6 | −0.67 |
| Bilberries | 85.1 | −1.11 |
| Bilberry juice | 89.5 | −1.11 |
| Carrots | 87.5 | −1.11 |
| Cherry juice | 86.7 | −1.44 |
| Grape juice | 84.7 | −1.78 |
| Onions | 85.5 | −1.44 |
| Orange juice | 89.0 | −1.17 |
| Peaches | 85.1 | −1.56 |
| Pears | 83.8 | −1.61 |
| Plums | 80.3 | −2.28 |
| Raspberries | 82.7 | −1.22 |
| Raspberry juice | 88.5 | −1.22 |
| Spinach | 90.2 | −0.56 |
| Strawberries | 89.3 | −0.89 |
| Strawberry juice | 91.7 | −0.89 |
| Sweet cherries | 77.0 | −2.61 |
| Tall peas | 75.8 | −1.83 |
| Tomato pulp | 92.9 | −0.72 |

TABLE A.13. HEAT TRANSFER COEFFICIENTS

| Condition | Heat Transfer Coefficient ($W/m^2K$) |
|---|---|
| Naturally circulating | 5 |
| Air blast | 22 |
| Plate contact freezer | 56 |
| Slowly circulating brine | 56 |
| Rapidly circulating brine | 85 |
| Liquid nitrogen | |
| low side of horizontal plate where gas blanket forms | 170 |
| upper side of horizontal plate | 425 |
| Boiling water | 568 |

**TABLE A.14. VARIOUS FORMS OF GAS CONSTANT**

| | |
|---|---|
| 0.0821 | atm liter/g-mole K |
| 1.987 | cal/g-mole °K |
| 1.987 | BTU/lb-mole R |
| 8.314 | joules/g-mole K |
| 1546 | ft lbs/lb-mole R |
| 10.73 | $(lb_f/in^2)\,ft^3$/lb-mole R |
| 18,510 | $(lb_f/in^2)\,in^3$/lb-mole R |
| 0.7302 | atm $ft^3$/lb-mole R |
| 848,000 | $(Kg/m^2)\,cm^3$/lb-mole K |
| 8,314.34 | $m^3$Pa/kg mole K |
| 62,363.32 | $cm^3$mm Hg/mole K |

**TABLE A.15. FREQUENTLY USED CONVERSION FACTORS FOR ENGLISH TO STANDARD INTERNATIONAL UNITS**

*Area*
$1\ ft^2 = 0.0929\ m^2$

*Density*
$1\ lb_m/ft^3 = 16.0185\ kg/m^3$

*Diffusivity*
$1\ ft^2/hr = 2.581 \times 10^{-5}\ m^2/s$

*Energy*
1 BTU = 1055 J = 1.055 kJ
1 kcal = 4.184 kJ

*Enthalpy*
$1\ BTU/lb_m = 2.3258$ kJ/kg

*Force*
$1\ lb_f = 4.4482N$
$1\ N = 1\ kg\ m/s^2$

*Heat flux*
1 BTU/hr = 0.29307 W
1 BTU/min = 17.58 W
$1\ kJ/hr = 2.778 \times 10^{-4} kW$
1 J/s = 1W

*Heat Transfer Coefficient*
$1\ BTU/hr\ ft^2F = 5.6783\ W/m^2K$

*Length*
1 ft = 0.3048 m
$1\ micron = 10^{-6}\ m$

*Mass*
$1\ lb_m = 0.4536$ kg

*Mass Transfer Coefficient*
$1$ lb mole/hr $ft^2$ mole fraction
$= 1.3562 \times 10^{-3}$ kg mole/s $m^2$mole fraction

*Pressure*
1 psia = 6.895 kPa
$1\ psia = 6.895 \times 10^3\ N/m^2$

*Specific Heat*
$1\ BTU/lb_mF = 4.1865$ J/gK

*Temperature*
1°F = 1.8°C

*Thermal Conductivity*
1 BTU/hr ft F = 1.731 W/mK

*Viscosity*
$1\ lb_m/ft\ h = 0.4134$ cp
$1\ lb_m ft\ s = 1488.16$ cp
$1\ cp = 10^{-3}$ Pa s
$1\ lb_f s/ft^2 = 4.7879 \times 10^4$ cp
$1\ N\ s/m^2 = 1$ Pa s
1 kg/m s = 1 Pa s

*Volume*
$1\ ft^3 = 0.02832\ m^3$
$1\ gal = 3.785 \times 10^{-3}\ m^3$

## BIBLIOGRAPHY

CHARM, S.E. 1978. The Fundamentals of Food Engineering, 3rd Edition. The AVI Publishing Co., Westport, Conn.

CLARY, B.L., G.L. NELSON, and R.E. SMITH. 1971. The application of the geometry analysis technique in determining the heat transfer rates from biological materials. ASAE Trans. *14*(3)386.

CRANK, J. and G. PARK. 1968. *Diffusion in Polymers*, Academic Press, New York, NY.

DICKERSON, R.W., Jr. 1981. Enthalpy of frozen foods. *In* Handbook and Product Directory Fundamentals. American Society of Heating, Refrigeration and Air Conditioning Engineers. New York.

DICKERSON, R.W., Jr. 1969. Thermal properties of foods. *In* The Freezing Preservation of Foods, 4th Ed., Vol. 2. D.K. Tressler, W.B. Van Arsdel and M.J. Copley. The AVI Publishing Co., Westport, Conn.

HARPER, J.C. 1960. Viscometric behavior in relation to evaporation of fruit purees. Food Technol. *14*:557.

HARPER, J.C. and A.F. EL SAHRIGI. 1965. Viscometric behavior of tomato concentrates. J. Food Sci. *30*:470.

HARPER, J.C. and K.W. LEBERMANN. 1964. Rheological behavior of pear purees. Proc. 1st Intern. Congr. Food Sci. & Technol. 719–728.

KEENAN, J.H., F.G. KEYES, P.G. HILL, and J.G. MOORE. 1969. Steam Tables - Metric Units. John Wiley & Sons, Inc., New York, NY.

MOHSENIN, N.N. 1978. Physical Properties of Plant and Animal Materials. Vol. I, Part II. Gordon and Breach Science Publishers. New York, NY.

RAZNJEVIC, K. 1978. Handbook of Thermodynamic Tables and Charts. McGraw-Hill Book Co., New York.

REIDY, G.A. 1968. Thermal Properties of Foods and Methods of their Determination. M.S. Thesis. Food Science Dept., Michigan State University. E. Lansing, MI.

SARAVACOS, G.D. 1968. Tube viscometry of fruit purees and juices. Food Technol. *22*(12)1585–1588.

WATSON, E.L. 1968. Rheological behavior of apricot purees and concentrates. Can. Agri. Engr. *10*(1)8–12.

## Answers to Selected Problems

### Chapter 1

1.1. 105.6 kj/kg
1.3. 0.288 mg/l

### Chapter 2

2.1. $m = 2.236$ kPa; $n = 0.295$
2.3. $m = 1.8676 \times 10^{-3}$ kPa;
$n = 0.399$
2.5. $f = \Delta PR/\rho \bar{u}^2 L$
2.7. $\Delta P = 17.7$ kPa (water)
$\Delta P = 468.9$ kPa (puree)
2.9. $\Delta P = 1.53$ ($\Delta P$ for water)
2.11. $D = 2.93$ cm

### Chapter 3

3.1. 23 cm
3.3. 6.18 $W/m^2C$
3.5. $1 - \exp(UA/Wc_p)$
3.7. 30.6 C
3.9. 60.5 C
3.11. 0.79 hr
3.13. 109.7 C

### Chapter 4

4.1. 86.5%
4.3. 350 kj/kg; 88.5%
4.5. $t_F = 3.49$ min
4.7. $t_F = 6.02$ hr; $d = 23.87$ m

### Chapter 5

5.1. 2689 $W/m^2K$
5.3. $F = 58.878$ kg/hr

### Chapter 6

6.2. 0.1166 kg$H_2O$/kg dry solids
6.4. 82.5 micron
6.6. 9.1% MC

### Chapter 7

7.1. 55.56 kg/min
7.3. 8.1 min
7.5. 2 stages

### Chapter 8

8.1. 36 hr
8.3. 10.03 cm

# Index

# Other AVI Books

AGRICULTURAL ENERGETICS
*Fluck and Baird*
DRYING CEREAL GRAINS
*Brooker, Bakker-Arkema, Hall*
ENCYCLOPEDIA OF FOOD ENGINEERING
*Hall, Farrall, Rippen*
ENCYCLOPEDIA OF FOOD TECHNOLOGY
Vol. 2 *Johnson and Peterson*
FOOD DEHYDRATION
Vol. 1 and 2 2nd Edition
*Van Arsdel, Copley, Morgan*
FOOD ENGINEERING SYSTEMS
Vol. 1 and 2 *Farrall*
FOOD PROCESSING WASTE MANAGEMENT
*Green and Kramer*
FUNDAMENTALS OF ELECTRICITY FOR AGRICULTURE
*Gustafson*
FUNDAMENTALS OF FOOD ENGINEERING
3rd Edition *Charm*
GRAIN STORAGE: PART OF A SYSTEM
*Sinha and Muir*
HANDLING, TRANSPORTATION AND STORAGE OF FRUITS AND VEGETABLES
Vol. 1 2nd Edition *Ryall and Lipton*
Vol. 2 *Ryall and Pentzer*
AN INTRODUCTION TO AGRICULTURAL ENGINEERING
*Roth, Crow, Mahoney*
NUTRITIONAL EVALUATION OF FOOD PROCESSING
2nd Edition *Harris and Karmas*
POSTHARVEST BIOLOGY AND HANDLING OF FRUITS AND VEGETABLES
*Haard and Salunkhe*
POTATOES: PRODUCTION, STORING, PROCESSING
2nd Edition *Smith*
PRINCIPLES OF FARM MACHINERY
3rd Edition *Kepner, Bainer, Barger*
PROCESSING EQUIPMENT FOR AGRICULTURAL PRODUCTS
2nd Edition *Hall and Davis*
SOYBEANS: CHEMISTRY AND TECHNOLOGY
Vol. 1 Revised Edition *Smith and Circle*
THE TECHNOLOGY OF FOOD PRESERVATION
4th Edition *Desrosier and Desrosier*

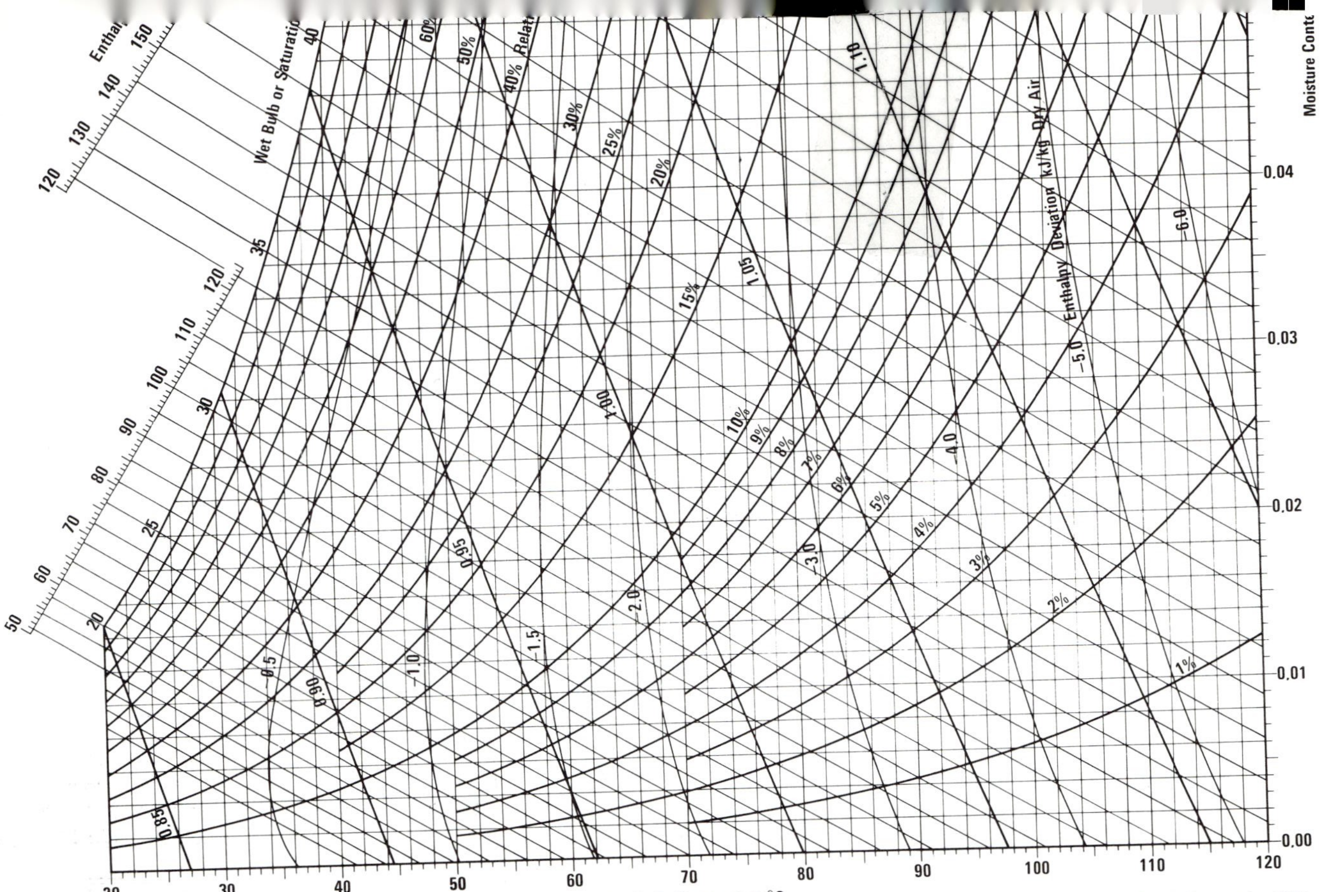

Enthalpy
150
140
130
120
Wet Bulb or Saturation
40
35
30
25
20
120
110
100
90
80
70
60
50
60%
50%
40% Relative
30%
25%
20%
15%
10%
9%
8%
7%
6%
5%
4%
3%
2%
1%
1.10
1.05
1.00
0.95
0.90
0.85
−0.5
−1.0
−1.5
−2.0
−3.0
−4.0
−5.0 Enthalpy Deviation kJ/kg Dry Air
−6.0
Moisture Content
0.04
0.03
0.02
0.01
0.00
20 30 40 50 60 70 80 90 100 110 120
Dry Bulb Temperature °C



FIG. A.5. PSYCHROMETRIC CHART

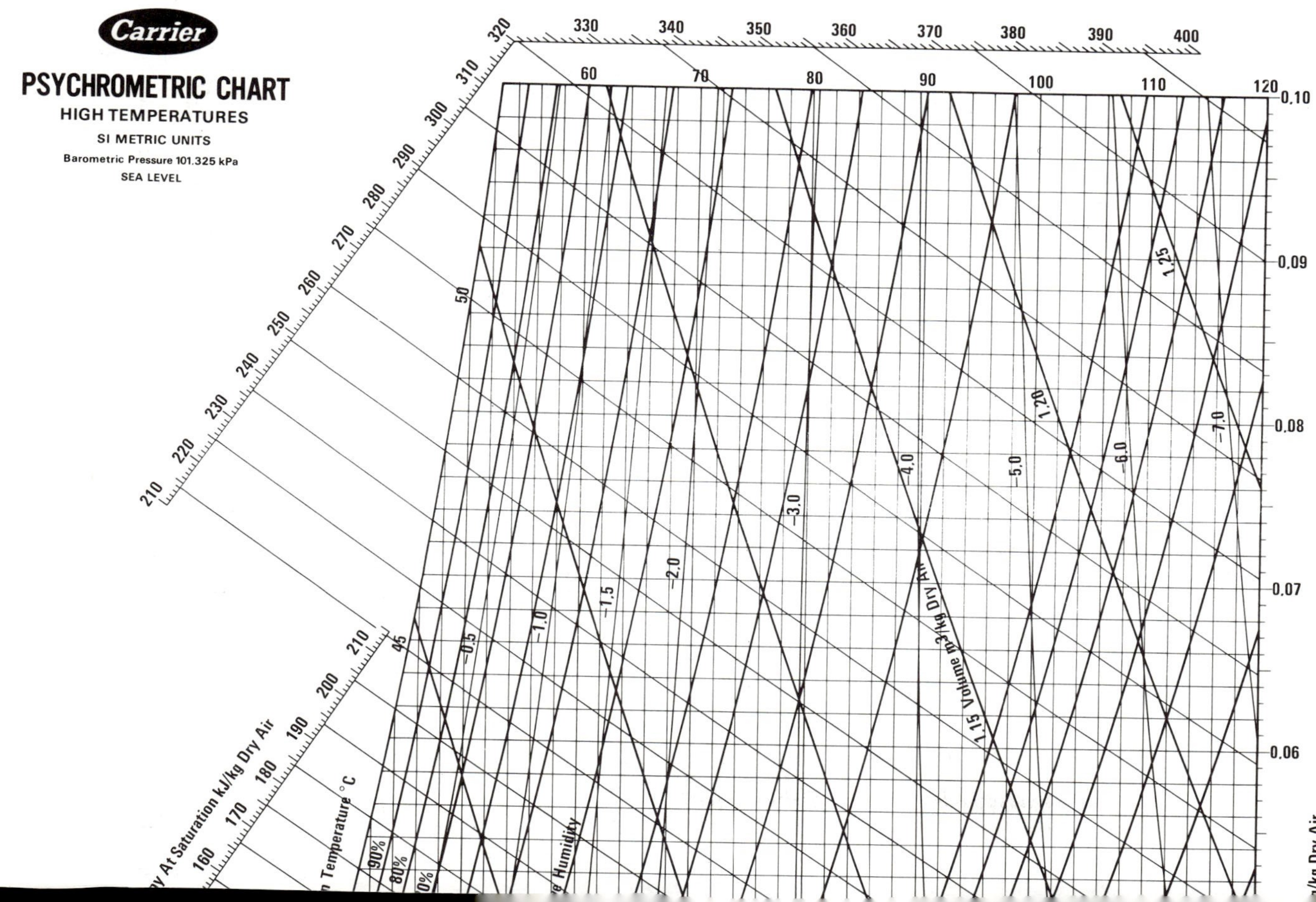
Carrier
PSYCHROMETRIC CHART
HIGH TEMPERATURES
SI METRIC UNITS
Barometric Pressure 101.325 kPa
SEA LEVEL
At Saturation kJ/kg Dry Air
Temperature °C
Humidity
1.15 Volume m3/kg Dry Air
1.20
1.25
g/kg Dry Air